NICOTINIC RECEPTORS IN THE NERVOUS SYSTEM

METHODS & NEW FRONTIERS IN NEUROSCIENCE

Series Editors
Sidney A. Simon, Ph.D.
Miguel A.L. Nicolelis, M.D., Ph.D.

Published Titles

Apoptosis in Neurobiology
Yusuf A. Hannun, M.D., Professor/Biomedical Research and Department Chairman/
Biochemistry and Molecular Biology, Medical University of South Carolina
Rose-Mary Boustany, M.D., tenured Associate Professor/Pediatrics and Neurobiology,
Duke University Medical Center

Methods for Neural Ensemble Recordings
Miguel A.L. Nicolelis, M.D., Ph.D., Associate Professor/Department of Neurobiology,
Duke University Medical Center

Methods of Behavioral Analysis in Neuroscience
Jerry J. Buccafusco, Ph.D., Professor/Pharmacology and Toxicology,
Professor/Psychiatry and Health Behavior, Medical College of Georgia

Neural Prostheses for Restoration of Sensory and Motor Function
John K. Chapin, Ph.D., MCP and Hahnemann School of Medicine
Karen A. Moxon, Ph.D., Department of Electrical and Computer Engineering,
Drexel University

Computational Neuroscience: Realistic Modeling for Experimentalists
Eric DeSchutter, M.D., Ph.D., Department of Medicine, University of Antwerp

Methods in Pain Research
Lawrence Kruger, Ph.D., Professor Emeritus/Neurobiology, UCLA School of Medicine

Motor Neurobiology of the Spinal Cord
Timothy C. Cope, Ph.D., Department of Physiology, Emory University School of Medicine

Dedication

This book is dedicated to my wife, Risa Hiller, and our daughters, Holly and Laura, for their love and support, which make endeavors such as this book possible.

Methods & New Frontiers in Neuroscience

Series Editors
Sidney A. Simon, Ph.D.
Miguel A. L. Nicolelis, M.D., Ph.D.

Our goal in creating the Methods & New Frontiers in Neuroscience Series is to present the insights of experts on emerging experimental techniques and theoretical concepts that are, or will be, at the vanguard of neuroscience. Books in the series cover topics ranging from methods to investigate apoptosis to modern techniques for neural ensemble recordings in behaving animals. The series also covers new and exciting multidisciplinary areas of brain research, such as computational neuroscience and neuroengineering, and describes breakthroughs in classical fields such as behavioral neuroscience. We want these to be the books every neuroscientist will use in order to get acquainted with new methodologies in brain research. These books can be given to graduate students and postdoctoral fellows when they are looking for guidance to start a new line of research.

Each book is edited by an expert and consists of chapters written by the leaders in a particular field. Books are richly illustrated and contain comprehensive bibliographies. Chapters provide substantial background material relevant to the particular subject. Hence, they are not just "methods books." They contain detailed "tricks of the trade" and information as to where these methods can be safely applied. In addition, they include information about where to buy equipment and Web sites helpful in solving both practical and theoretical problems.

We hope that as the volumes become available, the effort put in by us, the publisher, the book editors, and the individual authors will contribute to the further development of brain research. The extent to which we achieve this goal will be determined by the utility of these books.

Preface

Nicotine is a drug with a wide variety of effects — effects that are not black or white or even a shade of gray. Rather, the multifaceted effects of nicotine form a checkerboard pattern of effects, some of which are adverse, such as promoting tobacco smoking, and others that may provide therapeutic benefit, such as cognitive improvement and pain relief. To understand nicotine's effects, research must progress on a spectrum of levels from nicotinic receptors through complex neurobehavioral systems to clinical outcome. This book presents research approaches to all of these levels. Considerable progress can be made by integrating research across traditional disciplines so that novel nicotinic ligands discovered in receptor studies can be used in experimental animal models to gain information about specific nicotinic systems involved in behavioral function. This functional understanding can be used to develop better nicotinic-based therapeutic treatments.

Preface

The Editor

Edward D. Levin, is an associate professor in the departments of psychiatry and pharmacology at Duke University Medical Center. He earned his B.A. in psychology at the University of Rochester in 1976, his M.S. in physiological psychology in 1982, and his Ph.D. in environmental toxicology in 1984 at the University of Wisconsin-Madison. Dr. Levin held an a NIH-sponsored post-doctoral position in the psychology department at the University of California at Los Angeles and was a visiting scientist at Uppsala University in Sweden. Since 1989, he has conducted research at Duke University. Dr. Levin's research interests concern neurobehavioral pharmacology. He is particularly interested in the nicotinic effects on cognitive function. He has characterized attentional and memory improvements caused by nicotinic agonists and is currently conducting research on neural mechanisms and their possible clinical applications.

Contributors

Neural Studies

Amy Bradley
Department of Biochemistry
Bath University
Bath, UK

J. Brek Eaton
Barrow Neurological Institute
Phoenix, AZ

John D. Fryer
Barrow Neurological Institute
Phoenix, AZ

Cynthia L. Gentry
Barrow Neurological Institute
Phoenix, AZ

Ian W. Jones
Department of Biochemistry
Bath University
Bath, UK

Susan Jones
Department of Molecular
  Pharmacology, Physiology, and
  Biotechnology
Brown University
Providence, RI

Hao Lo
Department of Neurobiology
Duke University
Durham, NC

Ronald J. Lukas
Barrow Neurological Institute
Phoenix, AZ

Adrian Mogg
Department of Biochemistry
Bath University
Bath, UK

Peter P. Rowell
Department of Pharmacology
University of Louisville Medical Center
Louisville, KY

Sidney A. Simon
Department of Neurobiology
Duke University Medical Center
Durham, NC

Susan Wonnacott
Department of Biochemistry
Bath University
Bath, UK

Animal Studies

Jerry J. Buccafusco
Department of Pharmacology and
  Toxicology
Medical College of Georgia
Augusta, GA

Allan C. Collins
Institute for Behavioral Genetics
University of Colorado
Boulder, CO

William A. Corrigall
Addiction Research Foundation
Toronto, ON

M. Imad Damaj
Departments of Pharmacology and
  Toxicology
Medical College of Virginia
Richmond, VA

Christopher M. Flores
Departments of Endodontics and
  Pharmacology
University of Texas Health Science
  Center
San Antonio, TX

Edward D. Levin
Departments of Psychiatry and
  Pharmacology
Duke University Medical Center
Durham, NC

Michael J. Marks
Institute for Behavioral Genetics
University of Colorado
Boulder, CO

Amir H. Rezvani
Department of Psychiatry and
  Behavioral Sciences
Duke University Medical Center
Durham, NC

Alvin V. Terry, Jr.
College of Pharmacy
University of Georgia
Augusta, GA

Clinical Studies

Edward D. Levin
Departments of Psychiatry and
  Pharmacology
Duke University Medical Center
Durham, NC

Joseph P. McEvoy
Department of Psychiatry
Duke University Medical Center
Durham, NC

Paul A. Newhouse
Department of Psychiatry
University of Vermont
Burlington, VT

Mary Newman
Department of Psychology
College of Medicine
University of South Florida
Tampa, FL

Alexandra Potter
Department of Psychiatry
University of Vermont
Burlington, VT

Jed E. Rose
Nicotine Research Laboratory
VA Medical Center
Durham, NC

Paul R. Sanberg
Division of Neurosurgery
College of Medicine
University of South Florida
Tampa, FL

R. Douglas Shytle
Departments of Psychiatry and
  Behavioral Medicine, Neurosurgery,
  Psychology,
  and Neuroscience Program
College of Medicine
University of South Florida
Tampa, FL

Archie A. Silver
Departments of Psychiatry and
  Behavioral Medicine and
  Neuroscience Program
College of Medicine
University of South Florida
Tampa, FL

Berney J. Wilkinson
College of Medicine
University of South Florida
Tampa, FL

Table of Contents

Section 1

Section 2

Section 3

Section 4

Section 1

1 Some Methods for Studies of Nicotinic Acetylcholine Receptor Pharmacology

Ronald J. Lukas, John D. Fryer, J. Brek Eaton, and Cynthia L. Gentry

CONTENTS

1.1 INTRODUCTION

Nicotinic acetylcholine receptors (nAChRs) play many important physiological roles. They participate in classical excitatory neurotransmission to activate selected neuronal circuits, and their stimulation can modulate release of neurotransmitters or hormones that in turn can have additional actions.[1-7] Due to their wide dispersal and essential functional roles, nAChRs are ideal targets for the modulation by drug therapy of brain and body function in health and in disease.[2-4, 8-10] nAChR exist as a diverse collection of subtypes, each with a unique subunit composition.[3-4, 7, 9, 11]

0-8493-2386-X/02/$0.00+$1.50

Diversity of nAChR provides therapeutic opportunities, in that drugs might be designed to interact selectively at an nAChR subtypes, thus contributing to specific behaviors or functions without causing adverse or undesired side effects due to interactions at other nAChR subtypes. Even in an era dominated by molecular genetics, understanding the pharmacology of nAChR remains central to an improved understanding of the physiological roles of these receptors and how they can be manipulated to combat disease or, perhaps, to optimize human function.

Studies of nAChR pharmacology employ assays of nAChR function and ligand binding. Another chapter in this volume describes techniques used for electrophysiological studies of nAChR function, and Wonnacott's chapter elucidates nAChR functional assays based on measures of neurotransmitter release. The chapter by Marks refers to methods for isotopic ion flux assays of nAChR function in synaptosomal preparations; in contrast, isotopic ion flux assays of nAChR function using intact cells are described in this chapter. Techniques for ligand binding analyses of nAChR also are outlined here. The chapter begins with an overview of the use of cell lines as models for studies of nAChR pharmacology, including a brief exposition of strategies for generation of cell lines stably and heterologously expressing functional and ligand binding nAChR. Methods described are those used in our laboratory unless otherwise noted. Published work selectively cited in this overview provides additional examples of techniques outlined, but, with apologies to authors of such studies, it is not possible nor has it been our intent to cite all such useful articles.

1.2 MAINTENANCE OF CELL LINES NATURALLY EXPRESSING nAChR

Cell lines, naturally immortalized through carcinogenesis, have been useful factories expressing some of the same kinds of nAChR found in related, non-neoplastic cells (Table 1.1).[12] Limitations and attributes of cell lines in nAChR studies and examples of their uses in such work have already been described elsewhere in detail not to be duplicated here.[12] However, to summarize, cell lines allow functional and ligand binding pharmacological profiles to be derived and integrated in one experimental system, taking advantage of the potentially limitless quantities of easily manipulated and comparatively homogenous material to accommodate many kinds of assays and experimental strategies. Cell lines also are suitable for somewhat indirect, but still useful and informative, assessments of nAChR pharmacology based on how receptor activation or blockade affects phenomena including gene expression, cell shape and survival, intracellular message levels, release of other chemical messengers, and enzyme activity.

If a given cell line expresses just a single nAChR subtype, then pharmacological characterization of that nAChR subtype is reasonably straightforward and usually is much simpler than pharmacological characterization of the same subtype from a tissue or brain region in which additional nAChR subtypes may be expressed. Some cell lines, such as those corresponding to autonomic neurons (e.g., PC12, SH-SY5Y, IMR-32), also may express multiple nAChR subtypes, requiring identification of

TABLE 1.1
Cell Lines Naturally Expressing Specific nAChR Subtypes

nAChR Subtype	Cell Lines
$\alpha 1\beta 1\gamma\delta$-nAChR	TE671/RD human clonal line (rhabdomyosarcoma)
	RM0 rat muscle cell line
	BC_3.H-1 and C2 mouse muscle cell lines
$\alpha 3\beta 4^*$-nAChR	IMR-32 and SH-SY5Y human peripheral neuroblastoma lines
	PC12 rat pheochromocytoma cell line
$\alpha 7$-nAChR	IMR-32 and SH-SY5Y human peripheral neuroblastoma lines
	PC12 rat pheochromocytoma cell line

means for discriminating ligand binding and functional properties of those nAChR before pharmacological characterizations can be achieved confidently.

Maintenance of cell lines is both an art and a science. There are many monographs and instructional brochures pertaining to cell culture techniques that provide specific details that will not be repeated here. Use of sterile technique and manipulation of pathogen- or infectious agent-free cells in Class II Type A, laminar flow biosafety cabinets (recirculating all cabinet air through high efficiency particulate air (HEPA) filters and venting 30% of the filtered air back into the laboratory room) is central to handling the types of cell lines mentioned in this chapter that naturally express nAChR. These techniques help to ensure the safety of the investigator and also protect cells from investigator- or laboratory-introduced contamination. In cases where there are no assurances that cell lines being developed or used are pathogen-free, additional precautions, such as use of Class II Type B cabinets exhausting different proportions of cabinet air to the external (outside) environment, should be taken. Some cell lines can be safely manipulated under sterile conditions in open laboratory areas through the use of careful sterile technique and periodic ultraviolet illumination of the area, but costs of inevitable contaminations are likely to quickly surpass the investment in a suitable biosafety cabinet.

Conditions for cell incubation are quite standard, typically involving maintenance of cells at 37°C, at ~95% relative humidity to slow evaporation from media that would lead to concentration of salts and nutrients, and in 5% CO_2 in air to help maintain neutral pH in bicarbonate-buffered medium. However, some studies indicate that nAChR assembly and levels of expression are higher in cultures maintained at lower temperatures.[13,14] In addition, it has been found that elevated CO_2 levels facilitate proliferative effects of nAChR activation on pulmonary neuroendocrine carcinoma cells.[15] Thus, careful consideration and pilot studies may be warranted before adopting standard cell incubation conditions for studies of cells expressing nAChR.

Cell culture medium constituents can vary for maintenance of different cell lines and across laboratories. The laboratory has adapted a number of clonal cell lines to standard conditions of maintenance in Dulbecco's modified Eagle's medium (DMEM; high (4.5 mg/L) glucose, with 1 mM pyruvate, supplemented with 0.58 mg/ml (4 mM) L-glutamine), to which are added fetal calf serum (5% of DMEM volume) and horse serum (10% of DMEM volume; final osmolarity of serum-supplemented, "complete" medium is ~330 milliosmolar). However, studies have been performed to ensure nAChR expression by cells of the same passage is the same in this medium and in medium described in the initial report establishing the relevant cell line. In some cases, induction of nAChR expression requires change in growth medium. For example, serum deprivation and/or growth factor addition induces morphological differentiation and/or nAChR expression in some cell lines.[16,17] Cells are maintained on standard tissue culture plastic. Most often, 100-mm diameter dishes are used for cell maintenance, but growth in 25 cm² or 75 cm² flasks as a precaution against microbial contamination is an option particularly recommended in climates or periods of high humidity.

Antibacterial and antifungal agents are common additives to cell culture medium to help combat microbial contamination. For example, the laboratory routinely supplements medium with penicillin (200 µ/ml), streptomycin (200 µg/ml), and amphotericin B (2 µg/ml; ~2µM). Amphotericin B at ~120 µM is used for its pore forming abilities in perforated patch electrophysiological recording, but modest increases in basal $^{86}Rb^+$ efflux from cells in the presence of 3 µM amphotericin B have been observed. Consequently, effects of cell culture additives on function of nAChR and rates of recovery from any such effects should be documented if those additives are to be present during or shortly before studies of nAChR function.

Each cell line has its own characteristic doubling rate. For the purpose of expedience, cells are typically passaged weekly when they approach confluence. Loosely adherent cells can be dislodged by streams of medium applied tangentially to the plate surface, whereas more adherent cells can be dislodged in this way after mild trypsinization. For the latter approach, bulk medium is aspirated and a small volume (1.5–2 ml per 100-mm dish) of 0.25% trypsin in calcium-free Hank's balanced salt solution is applied to the dish for a few seconds. Typically, the enzyme solution is aspirated and the dish is placed in the incubator for a few minutes (i.e., a period empirically determined to yield rounded-up cells dislodged from one another). For especially adherent cells or for economy, dishes can be rinsed with phosphate-buffered saline to remove serum proteins prior to digestion in 0.25% or 0.025% trypsin, respectively. Enzymatic digestion is terminated and cells are harvested by addition of about 1^{-10} ml of complete medium per 100-mm diameter dish; this medium is applied from a pipette as a laminar stream across the dish surface to dislodge cells. Harvested cells are suspended to a specific volume in fresh medium, serially diluted (typically no more than 1:10 in a given dilution to minimize cell clumping), and plated at initial densities that should yield cultures near confluence within one week.

1.3   CREATION AND MAINTENANCE OF CELL LINES STABLY AND HETEROLOGOUSLY EXPRESSING nAChR OF DEFINED SUBUNIT COMPOSITION

Cell lines naturally expressing some important known nAChR subtypes have not yet been identified. In the case of nAChR subtypes found in the brain, this may be because mature, post-mitotic neurons have lost their susceptibility to neoplastic transformation, which may only occur in proliferating cells. This circumstance may change with advances in neuronal stem cell science and in creation of neuronal cell lines from transgenic animals in which oncogene expression is driven by neuron-specific promoter elements. Nevertheless, development of cell lines stably and heterologously expressing specific nAChR subtypes has evolved to address this limitation in experimental resources. (Transient expression will not be discussed here.) Generation of heterologous expression systems also offers advantages in that nAChR subtypes can be characterized pharmacologically in isolation, in contrast to studies using many tissues or brain regions or even some clonal cell lines that naturally express more than one nAChR subtype. Creation of cell lines stably and heterologously expressing nAChR has been fraught with failures, not in ability to express transgenes as message, but in ability of messages to be translated into nAChR protein subunits capable of assembling into ligand binding and/or surface-expressed, functional nAChR. Nevertheless, there have been some victories, and some of the details of strategy and technique in stable, heterologous expression of nAChR in mammalian cells are outlined here.

1.3.1 VECTOR SELECTION, TRANSFECTION, AND SUBCLONING

Initial success in stably and heterologously expressing α7-nAChR[18] can be attributed to selection of host cell (see next section) and selection of vector. The pCEP4 (Invitrogen) vector was chosen because it has the cytomegalovirus (CMV) promoter, which gives high constitutive expression of the downstream transgene of interest in human host cell lines. pCEP4 also contains genes coding for the Epstein–Barr virus nuclear antigen (EBNA) and origin of replication (ORI) allowing episomal replication of the vector in human cells. That is, a vector was chosen that would eliminate a requirement for insertion of the transgene into the host genome, where it might become silenced or subject to variable regulation of expression. This helped to ensure high copy number for the transgene and its stable expression under constant selective pressure via coordinate expression of the hygromycin resistance gene, all from the same plasmid.

Subsequently, there were succesful efforts in using integrating or episomal vectors for the stable expression of several different nAChR subunits as ligand binding and/or functional nAChR. (pcDNA3.1 (Invitrogen) is used as an integrating vector in our studies.)[19–30] Particularly when the objective is to express nAChR composed as binary (or higher order) complexes of subunits (e.g., binary α4β2-nAChR, ternary α3α5β4-nAChR), availability of vectors containing different selection markers is useful. (pcDNA3.1 comes in zeocin, hygromycin, or G418 resistance forms.) In one of the two initial reports of stable expression of neuronal nAChR as ligand binding and functional sites,[31,32] a vector harboring an inducible promoter was used.[31] We also have had reasons for expressing nAChR subunits from inducible promoters and have successfully used the Tet-on and Tet-off systems (Clontech) for this purpose.[27,28] These approaches require stable integration of both the pTet-off or pTet-on plasmids (containing tetracycline-sensitive regulatory elements and also conferring G418 resistance) and the pTRE plasmid (containing the gene of interest downstream from a tetracycline response element).

To introduce nAChR subunit cDNA(s) into host cells, no consensus exists in the literature as to whether electroporation, any of a number of homemade or commercially available lipofection aids, or calcium-phosphate precipitation is superior. The latter two techniques give higher initial survival of cells and are likely to be preferred for transient transfection studies (involving study of cells and nAChR that they make within a week of transfection). Electroporation is a more harsh approach (lower initial viability) that has to be custom-designed to optimize transfection efficiency balanced against cytotoxicity for every type of host cell, but requires an equipment investment rather than recurring costs for transfection reagents. In the laboratory, good and comparable success has been experienced using each of these approaches, but studies have not been done systematically. Typically, a 24- to 48-hour period of recovery from transfection is allowed, thus permitting transgenes of interest and genes conferring antibiotic resistance to be expressed before adding the appropriate antibiotic for positive selection of stable transfectants. Efficiency of transfection is then assessed based on numbers of cells and/or numbers of colonies of cells at different times after selection. Higher success in expression of multiple subunits occurs by isolating a stable mono- or poly-clone expressing one nAChR subunit and then using it as host for introduction of the second type of cDNA. Antibiotic kill curves (plots of untransfected cell survival as a function of time in the presence of selection antibiotic) are acquired for each prospective cell host before transfections are done; kill curves are also required to assess any synergistic effects on cell survival of treatment with more than one antibiotic.

Whether using episomal or integrating vectors, clones surviving selection are isolated while surviving cell densities are still low, either by using "ring cloning," "filter disc cloning," or a "stab-and-grab" technique. For ring cloning, a small amount of vaseline is applied to the bottom edge of 5-mm diameter cloning cylinders, which are placed to isolate targeted cell colonies. Medium within the cloning cylinders is removed, and 0.25% trypsin solution is applied until cells have lifted from the dish surface. Fresh medium is added, and the suspended cells are transferred to single wells in a 24-well tray. For filter disc cloning, medium is aspirated from the plate, and 5-mm diameter filter discs previously soaked in trypsin solution but wiped free of excess solvent are laid over visually identified colonies using flame-sterilized forceps. After 3 to 10 minutes, each filter disc is gently rubbed against the dish surface over the colony. Each disc with attached cells is then transferred to a well in a 24-well tray filled with fresh medium, shaken to dislodge cells, and removed if desired. The stab-and-grab technique simply involves aspiration to moistness of medium from a dish followed by positioning and manipulation of a pipette tip so that it wipes up some cells from a colony. The pipette tip is then dipped into fresh medium in a well in a 24-well tray and agitated to displace attached cells. Irrespective of the method for their physical isolation, cell clones are expanded for further subcloning as needed and screening for function and/or radioligand binding.

1.3.2 HOST CELL TYPES USED AND nAChR SUBTYPES EXPRESSED

Systematic studies are still needed to determine roles that host cell types play in the success of nAChR subunit transfection and heterologous expression. Indeed, such studies might progress to identify critical molecular and cellular bases for regulation of receptor expression, thereby providing insight into phenomena such as chaperone-assisted protein folding, assembly, and trafficking. These studies also may provide a fundamental understanding of how and why some cells differ in their ability to cope with expression of nAChR subtypes differing in kinetics of channel opening and/or ion permeability.

Cognizance of the unpublished difficulties of nAChR expression in common host cell lines such as CHO and COS cells resulted in initial reasoning that transfection and heterologous expression of nAChR should be perfected first using cells that naturally make at least some nAChR subtype. Hence, it was decided to over-express rat α7 subunits in the SH-SY5Y cell line known to naturally express human α7- and α3β4*-nAChR.[18,33] Bolstered by success with that approach and in expression of wild-type or mutant, human or chick α7 subunits in the same host cell line,[19] attention was then turned to the SH-EP1 human epithelial cell line. This cell line was initially isolated from the same tumor that yielded the SH-SY5Y clone, but early control studies indicated that it was native nAChR-null,[33] consistent with observations that the SH-EP1 and SH-SY5Y cell lines evolved divergent morphologies and chemical phenotypes as they were cloned.[34] Nevertheless, neuronal and epithelial cells have a shared embryological lineage, and both types of cells exhibit polar morphologies (dendrite-soma/axon compared to apical/basolateral or tissular/lumenal dispositions). Consequently, reasoning was that SH-EP1 epithelial cells might share abilities with neuronal cells to process and properly express complex transmembrane proteins. SH-EP1 cells have now been used successfully to stably and heterologously express functional nAChR composed of α7, α4 plus β2, or α4 plus β4 subunits;[19, 27–30] preliminary data indicate successful expression of other homomeric, binary, and even ternary complexes in these epithelial cells.

Success in heterologous expression of nAChR has also been achieved using at least some subclones of HEK-293 human embryonic kidney and other fibroblast cells.[20–22, 24–26, 31] Data are available showing expression of chick α7-nAChR (mutant form) from the pCEP4 vector in HEK-293 cells as well as in IMR-32 human neuroblastoma cells, PC12 rat pheochromocytoma cells, and CATH.a mouse neuronal cells. Stability of nAChR expression in each of these cells has not been systematically evaluated. Nevertheless, these studies clearly indicate that many cell lines have the capacity to express nAChR heterologously. How they differ from cell lines that lack such a capacity is yet to be determined.

The issue of "stability" of transfection and uniformity in transgene expression across and within passages requires clarification and warrants discussion. Primary cell cultures, which presumably have not become immortalized, are commonly defined as those passaged ten times or less from initial seeding. Continuous cell lines are commonly defined as those cultures surviving over 24 passages (leaving a gray area in nomenclature for cells carried through passages 11 to 23). Clonal cell lines are those derived from a single cell. Flow cytometric studies show that chromosomal makeup and numbers can vary wildly as a tumor-derived cell line becomes established. Continuous cell lines usually arise when a particular chromosomal makeup stabilizes growth and cell phenotype. However, recombinations still can occur to give cells a competitive growth advantage but could result in loss of genes or gene expression of interest. The policy is not to carry any cell line for more than 6 months or about 30 passages (realizing that cells in stocks are no less than 5 passages and may be as far as 20 passages removed from the true first passage). Even established cell lines, such as PC12 or SH-SY5Y, sporadically show loss of nAChR when carried for more than 20 passages from the stocks. Engineered cell lines are considered to be stably transfected if they have been carried for at least 24 passages without evidence of loss of transfected gene expression. However, this does not ensure that all passages from the same frozen stock, or cells carried from the same stock in different laboratories, will express the same quantities of nAChR or will do so through 20 to 30 passages every time. Even if a cell line expresses the same amounts of nAChR subunit message, expression of nAChR as functional or ligand binding sites may still differ because of differences in expression of some other gene critical to synthesis, assembly, and/or maturation of nAChR. Continuous quality assurance is required to monitor transgene expression. However, it is not necessarily catastrophic if a cell line loses expression while being maintained, so long as frozen stocks of low passage cells are maintained in abundance to begin a fresh passage of cells.

While not systematically evaluated, experience indicates that there may be variations in levels of nAChR expression even across cells within the same passage and on the same dish. Heterogeneity in levels of nAChR expression returns quickly, even after cells identified for their high level of expression have been subcloned. It is not clear whether variability in nAChR expression reflects dependence on position in the cell cycle, cell–cell contacts, or other cellular features.

Colleagues practicing electrophysiology have remarked about physical differences in membranes of cells expressing different nAChR subtypes from transfected genes but derived from the same host cell stocks. It should not be surprising, given choice of a single clone from dozens arising from a transfection of 10^5 to 10^6 cells, that two clones isolated from a master stock might evolve differently during transfection, selection, subcloning, and passage to exhibit different properties while retaining capacity to express nAChR.

1.4   ISOTOPIC ION FLUX ASSAYS FOR RAPID CHARACTERIZATION OF FUNCTIONAL nAChR ENSEMBLES

Isotopic ion flux assays are a proven and classical (over 25 years of use) means to characterize function of voltage- or ligand-gated ion channels in clonal cell lines.[35,36] Ion flux assays complement, and in many ways offer advantages over, more tedious electrophysiological analyses of channel function. Ion flux assays adapted for stopped-flow studies like those used in enzymology give temporal resolution comparable to that for the fastest electrophysiological studies and are fully suitable for intricate study of nAChR functional kinetics.[37] Membrane vesicles very rich in nAChR or other channels of interest and resistant to fluid pressures attained in stopped-flow studies, but not to whole cells, are preparations of choice for such studies. Nevertheless, ion flux assays using intact cells are ideally suited for high-throughput analyses of nAChR function using very simple techniques and common instrumentation. Ion flux assays integrate responses for the ensemble of nAChR from the entire population of cells in a cell culture dish or microwell, typically summed across >10^8 receptors (>10^5 cells per 15.5-mm diameter well (24-well tray) with ~10^3 surface receptors per cell).[38,39] Cellular nAChR responses can be determined using ion flux assays with a temporal resolution of seconds, and ion flux rates typically remain constant over a period of 45 to 60 seconds extrapolated through zero ion flux at time zero. When bi- or multiphasic kinetics of ion flux have been observed, it has reflected time-dependent inactivation of nAChR rather than exhaustion of accessible radiotracer.

Agonist dose-response profiles for test ligands can be obtained from studies of cells plated at equal density in wells in a multiwell array each incubated with a different dose of the test ligand. Positive control (total) responses are determined in samples exposed to a maximally efficacious dose of a standard agonist or containing both agonist and fully-blocking antagonist. Nonspecific ion flux is determined as a negative control in samples lacking agonist. Nonspecific ion flux is subtracted from positive control or test sample responses to yield specific ion flux for those samples. Specific ion flux is plotted as a function of concentration of test ligand, and the data are analyzed using nonlinear regression fits to the formula $F = F_{max} / (1 + (EC_{50}/[L])^n)$, where F is the measured specific ion flux and [L] is the molar ligand concentration, to yield the parameters F_{max} as the maximum ion flux, EC_{50} as the ligand concentration giving one-half of the maximal ion flux response to the standard agonist, and n (> 0) as the Hill coefficient for the process. Efficacy of test ligand can be determined relative to maximal response to standard agonist in positive control samples. Some test ligands may produce maximal responses at high concentrations that are less than the maximum response to the standard agonist. If agonist dose-response curves for these ligands plateau, then they act as partial agonists and their potencies can be estimated by the concentration giving one-half of their maximal effect. If agonist dose-response curves are bell-shaped, then the test ligand may be exhibiting self-inhibition of functional responses and/or inducing desensitization of nAChR at higher concentrations, but its potency also can be estimated as the dose giving one-half of its maximal effect. EC_{50} values also will differ from the concentration of a test ligand giving one-half of its maximal effect for the super efficacious drug having >100% of standard agonist efficacy. To provide valid measures of agonist activity, dose-response curves should include measurements at agonist concentrations at least ten times higher than the apparent EC_{50}, and responses at those concentrations should not be more than double responses obtained at the apparent EC_{50}.

Antagonist dose-response curves for test ligands can be obtained from studies of samples treated with different doses of the test ligand in the presence of a standard agonist at a constant concentration. Specific ion flux results are plotted as a function of concentration of test ligand, and the data are analyzed using the formula $F = F_{max} / (1 + (IC_{50}/[L])^n)$, where F is the measured specific ion flux and [L] is the molar ligand concentration, to yield the parameters F_{max} as the maximum ion flux in the absence of antagonist, IC_{50} as the ligand concentration giving a half-maximal inhibition of ion flux response, and n as the Hill coefficient for the process, which is < 0 for an antagonist in this formula. Tentatively, affinity of nAChR for the test ligand can be expressed as IC_{50} value (see below). Even if a ligand fails to produce blockade to negative control levels of ion flux, its affinity for nAChR can be tentatively expressed as the concentration giving one-half of that ligand's maximal degree of block.

Competitive or noncompetitive mechanisms of functional blockade can be distinguished based on studies of agonist dose-response profiles at zero and fixed antagonist concentrations. These curves shift to the right (to higher observed EC_{50} values, thus showing that functional block is surmountable) as competitive antagonist concentrations increase, but they shift downward (reflecting diminished agonist apparent efficacy in the face of insurmountable block) without substantially changing observed EC_{50} values as noncompetitive antagonist concentrations increase. Similarly, competitive antagonist dose-response profiles will shift to the right (giving increases in apparent IC_{50} values) as agonist concentrations within the maximally efficacious dose range increase, but noncompetitive antagonist dose-response curves will not shift appreciably left or right as agonist concentrations vary within the maximally efficacious range. IC_{50} values for noncompetitive antagonists are equal to K_i values (measures of functional nAChR affinity for the ligand; concentration at which there is half-maximal occupancy of nAChR) regardless of agonist concentration used (although agonist concentration should be equal to or greater than its EC_{50} value for practical reasons). However, determination of K_i values for competitive antagonists requires additional analysis. The competitive antagonist K_i value can be estimated as the concentration of antagonist that produces a doubling in apparent EC_{50} value for an agonist in an agonist dose-response profile relative to the EC_{50} value obtained from such a profile in the absence of antagonist. Competitive antagonist K_i values can be determined more precisely from nonlinear regression analysis of agonist dose-response curves and an expression describing receptor occupancy by agonist and competitive antagonist.[40]

Ion flux assays can also be used to derive additional information. Use dependence of blockade can be evaluated by testing for enhanced antagonism after short pretreatment with agonist. Insights into voltage sensitivity of functional blockade can be gained by studies of ion flux responses in the presence of extracellular medium containing different concentrations of potassium ion. Extracellular ion substitution experiments (e.g., N-methyl-D-glucamine exchanged for sodium, Ca^{2+} removal) can give insights into ion selectivity of channels under study. Spontaneous opening of channels can be assessed by comparing levels of ion flux in the absence of agonist to flux in samples treated with antagonist alone.

On balance, ion flux assays can give information comparable and complementary to that obtained from whole cell current and other methods of electrophysiological recording. Single channel analyses remain the purview of electrical recording. Whole cell current recording has advantages in studies of acute desensitization (occurring in seconds or less) and in some studies of very rapidly inactivating channels that are not open long enough to give significant, integrated ion flux signal above background, such as α7-nAChR. Theoretically, agonist dose-response profiles obtained from peak whole cell currents might differ from those obtained at later times in whole cell current records or from integrated ion flux responses if rates of desensitization of a given nAChR subtype differ across agonists. Similarly, dose-response profiles for a given agonist acting at different nAChR subtypes might differ when taken from peak whole cell currents or when derived from integrated ion flux responses if the two nAChR subtypes have different rates of desensitization. However, there is no reason why antagonist dose-response profiles determined at agonist EC_{50} values should differ when using, for example, peak whole cell current or integrating ion flux assays. In practice, very few studies have made direct comparisons between ion flux and whole cell current results. Experience shows that dose-response profiles differ more for analysis of a given nAChR subtype expressed in different systems (mammalian cell vs. *Xenopus* oocyte) than when analyzed using ion flux and electrophysiological techniques in the same expression system.

1.4.1 Efflux and Influx Assays

$^{86}Rb^+$ efflux assays have the highest sensitivity and resolution of any isotopic ion flux assay of nAChR function tested.[38, 41] They can be applied to nAChR because all indications are that nAChR channels are large and relatively nonspecific for size of monovalent cations. Even though the physiologically important current mediated by most nAChR is an inward current carried by Na^+, monovalent cations will flow down their concentration gradients, into or out of the cell, when nAChR are opened. $^{86}Rb^+$ efflux assays essentially establish an infinite gradient for $^{86}Rb^+$, which rapidly leaves the cell on nAChR stimulation. The specificity of $^{86}Rb^+$ efflux assays for nAChR responses has been demonstrated repeatedly. Nevertheless, pharmacological studies should be done to discount potential contributions of voltage- or Ca^{2+}-gated K^+ channels to nicotinic agonist-triggered $^{86}Rb^+$ efflux responses in a specific cell type. Note, however, that many different kinds of channel blockers, including voltage-gated K^+ channel blockers such as tetraethylammonium and 4-aminopyridine, voltage-gated Ca^{2+} channel blockers such as dihydropyridines, ionotropic glutamate receptor blockers such as MK-801, etc., also block nAChR channels, sometimes with comparable affinities.

In a typical protocol, cells of interest harvested from master plates via trypsinization (see above) are seeded into wells of multiwell trays and maintained overnight in the incubator. Cells are typically seeded in 0.5 to 1 ml of growth medium to achieve confluence at the time of assay (2–5 • 10^5 cells per 15.5-mm diameter well in a 24-well tray; 150 to 300 μg of total cell protein per well, depending on cell type). If cell plating density on the planned day of assay is too low, then ion flux assay signal will be suboptimal, and delaying the assay until cells achieve confluence is recommended. If cells are overconfluent, then there is a risk that cells will lift as a sheet during sample processing, most notably partially around the edges of the plate, causing loss of cells in initial rinses and/or transfer of cells to the efflux sample. Enhanced attachment of some cell types can be effected using specially treated plates (e.g., Falcon Primaria plates for TE671/RD cells). A more standard routine to prevent cell lifting during the assay is to treat any kind of plate momentarily with 100 μg/ml poly-D-lysine (70,000 to 150,000 Da average size; 1 ml per 15.5-mm diameter well) before seeding cells. Function of some nAChR subtypes is compromised when cells expressing those nAChR are seeded onto plates treated with polyethyleneimine, so pilot studies are recommended to ensure that poly-D-lysine, polyethyleneimine, or polyornithine treatment of assay plates does not cause inhibition of nAChR function. Note that cells will be rinsed several times before assay of nAChR function commences, so any effects of residual fluid phase attachment factor should be minimized in most instances; plates are not routinely rinsed after poly-D-lysine treatment before seeding cells. For some cell types (e.g., BC_3H-1), cell lifting occurs unless cells are seeded into 35-mm diameter dishes or into wells of 6-well trays. Probably because of the sheet-like nature of growth of those cells, they lift easily from smaller diameter wells that have larger ratios of circumference to surface area. For cell types that tend to clump rather than adhere to substratum, such as PC12, SH-SY5Y or IMR-32 cells, creating a fine, single-cell suspension of adequate concentration to ensure broad and uniform dispersal of seeded cells gives optimal results. For cell types like IMR-32 cells that tend to lift even when seeded appropriately and onto a good substratum, placement of multiwell trays on a warming plate at 37°C and use of warm instead of room-temperature rinsing media during processing (see below) helps to prevent cell lifting.

Plated cells are loaded with $^{86}Rb^+$ by aspirating seeding medium (typically serum-supplemented DMEM) and replacing it with otherwise identical medium typically supplemented with 1 to 2 μCi/ml of isotope (250 μl per 15.5-mm diameter well). Because $^{86}Rb^+$ has such a short half-life, preparation of loading medium from isotope stocks to achieve such a level of activity is empirical, and the amount of $^{86}Rb^+$ in loading medium for each experiment should be noted. Obviously, the amount of $^{86}Rb^+$ in loading medium can be decreased to conserve resources if the nAChR function being assessed is spectacular or can be increased to enhance signal in studies of nAChR with low functional activity and/or if cells are expressing low numbers of functional nAChR. Samples are returned to the incubator for at least 4 hours to allow activity of the Na-K-ATPase to concentrate $^{86}Rb^+$ (as a K^+ analogue) in intracellular medium. Studies of the kinetics of $^{86}Rb^+$ loading should be done for every cell type being assayed, but experience indicates that loading is >90% complete over 4 hours. It is recommended that some cell samples be used to determine the amount of $^{86}Rb^+$ loaded and the amount of $^{86}Rb^+$ remaining in the extracellular fluid. Specific radioactivity of $^{86}Rb^+$ can be calculated from the latter value and from knowing the extracellular volume and the concentration of extracellular K^+ and analogues. Based on the calculated specific activity of $^{86}Rb^+$, the determined amount of $^{86}Rb^+$ loaded into cells, and an assumed or determined intracellular concentration of K^+ and analogues (typically ~120 mM), intracellular volume accessible to $^{86}Rb^+$ can be calculated.

Once loaded with $^{86}Rb^+$, plated cells are removed from the incubator, and subsequent procedures are conducted behind a Lucite shield at room temperature (with exceptions noted above), in part to slow functional desensitization. Loading is terminated by aspiration of medium into a shielded collection flask and application of ion efflux buffer (130 mM NaCl, 5.4 mM KCl, 2 mM $CaCl_2$, 5 mM glucose, 50 mM HEPES, pH 7.4, ~300 milliosmolar; typically 3, 2, or 1 ml, respectively, per well for 12-, 24-, or 48-well trays). In some early studies, cell culture medium was used as the efflux buffer. However, phenol red can have activity as an nAChR antagonist, and serum components can include esterases that can cleave ester bond-containing nicotinic ligands like acetylcholine, suberyldicholine, or succinylcholine, so it is better to avoid using medium and serum in efflux assays. Buffer aspiration and cell rinsing are repeated two more times for a total time of 20 seconds to 2 minutes. Rinse buffer is then replaced with fresh ion efflux buffer usually containing nicotinic ligands of choice. Whereas stock solutions of efflux buffer are made up and stored at 4°C until use, drugs are made fresh daily from powder unless control studies show that frozen aqueous stocks or stocks dissolved in other media (e.g., ethanol or dimethylsulfoxide) give the same results as freshly prepared ligand. An option is to create high concentration stock solutions of more expensive ligands sold in small quantities or of commonly used ligands. Stocks of 1 M carbamylcholine in water, 10 mM d-tubocurarine in efflux buffer, 1 M nicotine in water, and 100 mM methyllycaconitine in ethanol can be stored at –20°C without loss of drug activity. After incubation for a prescribed period (1 to 5 minutes for studies in the laboratory), efflux medium is collected for counting, and cells are dissolved in 0.01 N NaOH, 0.1% sodium dodecyl sulfate for further analysis. $^{86}Rb^+$ can be quantified by Cerenkov counting of aqueous samples in liquid scintillation counters at ~25%/45% efficiency (glass/plastic vials) or by scintillation counting in scintillation fluid at ~95% efficiency.

$^{22}Na^+$ influx assays of nAChR (and other channel) function have been described in detail elsewhere and will not be repeated here.[35,36] $^{86}Rb^+$ influx assays, designed as variations on methods established for $^{22}Na^+$ influx assays, also have seen substantial use in studies of nAChR.[42–44] Cells are plated and processed and solution volumes are used as for $^{86}Rb^+$ efflux assays. Sample processing is initiated by removal of cell growth medium by rinsing cells and incubation for up to 30 min in rinse/equilibration medium. For cell rinses and equilibration, some published methods used complete or serum-free cell culture medium (~330 milliosmolar),[41,44] sometimes because cells were preincubated for long times with nicotinic ligands. For simpler pharmacological characterization, HEPES-buffered salt solutions (e.g., 150 mM NaCl, 5 mM KCl, 1.8 mM $CaCl_2$, 1.2 mM $MgCl_2$, 0.8 mM NaH_2PO_4, 10 mM glucose, 15 mM HEPES, adjusted to pH 7.4 with NaOH, ~310 milliosmolar)[42] are used for rinses and equilibration. For ion flux assays, Na^+-free influx buffer (replacing NaCl in whole or in part with sucrose, e.g., 0.25 M sucrose, 5 mM KCl, 1.8 mM $CaCl_2$, 10 mM glucose, 15 mM HEPES, pH 7.4, ~325 milliosmolar)[44] is substituted. Use of Na^+-free medium for influx assay reduces extracellular monovalent cations that could compete for influx of radiotracer and minimizes changes in membrane potential during nAChR activation. Just before starting the influx assay, all samples are subjected to 1 minute of incubation in Na^+-free influx buffer supplemented with 1 to 2 mM ouabain to inhibit action of the $Na^+K^+ATPase$. To initiate influx, this medium is then simply removed by aspiration, and cells are bathed in influx buffer supplemented with 1 mM ouabain (to continue inhibition of $^{86}Rb^+$ uptake via Na-K-ATPase), $^{86}Rb^+$ (about 5 μCi/ml), and nicotinic ligands of choice. After a prescribed period of time (20 sec to 2 min for studies in the laboratory), the influx period is terminated as cells are rinsed 3 to 4 times over ~0.5 to 3 minutes to remove extracellular $^{86}Rb^+$. $^{86}Rb^+$ uptake is quantified by Cerenkov or liquid scintillation counting of cellular samples dissolved in 0.01 N NaOH, 0.1% sodium dodecyl sulfate.

$^{86}Rb^+$ or $^{22}Na^+$ influx assays can and do provide essentially equivalent information to that obtained through $^{86}Rb^+$ efflux assays. Influx assays have advantages for some types of studies. For example, studies of nAChR desensitization involving pretreatment of cells with nicotinic ligands are simpler to interpret initially if influx assays are used to monitor nAChR function. Use of $^{86}Rb^+$ efflux assays for such studies is potentially complicated if ligand pretreatment causes transient activation of nAChR and loss of loaded $^{86}Rb^+$ from cells before challenge doses of ligand are applied to initiate nAChR functional analysis. However, it has been demonstrated that normalization of $^{86}Rb^+$ efflux data to the amount present in cells at the time of efflux assay initiation, even when used in studies involving pretreatment of cells with nicotinic ligands, adequately accounts for any loss of loaded $^{86}Rb^+$. $^{86}Rb^+$ efflux assays are superior to either kind of influx assay in terms of resolution (signal:noise) and sensitivity (signal when using the same amount of isotope and/or biological material), which translates into lower cost. Moreover, emissions from $^{86}Rb^+$ are less energetic than those from $^{22}Na^+$, which translates into improved safety. In addition, $^{86}Rb^+$ can be detected using Cerenkov counting without use of scintillation vials or fluid, again making them advantageous economically and in terms of safety (no need to purchase and dispose of organic scintillants). On balance, $^{86}Rb^+$ efflux assays are the clear choice for ion flux analyses of nAChR function.

1.4.2 Techniques for Cell Manipulation

Sample handling in isotopic ion flux assays most commonly involves using conventional or (for some steps in the method) repeating pipettes for solution application or aspiration. In typical $^{86}Rb^+$ efflux assays, the sample processing interval for a 24-well tray is 12 seconds, and sample wells are processed sequentially one at a time. For a typical efflux period of 5 minutes, 10 minutes pass while processing a single plate. Plates are tilted during solution exchange so that fluids are removed or applied more gently via laminar flow to the cell plating surface. Pipette tips for solution aspiration and application also are positioned consistently at one point inside the bottom edge of the well, and care must be taken to prevent or minimize displacement of some cells or a spot of cells to minimize data scatter.

A recently developed novel approach (the "flip-plate" method) for conduct of ion flux assays provides higher throughput, gentler sample handling, more flexibility in experimental design, superior temporal accuracy, and improved sample-to-sample reproducibility than sequential pipetting methods. Cells are seeded into poly-D-lysine-coated wells and loaded with $^{86}Rb^+$ (in this example for efflux assays) as usual on Falcon or Corning multiwell plates. Only Falcon or Corning brands of multiwell plates are currently manufactured so that the lips at the top of each well are elevated/level relative to the lips at the edge of the plate. This allows formation of a tight seal between well lips when two plates are opposed top to top and pressed together under uniform finger pressure. $^{86}Rb^+$ loading medium is aspirated from wells of "cell plates" as usual into a shielded collection flask, but then the cell plate is inverted and aligned top to top over the first of two "rinse plates" set up in advance to contain fresh efflux buffer (2 ml per 15.5-mm diameter well). The plates are held together firmly, and the ensemble is gently and slowly flipped (rotated) so that rinse buffer bathes the cells. After a few seconds, the plates are then gently flipped back to allow the rinse solution to fall back into the first rinse plate. The cell plate is lifted off, inverted over the second rinse plate, and rinsed in the same fashion. (Rinse plates can be washed and reused; the bulk of isotope is safely removed during the initial aspiration step.) There is a small, consistent, residual volume of buffer that remains in the cell plate (typically 30 to 40 μL for a 24-well plate) after removal of extracellular ^{86}Rb$^+$ and rinses. Therefore, to maintain exact test concentrations, the residual rinse solution is aspirated from each well as quickly as possible (usually about 10 seconds total time per plate). The cell plate is then inverted and aligned over an "efflux plate" set up in advance to contain 2 ml of nicotinic ligands of choice in efflux buffer. Efflux is initiated by again holding the plates firmly together and gently flipping them, allowing the test concentrations of drug to fall into and bathe cells in wells of the cell plate. After a prescribed incubation period, the plates are again held firmly together (sometimes gently swirled once to displace pericellular ^{86}Rb$^+$) and gently flipped, allowing assay solution containing effluxed ^{86}Rb$^+$ to fall back into the efflux plate. Although a small amount of residual volume remains in the wells of the cell plate, this quantity is uniform and can be determined to allow data normalization. The wells of the cell plate are then filled with the same volume of solution as was used in the efflux plate but containing 0.01 N NaOH, 0.1% sodium dodecyl sulfate to dissolve the cells and their contents. Samples from each well of the efflux plate and the cell plate can then be used to determine levels of effluxed and remaining intracellular isotope, respectively, for each sample after transfer to scintillation vials for Cerenkov or liquid scintillation counting. It has been found to be more convenient, more economical, and less labor intensive simply to prepare the plates for Cerenkov counting using an E&G Wallac (now Perkin-Elmer) Tri-Lux 1450 Microbeta plate-reading liquid scintillation and luminescence detector. The flip-plate method has been used routinely in 6-, 12-, 24- and 48-well formats. Use of a 96-well format is, however, not recommended, as surface tension and the requirement for higher tolerance in well dimensions and orientation make assays problematic. The flip-plate method has also been used successfully after appropriate adaptation for influx assays. Variants on the flip-plate approach have potential applicability in virtually any assay using cells or substrate adherent to or immobilized on a test plate.

1.5 RADIOLIGAND BINDING ASSAYS

Pharmacological characterization of nAChR can be assessed based on specific binding of radiolabeled nicotinic ligands and on competition by unlabeled compounds for specific radioligand binding. Central to this approach is identification of a suitable radioligand acting with reasonable selectivity (binding with much lower affinity to other nAChR subtypes) or specificity (showing no binding to other nAChR subtypes) at a given nAChR subtype(s). Most radioligands for nAChR are agonists (e.g., ^{3}H-labeled epibatidine, nicotine, acetylcholine, or cytisine) or competitive antagonists (e.g., ^{125}I-labeled α-bungarotoxin or α-cobratoxin; ^{3}H-labeled methyllycaconitine) interacting at overlapping sites on the extracellular face of nAChR. However, high affinity radioligands targeting the channel domain (e.g., ^{3}H-labeled histrionicotoxin or phencyclidine) have been used in studies of preparations highly enriched in nAChR to characterize binding sites for noncompetitive functional antagonists. Radioligand binding studies can be used to make predictions about, or to help confirm, sites of action of functionally potent compounds. For example, ligands acting as competitive antagonists or as agonists in functional assays should also inhibit binding of radiolabeled agonists or competitive antagonists to nAChR with comparable affinities. By contrast, ligands acting as noncompetitive antagonists in functional assays might act with equal potency as competitive inhibitors for radio-ligands binding in the ion channel, but should display no or lower affinity as inhibitors of radiolabeled agonist or competitive antagonist binding.

Essential to radioligand-based characterization of nAChR is determination of numbers of sites and affinities for specific radioligands. Radioligand binding satu-ration curves are determined for fixed amounts of intact cells or cellular membrane fragments incubated with different concentrations of radioligand. Positive control (total) binding is determined in samples exposed to radioligand alone. Nonspecific binding is determined as a negative control in samples exposed to radioligand and to an at least 100-fold excess of nonradiolabeled probe. Nonspecific binding is subtracted from total binding to yield specific binding for those samples. Plots of specific binding as a function of radioligand concentration are analyzed using non-linear regression fits commonly found now in data analysis software packages. The simplest analyses assume achievement of equilibrium conditions (see below) and use the formula $B = B_{max} / (1 + (K_D/[L])^n)$, where B is the observed amount of binding and [L] is the radioligand concentration to yield parameters B_{max} for the number of specific binding sites, K_D for the dissociation constant (a measure of affinity of nAChR for the radioligand), and n as the Hill coefficient for radioligand binding (> 0). Analyses can be extended to assess whether or not more than one class of specific binding sites displaying different affinities for the radioligand exists in a particular preparation and how many receptors exist in each class of binding sites.

Many monographs, review articles, or data analysis software packages explain in further detail transforms for analysis of radioligand binding assays and features of appropriate experimental design and interpretation. Only some of the more salient points will be summarized here. Binding saturation curves should be conducted under conditions where fixed receptor concentration in assay mixtures is less than the K_D for the radioligand under use and less than the radioligand concentration. The highest amount of bound radioligand in a saturation assay should be less than 10% of the free radioligand in that sample. This is particularly critical in binding studies using ^{3}H-epibatidine, which can have K_D values for binding to some nAChR subtypes of 10 pM or less. Such a high affinity for ^{3}H-epibatidine means that increased reaction volumes are required to lower receptor concentration while main-taining enough receptor to give significant levels of radioligand binding. The highest radioligand concentration in a saturation curve should be ten times higher than K_D to help ensure approach to B_{max} values. Kinetics of ligand binding should be assessed in concert with saturation binding analyses. Association rate constants (k_{on}) for radioligand binding are typically diffusion controlled yielding values of ~10^8/mol · min, but slower rates of association are observed for some of the more bulky radioligands, such as ^{125}I-labeled α-bungarotoxin. Observed association rates constants (k_{on}^{obs}) are related to true k_{on} by the formula $k_{on}^{obs} = k_{off} + k_{on} [L]$, where k_{off} is the dissociation rate constant and [L] is the concentration of radioligand used. There are several ways to determine k_{off} that will not be discussed in detail here. However, K_D values for radioligand binding obtained from saturation curves should be consistent with kinetic determinations because of the relationship $K_D = k_{off}/k_{on}$. Moreover, if k_{off} values are greater than 1/min, then dissociation is too fast for most means of sample processing to give reliable data. K_D values >10 nM are likely to result. If k_{off} is less than 0.001/min, then half times for dissociation ($0.693/k_{off}$) will be ~12 hours or more, and K_D values <1 pM would be expected. This also means that saturation analyses are complicated, because true equilibrium conditions are not achieved unless samples are incubated for periods equal to four- to five times the half time for radioligand dissociation, and receptor concentrations will need to be lowered in reaction mixtures as discussed previously.

Radioligand binding studies can also be used to derive information about non-radiolabeled ligand interactions at nAChR. Dose-response profiles for test ligands can be obtained from studies of intact cell or cell membrane fractions incubated with different doses of test ligand (preferably covering at least six orders of magnitude) and a constant concentration of radioligand. Positive control (total) and negative control (nonspecific) binding are defined as for saturation curves but at the single concentration of radioligand used. Nonspecific binding is subtracted from positive control or test samples to yield specific binding for those samples. Whether or not equilibrium conditions are achieved,[45] formulas describing the competition process have the general structure $B = B_0 / (1 + (IC_{50}/[L])^n)$, where B is the observed amount of binding, [L] is the test ligand concentration, and B_0 is the number of specific binding sites in the absence of competitor, to yield the parameters IC_{50} as the ligand concentration giving half-maximal blockade of radioligand binding and n as the Hill coefficient for the process (< 0). Affinity of test ligand for nAChR can be expressed tentatively in terms of IC_{50} value (the test ligand dose giving one-half of the binding in the positive control sample). However, some ligands may not occupy all nAChR interacting with the radioligand and/or will not inhibit more than 50% of radioligand binding. Apparent affinities for the affected subset of nAChR are then based on the test ligand concentration giving one-half of its maximal extent of inhibition of radioligand binding. K_i values (the inhibition constant or the concentration of ligand giving half-maximal occupancy of nAChR) are calculated from the Cheng-Prusoff conversion, $K_i = IC_{50}/(1+[L]/K_D)$, where [L] is the concentration of radioligand and K_D is its dissociation constant determined kinetically and/or by saturation analysis.

In principle, competitive or noncompetitive mechanisms of inhibition can be distinguished based on studies of radioligand saturation curves at zero and fixed competitor concentrations or on studies of competitor dose-response profiles at different concentrations of radioligand as described above for ion flux analyses. However, these types of studies are often confounded because ligand binding may induce conformational changes in receptors and because binding sites for ligands are not always totally and exclusively overlapping. Moreover, very attractive allosteric models of nAChR hold that binding of a particular ligand stabilizes specific states of nAChR,[46] perhaps making those states refractory to binding of other ligands, and giving experimental results indistinguishable from predictions based on non-competitive mechanisms of radioligand binding block.

For ^{125}I-labeled α-bungarotoxin binding assays conducted in the laboratory, radiolabeled toxin stocks are supplemented with 1 mg/ml of bovine serum albumin, and reaction mixtures contain at least 0.1 mg/ml of bovine serum albumin. This precaution helps to prevent adsorption of radiotoxin to stock or reaction tubes and slows irradiation decomposition of probe. Stock samples are maintained in aqueous solution at –20°C. Nonspecific binding is defined in samples containing 2 μM unlabeled α-bungarotoxin. ^{3}H-labeled agonists are maintained at –20°C in solvents suggested by the manufacturer. For ^{3}H-labeled epibatidine binding assays, we have been able to define nonspecific binding using samples containing 100 μM nicotine in Ringer's buffers that do not require supplementation with bovine serum albumin.

1.5.1 INTACT CELLS — SUSPENSION AND *IN SITU*

Radioligand binding assays of nAChR on the cell surface require use of intact cells assayed either while in suspension or seeded on assay plates.[41, 47] To initiate binding assays using plated cells *in situ*, cell culture growth medium is aspirated. Cells typically prepared on multiwell trays, such as for ion flux assays, are rinsed and equilibrated in ice-cold Ringer's buffer (150 mM NaCl, 5 mM KCl, 1.8 mM CaCl$_2$, 1.3 mM MgCl$_2$, 33 mM Tris, pH 7.4, ~310 milliosmolar). The final rinse buffer is aspirated and replaced with assay (Ringer's) buffer containing the radioligand with or without unlabeled homologous ligand or test ligands of choice and at the appropriate single concentration or range of concentrations. Reaction volumes are tailored to the size of the cell culture plate or well. After a prescribed period, the incubation mixture is removed by aspiration. Samples are then rinsed two to three times in ice-cold, fresh Ringer's buffer over a period of minutes (either by sequential aspiration and rinse buffer addition or by application of the flip-plate technique). Rinsed cells are dissolved in 0.01 N NaOH, 0.1% sodium dodecyl sulfate. Dissolved samples are then transferred to appropriate vessels for γ-counting of ^{125}I-labeled radioligand bound or to scintillation vials to which liquid scintillation fluid is added for quantitation of ^{3}H-labeled radioligands.

Cells for radioligand binding assays in suspension are prepared first by aspirating medium from large plates (100-mm diameter in this example) seeded with cells, rinsing cells with ice-cold Ringer's buffer, and adding ~1 ml of ice-cold Ringer's buffer to each dish. Cells are then gently harvested mechanically by scraping the dish bottom with an angled rubber or polypropylene policeman and collecting the cells dislodged by pipette-delivered streams of tangentially applied buffer as a suspension. Cells are gently centrifuged for 3 to 5 min at 500 g, excess buffer is withdrawn, the supernatant is discarded, and the cell pellet is suspended again in Ringer's buffer. The sample is centrifuged, the supernatant withdrawn and discarded, and the pelleted cells are resuspended in fresh Ringer's buffer to a density suitable for the intended assay. Sample aliquots placed in centrifuge tubes (16 mm × 100 mm polycarbonate) are supplemented with radioligand and/or unlabeled ligands of choice and incubated for a prescribed period. Sample volumes are typically 200 µl for ^{125}I-labeled α-bungarotoxin binding assays and 600 µl for ^{3}H-epibatidine binding assays. To end the reaction, samples are diluted in 3 to 4 ml of ice-cold Ringer's buffer, and centrifuged only for the time it takes to accelerate to 5000 rpm. Supernatants are aspirated and discarded, and cell pellets are rapidly but gently resuspended again in 3 to 4 ml of buffer and centrifuged. The process is repeated for a third cycle of dilute suspension and centrifugation, and the cell pellets are dissolved in 0.01 N NaOH, 0.1% sodium dodecyl sulfate before being processed to quantitate bound radioligand as described previously. Because longer times are needed to process samples using the cell suspension protocol, it is recommended that binding studies using radioligands with fast dissociation rates follow the protocol for intact cells *in situ*. However, the cell suspension protocol is more economical in terms of reaction volumes and quantities of reagents used.

^{125}I-labeled α–bungarotoxin, which has K_D values of ~1 nM for α1*-nAChR or α7-nAChR, can be used in assays with intact cells directly to quantify those nAChR subtypes on the cell surface. In principle, it is also possible to use assays with intact cells to quantify surface and intracellular binding sites for ^{3}H-labeled epibatidine, including those on α4β2-nAChR (typical K_D of ~10 pM), α3β4*-nAChR (typical K_D of ~100 pM) or α7-nAChR (typical K_D of ~1 nM). Otherwise equivalent samples containing ^{3}H-epibatidine only or ^{3}H-epibatidine plus 10 µM nicotine are used to define total and nonspecific binding, respectively, to cell surface plus intracellular pools of nAChR. Our studies have shown that levels of specific ^{3}H-epibatidine binding to intact cells are indistinguishable from levels of binding to membrane fractions from the same number of cells, showing access of ^{3}H-epibatidine and nicotine to all nAChR pools, even in unbroken cells. Theoretically, positively charged and relatively membrane-impermeant nicotinic ligands such as carbamylcholine should not enter the cell and block intracellular nAChR. Therefore, samples containing ^{3}H-epibatidine plus carbamylcholine should display ^{3}H-epibatidine binding to intracellular and nonspecific sites only, allowing specific ^{3}H-epibatidine binding to cell surface sites to be calculated. However, it is recommended that full carbamylcholine competition dose-response profiles be obtained using both intact cell and membrane preparations (see below). This will help determine whether there is any movement of carbamylcholine at higher concentrations into cells. If biphasic competition profiles are obtained when using intact cells, then this indicates that the fraction of binding sites corresponding to surface receptors is blocked by carbamylcholine with high affinity at lower concentrations. The second phase of such a competition profile would indicate that carbamylcholine is entering the cell at higher concentrations and gaining access to intracellular receptors, which would appear to be blocked by carbamylcholine with lower affinity. At a minimum, these studies would help identify a concentration of carbamylcholine adequate to block only surface receptors.

Assays using intact cells should involve incubation at 0 to 4°C to prevent internalization of surface nAChR complexed with radioligand; assay conditions should be chosen to ensure that ligand and receptor concentrations and incubation periods are appropriate to achieve binding equilibrium at these lower temperatures.

1.5.2 MEMBRANES AND DETERGENT-SOLUBILIZED PREPARATIONS

Preparation of membrane fragments for radioligand binding assays begins with medium removal, rinsing, and mechanical harvesting of cells as described previously. Cell suspensions are gently centrifuged for 3 to 5 min at 500 g, and the supernatant is withdrawn and discarded. The cell pellet is suspended again either in ice-cold, hypo-osmotic 5 mM Tris, pH 7.4 (to help ensure swelling of cells and maximal yield of membranes from small diameter cells such as PC12 and SH-SY5Y) or in ice-cold Ringer's buffer. Samples are subjected to homogenization for 45 seconds using a Polytron at setting 65, using a probe tip, suspension volume, and vessel size to minimize foaming of the suspension. The homogenate is transferred to centrifuge tubes (16 mm × 100 mm polycarbonate) and sedimented at ~40,000 g for 10 minutes. The supernatant is withdrawn and discarded, and the membrane pellet is suspended in fresh Ringer's buffer supplemented with 0.4 mg/ml of sodium azide to a density suitable for the intended assay. If obtained by homogenization in hypotonic buffer, the sample is centrifuged and resuspended in fresh Ringer's buffer one additional time. Brief sonication can be used at this point to aid in obtaining a uniform suspension of membranes. It has been found that resuspension in Ringer's buffer supplemented with 0.4 mg/ml sodium azide allows preservation of membranes in sealed tubes maintained at 4°C for many months without loss of ^{125}I-labeled α-bungarotoxin or ^{3}H-epibatidine binding capacity. Cell pellets can be frozen and stored at –80°C — as can tissues (brain, muscle) — and still yield membrane preparations with preserved nAChR radioligand binding sites. However, nAChR radioligand binding sites are not well preserved by frozen storage of membrane preparations. Membrane fractions suspended in hypo-osmotic buffers (e.g., 25 mM Tris, pH 7.4) have been used by others in nAChR radioligand binding assays, but experience suggests that nonspecific binding is increased substantially under these conditions compared to assays done in Ringer's or other physiological, extracellular salt solutions.

For centrifugation-based assays, membrane sample aliquots, unlabeled ligands of choice, and radioligand are placed in centrifuge tubes (16 mm × 100 mm polycarbonate). Reaction mixtures are gently flicked to ensure mixing after each addition, and reaction tubes are placed on an orbital shaker throughout incubation for the prescribed period. To end the reaction, samples are diluted in 3 to 4 ml of ice-cold Ringer's buffer supplemented with 0.1% bovine serum albumin and centrifuged at ~40,000 g for 10 minutes. Supernatants are aspirated and discarded, and membrane pellets are resuspended in 3 to 4 ml of buffer and centrifuged. Resuspension is most efficient if ~250 μl of buffer is added to the tube, allowing the sample to be blended to a fine paste during vigorous vortex mixing, before the bulk of the buffer is added to dilute the suspension. The process is repeated for a third cycle of dilute suspension and centrifugation, and the final cell pellets are dissolved in 0.01 N NaOH, 0.1% sodium dodecyl sulfate before being processed to quantitate bound radioligand as described previously. These assays can be used with radioligands that have slower dissociation constants (e.g., ^{125}I-labeled α-bungarotoxin), but they would underestimate numbers of nAChR if assayed with quickly dissociating radioligands. Centrifugation assays of ^{125}I-labeled α-bungarotoxin binding give lower nonspecific binding levels (hence, more resolution) than most filtration assays (but, see below).

For filtration-based assays, sample aliquots are prepared as for centrifugation-based assays but in borosilicate glass tubes (typically 12 mm × 75 mm). Also, Whatman GF/C filters are presoaked in ~0.1 mg/ml of polyethyleneimine before being rinsed with 3 ml of Ringer's buffer just prior to application of reaction samples. After a prescribed period of incubation with orbital shaking, reaction samples of 200 to 600 µl volume are diluted in 3 ml of ice-cold Ringer's buffer. Reaction samples of 6 ml volume can be processed directly. Each diluted suspension is applied to a polyethyleneimine-coated filter. Vacuum is then applied to draw buffer and unbound radioligand through the filter. Three to four ml of Ringer's buffer is added again to each reaction tube, and the contents are transferred to the filter under vacuum, preferably before the filter has dried. The rinse process is repeated twice; 25-mm diameter filter disks or filter pads used in semiautomated sample processors capture comparable quantities of membrane sites. If discs have been used to capture ^{125}I-labeled radioligand binding sites, they can be inserted into test tubes for γ counting immediately. Tests should be conducted to ensure that γ counting well geometry and sample placement in the counting tube are compatible with maximal detection of isotope. If discs have been used to capture ^{3}H-labeled radioligand binding sites, then they should be dried, placed in vials containing liquid scintillation medium, and left on a shaker overnight to ensure suspension of radiolabel before initiating liquid scintillation counting. Uniformity in efficiency of isotope detection across samples should be ascertained using a liquid scintillation counting internal standard. This is the method of choice for ^{3}H-labeled nAChR agonist binding assays. GF/C filtration-based assays for ^{125}I-labeled α-bungarotoxin binding can be done with greater ease and reproducibility than centrifugation-based assays and with comparable resolution if Ringer's buffer supplemented with 0.1% bovine serum albumin is used to rinse the polyethyleneimine-coated filters prior to and after sample application, as well as in reaction mixtures.

If levels of purity of nAChR allow, ion exchange techniques can be used to capture anionic nAChR-radioligand complexes. The version of the Whatman DE81 disc assay developed by Schmidt and Raftery[47] involves application of reaction mixtures to the center of dry, 25-mm diameter DE81 discs placed in ~40-mm diameter, round-bottom wells of a camper's egg carton. Three to four ml of wash solution (50 mM NaCl, 5 mM $NaPO_4$, 0.1% Triton X-100, pH 7.4) are added, and wash solution is removed after 2 to 3 min by aspiration, taking care to flip the filter disc over in the well. This process is repeated twice before discs are processed for determination of bound radioligand as for vacuum-filtered samples.

Specific ^{3}H-epibatidine binding to nonionic detergent-solubilized nAChR (e.g., used in immunoprecipitation assays or to quantitate nAChR fractionated on density gradients) is quantified using GF/C filtration to resolve free ^{3}H-epibatidine from nAChR and bound ^{3}H-epibatidine after the latter have been reprecipitated by addition to a final concentration of 20% polyethylene glycol-8000.

[125]I-labeled α-bungarotoxin binding assays are typically done in a 200 µl reaction volume containing 100 µl of membranes and 50 µl of radiotoxin at four times the final concentration. Reaction times are typically one hour at room temperature for

competition assays, to set up dissociation studies, and for pre-equilibration saturation analyses. ^{3}H-epibatidine binding competition and kinetics assays are typically done using 600 µl reaction volumes containing 100 µl of membranes and 100 µl of ^{3}H-epibatidine at six times the final concentration. ^{3}H-epibatidine saturation assays are done in 6 ml reaction volumes to ensure adequately low concentration of nAChR for assays in the presence of 1–100 pM radioligand. ^{3}H-epibatidine binding reaction times are typically 1 hour at room temperature, which is usually adequate to achieve equilibrium. For most studies, the sequence of reagent addition is dilution buffer, ligand to define nonspecific binding if needed, unlabeled ligand for competition studies if needed, membrane suspension, and then radioligand. If studies are designed to assess unlabeled ligand competition toward initial rates of radioligand binding, then membrane suspension should be added last. For samples containing unlabeled ligand, it should be added in a volume no less than one tenth of the final reaction volume to ensure that there is no more than a ten-fold dilution from the unlabeled ligand stock solution.

REFERENCES

1. Nordberg, A., Fuxe, K., Holmstedt, B., and Sundwall, A., Eds., *Nicotinic Receptors in the CNS: Their Role in Synaptic Transmission*, Elsevier, Amsterdam, Vol. 79, 1989.
2. Clarke, P. B. S., Quik, M., Adlkofer, F., and Thurau, K., Eds., *Effects of Nicotine on Biological Systems II*, Birkhauser Verlag, Basel, 1995.
3. Lindstrom, J., Neuronal nicotinic acetylcholine receptors, in *Ion Channels*, Narahashi, T., Ed., Plenum Press, New York, Vol. 4, 1996, 377–450.
4. Lindstrom, J., Nicotinic acetylcholine receptors in health and disease, *Mol. Neurobiol.* 15, 193–222, 1997.
5. Gotti, C., Fornasari, D., and Clementi, F., Human neuronal nicotinic receptors, *Prog. Brain Res.,* 53, 199–237, 1997.
6. Barrantes, F. J., Ed., *The Nicotinic Acetylcholine Receptor: Current Views and Future Trends*, Springer Verlag, Berlin/Heidelberg and Landes Publishing Co., Georgetown, TX, 1998.
7. Le Novère, N. and Changeux, J.-P., The ligand-gated ion channel database, *Nucleic Acids Res.* 27, 340–342, 1999 (http://www.pasteur.fr/recherche/banques/LGIC).
8. Levin, E. D., Rose, J. E., Lippiello, P., and Robinson, J., Eds., Nicotine addiction, *Drug Devel. Res.,* 38, 135–304, 1996.
9. Lukas, R. J., Neuronal nicotinic acetylcholine receptors, in *The Nicotinic Acetylcholine Receptor: Current Views and Future Trends,* Barrantes, F. J., Ed., Springer Verlag, Berlin/Heidelberg and Landes Publishing Co., Georgetown, TX, 1998, 145-173.
10. Arneric, S. P. and Brioni, J. D., Eds., *Neuronal Nicotinic Receptors: Pharmacology and Therapeutic Opportunities*, Wiley-Liss, Inc., New York, 1999.
11. Lukas, R., Changeux, J.-P., Le Novère, N., Albuquerque, E. X., Balfour, D. J. K., Berg, D. K., Bertrand, D., Chiappinelli, V. A., Clarke, P. B. S., Collins, A. C., Dani, J. A., Grady, S. R., Kellar, K. J., Lindstrom, J. M., Marks, M. J., Quik, M., Taylor, P. W., and Wonnacott, S., International Union of Pharmacology. XX. Current status of the nomenclature for nicotinic acetylcholine receptors and their subunits, *Pharmacol. Rev.,* 51, 397–401, 1999.

12. Lukas, R. J., Cell lines as models for studies of nicotinic acetylcholine receptors, in *Neuronal Nicotinic Receptors: Pharmacology and Therapeutic Opportunities,* Arneric, S. P. and Brioni, J. D., Eds., Wiley-Liss, Inc., New York, 1999.

13. Green, W. N. and Claudio, T., Acetylcholine receptor assembly: subunit folding and oligomerization occur sequentially, *Cell,* 74, 57–69, 1993.

14. Cooper, S. T., Harkness, P. C., Baker, E. R., and Millar, N. S., Up-regulation of cell-surface alpha4 beta2 neuronal nicotinic receptors by lower temperature and expression of chimeric subunits, *J. Biol. Chem.,* 274, 27145–52, 1999.

15. Schuller, H M., Carbon dioxide potentiates the mitogenic effects of nicotine and its carcinogenic derivative, NNK, in normal and neoplastic neuroendocrine lung cells via stimulation of autocrine and protein kinase C-dependent mitogenic pathways, *Neurotoxicology,* 15, 877–86, 1994.

16. Patrick, J., McMillan, J., Wolfson, H., and O'Brien, J. C., Acetylcholine metabolism in a nonfusing muscle cell line, *J. Biol. Chem.,* 252, 2143–53, 1977.

17. Rogers, S. W., Mandelzys, A., Deneris, E. S., Cooper, E., and Heinemann, S., The expression of nicotinic acetylcholine receptors by PC12 cells treated with NGF, *J. Neurosci.,* 12, 4611–23, 1992.

18. Puchacz, E., Buisson, B., Bertrand, D., and Lukas, R. J., Functional expression of nicotinic acetylcholine receptors containing rat α7 subunits in human neuroblastoma cells, *F. E. B. S. Letters,* 354, 155–159, 1994.

19. Puchacz, E., Galzi, J.-L., Buisson, B., Bertrand, D., Changeux, J.-P., and Lukas, R.J., Properties of transgenic nicotinic acetylcholine receptors stably expressed in human cells as homo-oligomers of wild-type or mutant rat or chick α7 subunits, *Soc. Neurosci. Abst.,* 21, 1335, 1995.

20. Gopalakrishnan, M., Buisson, B., Touma, E., Giordano, T., Campbell, J. E., Hi, I. C., Donnelly-Roberts, D., Arneric, S. P., Bertrand, D., and Sullivan, J. P., Stable expression and pharmacological properties of the human α7 nicotinic acetylcholine receptor, *Eur. J. Pharm.,* 290, 237-246, 1995.

21. Gopalakrishnan, M., Monteggia, L.M., Anderson, D. J., Molinari, E. J., Piattoni-Kaplan, M., Donnelly-Roberts, D., Arneric, S. P., and Sullivan, J. P., Stable expression, pharmacologic properties, and regulation of the human neuronal nicotinic acetylcholine α4β2 receptor, *J. Pharm. Exper. Thera.,* 276, 289–297, 1996.

22. Stetzer, E., Ebbinghaus, U., Storch, A., Poteur, L., Schrattenholz, A., Kramer, G., Methfessel, C., and Maelicke, A., Stable expression in HEK-293 cells of the rat α3/β4 subtype of neuronal nicotinic acetylcholine receptor, *F. E. B. S. Letters,* 397, 39–44, 1996.

23. Cooper, S. T. and Millar, N. S., Host cell-specific folding and assembly of the neuronal nicotinic acetylcholine receptor alpha7 subunit, *J. Neurochem.,* 68, 2140–51, 1997.

24. Stauderman, K. A., Mahaffy, L. S., Akong, M., Velicelebi, G., Chavez-Noriega, L. E., Crona, J. H., Johnson, E. C., Elliott, K. J., Gillespie, A., Reid, R. T., Adams, P., Harpold, M. M., and Corey-Naeve, J., Characterization of human recombinant neuronal nicotinic acetylcholine receptor subunit combinations α2β4, α3β4 and α4β4 stably expressed in HEK293 cells, *J. Pharm. Exper. Thera.,* 284, 777–789, 1998.

25. Wang, F., Nelson, M. E., Kuryatov, A., Olale, F., Cooper, J., Keyser, K., and Lindstrom, J., Chronic nicotine treatment up-regulates human alpha3 beta2 but not alpha3 beta4 acetylcholine receptors stably transfected in human embryonic kidney cells, *J. Biol. Chem.,* 273, 28721–32, 1998.

26. Xiao Y., Meyer, E. L., Thompson, J. M., Surin, A., Wroblewski, J., and Kellar, K. J., Rat $\alpha3/\beta4$ subtype of neuronal nicotinic acetylcholine receptor stably expressed in a transfected cell line: pharmacology of ligand binding and function, *Molec. Pharm.*, 54, 322-33, 1998.

27. Peng, J.-H. and Lukas, R. J., Heterologous expression of epibatidine- and α-bungarotoxin-binding human $\alpha7$-nicotinic acetylcholine receptor in a native receptor-null human epithelial cell line, *Soc. Neurosci. Abst.*, 24, 831, 1998.

28. Peng, J.-H., Lucero, L., Fryer, J., Herl, J., Leonard, S. S., and Lukas R. J., Inducible, heterologous expression of human $\alpha7$-neuronal nicotinic acetylcholine receptors in a native nicotinic receptor-null human clonal line, *Brain Res.*, 825, 172–179, 1999.

29. Peng, J.-H., Eaton, J. B., Eisenhour, C. M., Fryer, J. D., Lucero, L., and Lukas, R. J., Properties of stably and heterologously-expressed human $\alpha4\beta2$-nicotinic acetylcholine receptors (nAChR), *Soc. Neurosci. Abst.*, 25, 1723, 1999.

30. Eaton, J. B., Kuo, Y.-P., Fuh, L. P.-T., Krishnan, C., Steinlein, O., Lindstrom, J. M., and Lukas, R. J., Properties of stably and heterologously-expressed human $\alpha4\beta4$-nicotinic acetylcholine receptors (nAChR), *Soc. Neurosci. Abst.*, 26, in press, 2000.

31. Whiting, P. J., Schoepfer, R., Lindstrom, J. M., and Priestly, T., Structural and pharmacological characterization of the major brain nicotinic acetylcholine receptor subtype stably expressed in mouse fibroblasts, *Mol. Pharm.*, 40, 463–472, 1991.

32. Rogers, S.W., Gahring, L.C., Papke, R.L., and Heinemann, S., Identification of cultured cells expressing ligand-gated cation channels, *Protein Express. Purif.*, 2, 108–116, 1991.

33. Lukas, R.J., Norman, S.A., and Lucero, L., Characterization of nicotinic acetylcholine receptors expressed by cells of the SH-SY5Y human neuroblastoma clonal line, *Mol. Cell. Neurosci.*, 4, 1–12, 1993.

34. Ross, R. A., Spengler, B. A., and Biedler, J. L., Coordinate morphological and biochemical interconversion of human neuroblastoma cells, *J. Natl. Cancer Inst.*, 71, 741–747, 1983.

35. Catterall, W. A., Sodium transport by the acetylcholine receptor of cultured muscle cells, *J. Biol. Chem.*, 250, 1776–1781.

36. Stallcup, W. B., Sodium and calcium fluxes in a clonal nerve cell line, *J. Physiol.* (Lond.), 286, 525-540, 1979.

37. Hess, G. P., Cash, D. J., and Aoshima, H., Acetylcholine receptor-controlled ion fluxes in membrane vesicles investigated by fast reaction techniques, *Nature*, 282, 329–331, 1979.

38. Lukas, R. J. and Cullen, M. J., An isotopic rubidium ion efflux assay for the functional characterization of nicotinic acetylcholine receptors on clonal cell lines, *Anal. Biochem.*, 175, 212–218, 1988.

39. Lukas, R. J., Pharmacological distinctions between functional nicotinic acetylcholine receptors on the PC12 rat pheochromocytoma and the TE671 human medulloblastoma, *J. Pharm. Exper. Thera.*, 251, 175-182, 1989.

40. Lew, M. J. and Angus, J. A., Analysis of competitive agonist-antagonist interactions by nonlinear regression, *Trends Pharm. Sci.*, 16, 328-337.

41. Ke, L., Eisenhour, C. E., Bencherif, M., and Lukas, R. J., Effects of chronic nicotine treatment on expression and function of diverse nicotinic receptor subtypes. I. Dose- and time dependent effects of nicotine treatment, *J. Pharm. Exper. Thera.*, 286, 825–840, 1998.

42. Robinson, D. and McGee, R., Jr., Agonist-induced regulation of the neuronal nicotinic acetylcholine receptor of PC12 cells, *Mol. Pharm.*, 27, 409-417, 1985.

43. Boyd, N. D., Two distinct phases of desensitization of acetylcholine receptors of clonal rat PC12 cells, *J. Physiol.*, 389, 45–67, 1987.
44. Bencherif, M., Eisenhour, C. M., Prince, R. J., Lippiello, P. M., and Lukas, R. J., The "calcium antagonist" TMB-8 [3,4,5-trimethoxybenzoic acid 8-(diethylamino)octyl ester] is a potent, non-competitive, functional antagonist at diverse nicotinic acetylcholine receptors subtypes, *J. Pharm. Exper. Thera.*, 275, 1418–1426, 1995.
45. Lukas, R. J., Morimoto, H., and Bennett, E. L., Effects of thio-group modification and Ca^{2+} on agonist-specific state transitions of a central nicotinic acetylcholine receptor, *Biochemistry,* 18, 2384–2395, 1979.
46. Galzi, J.-L., Edelstein, S. J., and Changeux, J.-P., The multiple phenotypes of allosteric receptor mutants, *Proc. Natl. Acad. Sci., U.S.A.* 93, 1853–1858, 1996.
47. Bencherif, M. and Lukas, R. J., Cytochalasin modulation of nicotinic cholinergic receptor expression and muscarinic receptor function in human TE671/RD cells: a possible functional role of the cytoskeleton, *J. Neurochem.,* 61, 852–864, 1993.
48. Schmidt, J. and Raftery, M. A., A simple assay for the study of solubilized acetylcholine receptors, *Anal. Biochem.,* 52, 319–354.

2 Presynaptic Nicotinic Acetylcholine Receptors: Subtypes Mediating Neurotransmitter Release

Susan Wonnacott, Adrian Mogg, Amy Bradley, and Ian W. Jones

CONTENTS

0-8493-2386-X/02/$0.00+$1.50
© 2002 by CRC Press LLC

2.1 INTRODUCTION

There is substantial evidence for the occurrence of presynaptic nAChR on CNS neurones. They can increase (and under some circumstances decrease) transmitter release, a function compatible with a major modulatory role proposed for nAChR in the brain.[1] Presynaptic nAChR have been the subject of extensive recent reviews[2-5] and the reader is referred to these sources for a comprehensive account. Instead, this chapter will focus on strategies adopted to monitor nAChR-evoked transmitter release and to identify the receptor subtypes responsible. *In vitro* neurochemical and electrophysiological methods have provided valuable insights into presynaptic nAChR. Transmitter release measured by *in vivo* neurochemical methods has been mainly limited to examination of the effects of systemically administered nicotine, and both somato-dendritic and terminal nAChR have been implicated. An adjunct to these "functional pharmacology" approaches is the deployment of autoradiographic and immunocytochemical techniques to ascertain the cellular and subcellular localization of specific nAChR subunits and/or subtypes. We will attempt to review these approaches briefly, summarizing key findings from studies of mammalian CNS and highlighting advantages and limitations.

2.2 *IN VITRO* NEUROCHEMICAL TECHNIQUES

2.2.1 SUPERFUSION

Transmitter release from brain preparations (synaptosomes or slices) *in vitro* is typically measured by superfusion (Figure 2.1). The term "superfusion" was coined by H.E. Gaddum in 1953,[6] although the principle had been applied in earlier studies. It refers to the flow of liquid, typically physiological buffer, **over** tissue (as opposed to perfusion in which liquid flows **through** the tissue). Usually several samples are superfused in parallel, and serial fractions of superfusate are collected for analysis of released transmitter. After establishing a stable baseline, transmitter release can be evoked by drug application or electrical stimulation in the presence or absence of antagonists or other drugs. Automated superfusion systems, designed for brain slices, are produced commercially. A customized system, with subsecond time resolution, has been developed[7] but there have been no published reports yet of nicotine-evoked transmitter release measured over such brief timescales. Recently a higher throughput release assay has been described.[8] This is a static release system carried out in 96 well filter plates equipped with a support membrane to separate tissue slices from the bathing medium (containing the released transmitter) that can be removed by vacuum filtration. The pharmacological data presented for [³H]dopamine[9] and [³H]noradrenaline release[8] are comparable to results from conventional superfusion systems. This method may be useful for rapid screening of novel ligands.

Convenient assay methods for endogenous transmitter are generally too insensitive to detect amounts in the individual fractions of superfusate, although amperometry with carbon fiber microelectrodes can detect small changes in chemical concentration at the surface of cells *in vitro* (e.g., nicotine-evoked catecholamine

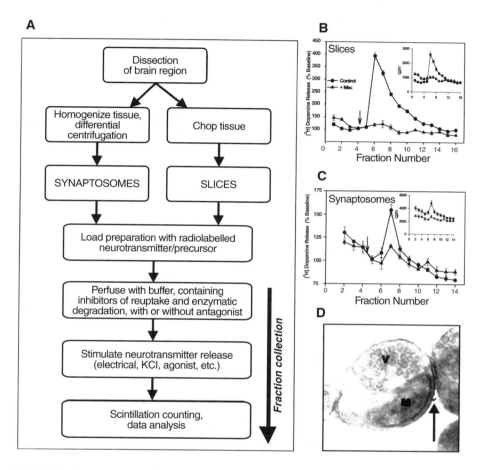

FIGURE 2.1 *In vitro* neurochemical techniques. A. Scheme summarizing the steps involved in superfusion experiments with either slices or synaptosomes; B. Typical profile for the release of [³H]dopamine from rat striatal slices. Slices (250 μm prisms) were prepared and loaded with [³H]dopamine and superfused in closed chambers.[22] A 40 sec pulse of 10 μM nicotine was delivered at the point shown by the arrow, in the presence (▲) or absence (●) of 10 μM mecamylamine. The insert panel shows the raw data, as cpm per fraction. In the main panel, release has been converted to percent baseline (to compensate for differences in amounts of slices or [³H]dopamine loaded between chambers); C. Typical profile for the release of [³H]dopamine from rat striatal synaptosomes. Crude synaptosomes were prepared and loaded with [³H]dopamine, and superfused in open chambers.[22] A 40 sec pulse of 10 μM nicotine was delivered at the point shown by the arrow, in the presence (▲) or absence (●) of 10 μM mecamylamine; D. Electron micrograph of a striatal synaptosome purified on a Percoll gradient.[22] Synaptic vesicles (V) and a mitochondrion (M) can be clearly seen. Two gold particles (arrow) immunolabel the nAChR β2 subunit on the surface on the synaptosome (that has not been permeabilized). Magnification × 10,000.

secretion from individual bovine adrenal chromaffin cells has been detected by this method[10]). Most commonly, tissue is loaded with radiolabelled transmitter prior to superfusion, and the release of radioactivity is measured. Thus, nicotinic agonists have been shown to stimulate the release of [³H]dopamine (see the chapter by Rowell), [³H]noradrenaline,[11,12] [³H]GABA,[13] and [³H]5HT.[14] Alternatively, radiolabelled precursors may be used, allowing the tissue to convert this into transmitter. It has been argued that the latter method ensures that physiological pools of releasable radiolabelled transmitter will be formed. This approach is necessary for ACh, as there is no uptake mechanism for the transmitter itself; most of the [³H]choline taken up is converted into [³H]ACh and the release of tritium evoked by nicotinic agonists can be attributed to the transmitter.[15,16] Similarly, [³H]tyrosine has been used to label dopamine in studies of presynaptic nAChR function.[17,18] In this case the preparation is continuously superfused with the [³H]tyrosine; hence it is necessary to chromatographically separate the precursor from the product in each fraction of superfusate collected. In general there is good accord between studies using [³H]tyrosine and [³H]dopamine to examine the effects of nicotinic drugs on striatal dopamine release (c.f. Cheramy *et al.*[18] and Kaiser and Wonnacott[19]).

Absent from this list is the measurement of nicotine-evoked glutamate release: this has proved refractory using [³H]glutamate or [³H]aspartate. Fluorimetric[20] and bioluminescent methods[21] exist but have apparently not been applied to study nicotine-evoked responses. If glutamate release is primarily regulated by α7* nAChR (see Section 2.3), a superfusion system with fast application and high temporal resolution may be crucial for discerning responses. A component of nAChR-evoked dopamine release elicited by α7-mediated glutamate release has been inferred *in vitro*[19,22] and *in vivo*[23,24] (see Section 2.4).

2.2.2 SLICES OR SYNAPTOSOMES?

Slices or minces (about 250 µm) are conventionally prepared using an automated device, such as a McIlwain tissue chopper. The advantages of such preparations are their speed and ease of preparation, and the preservation of local connectivity, making the preparation more representative of the intact brain. The latter aspect may also be a disadvantage, as indirect effects on transmitter release, via synaptic connections, can complicate the interpretation of results. The protocol typically used to study the effects of presynaptic **metabotropic** receptors on transmitter release from slices is to determine S2/S1 ratios. In this paradigm a stimulus (e.g., electrical or KCl depolarization) is given first with standard buffer (S1) and then repeated in the presence of test agonist or antagonist (S2). The S2/S1 ratio of amounts of transmitter released gives the degree of stimulation or antagonism. Thus each individual experiment has its own internal standard (S1). This protocol is unsuitable for studying nicotinic effects on transmitter release because (a) nAChR modulate **basal** release, and (b) nAChR desensitize so readily that successive stimulations with a nicotinic agonist can produce large decrements in responses that compromise the comparison of ratios (e.g., see Wilkie et al.[15]). Instead, absolute values of transmitter release (in cpm or fmoles, typically expressed as a percent of baseline or of total radiolabel originally accumulated by the preparation; see Figure 2.1B) are determined and comparisons made between parallel chambers with and without various concentrations of the drug of interest. In particular, slice studies have been used for the pharmacological characterization of nAChR-mediated [³H]dopamine release in striatum (e.g., Dwoskin et al.[25] and Marshall et al.[26]) and [³H]noradrenaline release in hippocampus.[11,27]

Isolated nerve terminals (synaptosomes; Figure 2.1D) are generated when brain tissue is homogenized under appropriate conditions.[28] These structures retain all the features of the intact nerve ending and enable aspects of nerve terminal function to be studied in isolation. Crude synaptosome preparations, produced by differential centrifugation, include free mitochondria and plasma membranes. Highly purified preparations can be prepared by density gradient centrifugation, for example, four-step Percoll gradients[29] separate synaptosomes from membrane fragments and mitochondria. Superfusion of synaptosomes avoids cross talk between nerve terminals, and modulation of transmitter release reflects a direct effect via presynaptic receptors. However, preparations are rather frail, being prone to osmotic or mechanical damage. Nevertheless, synaptosomes have been somewhat more popular than slices for the neurochemical examination of presynaptic nAChR mediating the release of various neurotransmitters from different brain regions.[5] Typically, a single stimulation with nicotinic agonist is given, and different agonist concentrations or the effect of antagonists or other drugs compared in parallel chambers, as for slices (Figure 2.1C).

2.2.3 SUBTYPES OF PRESYNAPTIC nAChR

While the nicotinic stimulation of various transmitters is well documented,[2,4] evidence in favor of particular subtypes is more limited, reflecting the dearth of subtype-selective tools. The current perspective on subtypes of presynaptic nAChR in the rodent CNS is summarized in Figure 2.2. The presynaptic nicotinic modulation of dopamine release in the striatum is the best characterized example, in large part due to the robust, readily quantifiable response to nicotinic agonists, illustrated in Figure 2.1B,C. Synaptosome studies have excluded the direct involvement of α7-type nAChR through the insensitivity of the response to the α7-selective antagonists αbungarotoxin (αBgt)[30] and αconotoxin IMI.[31] Inhibition by αconotoxin MII[31,32] implicated a nAChR subtype containing an α3β2 interface. The high levels of expression of the α6 nAChR subunit (which is most homologous to α3) in dopaminergic neurones[33,34] and sensitivity of chicken α6-containing nAChR to αconotoxin MII[35] has led to the current view that αconotoxin MII may recognize α3/α6 + β2 containing nAChR.[36] As α6 appears to require additional α subunits for assembly into functional nAChR,[37] it is conceivable that both α3 and α6 subunits may reside in the same pentameric receptor.

αconotoxin MII only partially inhibits nicotinic agonist-evoked dopamine release from striatal synaptosomes,[32] suggesting the presence of a second subtype of nAChR on these nerve terminals. α4β2* nAChR have been proposed as a candidate, based on pharmacological evidence;[38] and this is consistent with the loss of [³H]nicotine binding sites, following lesioning of the nigrostriatal pathway (see Section 2.5.1). The requirement for the β2 subunit for the formation of the two

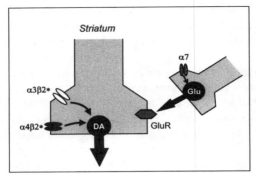
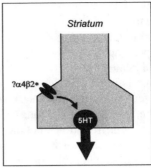
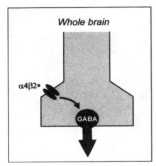
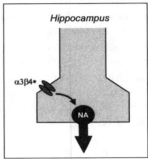
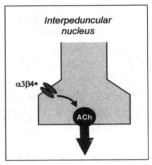

FIGURE 2.2 Putative subtypes of presynaptic nAChR in the rodent brain. Subtypes proposed to modulate the release of various transmitters, based largely on *in vitro* neurochemical experiments, are illustrated. See Section 2.2 for details.

putative subtypes of presynaptic nAChR on striatal dopamine terminals is consistent with the absence of nicotine-evoked [^{3}H]dopamine release from synaptosomes prepared from the striata of β2-null mutant mice.[16]

Recent studies on striatal slices have disclosed an additional component blocked by the α7-selective antagonists αBgt, αconotoxin IMI and MLA, and by glutamate receptor antagonists.[19,22] This component is not seen in synaptosome experiments, and is interpreted as an indirect nicotinic effect, whereby α7 nAChR on glutamate terminals cause the release of glutamate that in turn promotes the release of dopamine. The modulation of glutamate release by α7 nAChR is consistent with electrophysiological data from several brain regions (see Section 2.3).

Comparison of nAChR-evoked striatal dopamine release and hippocampal noradrenaline release[11,12] reveals marked pharmacological differences consistent with the involvement of different nAChR subtypes. The ability of αconotoxin AuIB to inhibit hippocampal noradrenaline release, but not striatal dopamine release, implicates the α3β4* nAChR subtype in the former response.[39] αconotoxin AuIB also partially inhibited the nicotinic stimulation of [^{3}H]ACh release from mouse interpeduncular nucleus synaptosomes.[16] The absence of the β2 subunit in null mutant mice did not diminish the response, consistent with an α3β4* subtype of nAChR. Previously, the nicotinic stimulation of rat hippocampal ACh release was tentatively attributed to the α4β2 nAChR subtype.[15] Definitive subtype-selective tools were not available in this study, although it is also possible that different presynaptic nAChR subtypes operate in different brain regions to modulate release of the same transmitter.

Nicotinic agonist-evoked [3H]GABA release has also been studied in wildtype and transgenic mice lacking the nAChR β2 subunit.[13] In this case, [3H]GABA release and [3H]nicotine binding decreased in parallel with the number of copies of the functional β2 gene, implicating the α4β2 subtype as a candidate for the presynaptic nAChR regulating GABA release. Nicotine-evoked [3H]5HT release from rat striatal synaptosomes has recently been characterized.[14] Sensitivity to low concentrations of dihydroβerythroidine and to higher concentrations of MLA exclude the α7 nAChR as a candidate receptor; the α4β2 subtype is consistent with some (but not all) of the pharmacological data,[14] and also with evidence from a lesioning study[40] (see Section 2.5.1).

2.3 *IN VITRO* ELECTROPHYSIOLOGICAL APPROACHES

Discussion of electrophysiological methods is presented elsewhere in this volume. In this section our discussion is limited to some general comments on the approach and a summary of major findings.

Because nerve endings are too tiny to record from (with the exception of certain calyces[41]) presynaptic nAChR activity is inferred from a modulation of transmitter release detected as postsynaptic currents. Nicotinic modulation at glutamate and GABA synapses has been the major focus of this approach.[42,43] Nicotine increases the frequency of miniature postsynaptic currents without affecting their amplitude, consistent with an enhancement of transmitter release rather than direct activation of a distinct population of postsynaptic nAChR. This is confirmed by blockade of currents by glutamate or GABA receptor antagonists. A presynaptic locus of action requires that the enhancement by nicotine should be Ca^{2+}-dependent (consistent with exocytosis) but unaffected by TTX (i.e., independent of axonal firing). Where TTX-sensitivity has been demonstrated in the absence of relevant cell bodies (e.g., slices and dissociated neurones from rat interpeduncular nucleus,[44]) a preterminal localization of nAChR has been proposed.[2]

The high temporal resolution of electrophysiological recording, coupled with a fast perfusion system for agonist application,[43] has enabled a nicotinic modulation of glutamate release via rapidly desensitizing α7 nAChR to be discerned. Examples of presynaptic α7 nAChR capable of modulating glutamate release have been described in rat hippocampus,[45] olfactory bulb,[46] and VTA.[47] The unambiguous assignment of the α7 subtype as the nAChR responsible is due to the availability of several α7-selective antagonists (αBgt, MLA, and αconotoxin IMI), plus the recognition that choline preferentially activates α7 nAChR.[48] More direct evidence to support the role of presynaptic α7 nAChR in hippocampal neurones was provided by Gray et al.[45] In parallel with electrophysiological recordings of postsynaptic currents, increases in intracellular Ca^{2+} in response to nicotine were demonstrated in large mossy fiber terminals loaded with Fura 2. Thus nAChR activation could increase the probability of transmitter release (and hence increase the rate of quantal release, reflected in an increase in the frequency of postsynaptic currents).

Radcliffe and Dani[49] observed a second type of nicotinic modulation of glutamate release in hippocampal cultures. This was also attributed to $\alpha 7$ nAChR, but in this case the effect lasted for minutes rather than seconds, persisting after the nicotinic stimulus had been removed. These properties prompted the suggestion that presynaptic $\alpha 7$ nAChR may contribute to glutamatergic synaptic plasticity. This concept has been advanced recently by Mansvelder and McGehee[47] who showed that, under appropriate conditions of repetitive postsynaptic electrical stimulation, coapplication of nicotine could induce a long lasting increase in the amplitude of postsynaptic currents, in response to subsequent standard stimulation conditions. The increased amplitude reflects an enhanced amount of transmitter release in response to a given stimulus that persisted for at least 40 minutes, consistent with the induction of an LTP-like phenomenon. The most parsimonious interpretation of the data is that nicotine acts at presynaptic $\alpha 7$ nAChR on glutamatergic terminals; pairing nicotine application with presynaptic electrical stimulation did not result in any change in amplitude of postsynaptic currents.[47]

2.4 *IN VIVO* NEUROCHEMICAL METHODS

In vivo microdialysis has been widely used to measure transmitter release; with respect to nicotine, the majority of studies has focused on monitoring dopamine overflow. This technique has the benefit that it can provide results from the functioning brain in conscious, freely moving animals. It has high anatomical precision, enabling small, adjacent structures (such as the core and the shell of the accumbens) to be examined independently. On the negative side, *in vivo* microdialysis has poor temporal resolution (*in vivo* voltametry offers higher temporal resolution[50]) and it is impractical for detailed dose-response relationships. Moreover, drug concentrations reaching the nAChR can only be estimated. Most applications of *in vivo* microdialysis to study nicotinic mechanisms in the brain have employed systemic administration of nicotine. However, local delivery of antagonists (into cell body areas via a cannula, or into terminal fields via the dialysis probe) can define the site of action of nicotine. Thus dopamine overflow in the nucleus accumbens in response to systemic nicotine is largely abolished by infusion of mecamylamine into the VTA,[51] interpreted in terms of somato-dendritic nAChR having a pre-eminent role with respect to nicotine-evoked dopamine release in the terminal field. However, more recent experiments implicate presynaptic $\alpha 7$ nAChR in the VTA in the indirect control of accumbens dopamine release: intrategmental infusion of NMDA receptor antagonists[24] or MLA[23] abolished nicotine-evoked dopamine release in the nucleus accumbens. Thus nicotine may exert some of its effects in the VTA through stimulation of $\alpha 7^*$ nAChR on the terminals of glutamatergic afferents. This view is reinforced by the observation that local infusion of MLA blocks the increase in overflow of glutamate and aspartate in the VTA in response to systemic nicotine.[52] The concept of presynaptic $\alpha 7$ nAChR on glutamate terminals in the VTA is in accord with the *in vitro* electrophysiological experiments of Mansvelder and McGehee[47] (see Section 2.3).

Infusion of nicotine directly into the terminal field (striatum, nucleus accumbens or frontal cortex) elicited dopamine overflow detected by microdialysis.[53] This was blocked by mecamylamine but no further pharmacological analysis was carried out to determine which nAChR subtypes are responsible. However, the data are consistent with the ability of local stimulation of nAChR (on presynaptic terminals or interneurons) to elicit dopamine release *in vivo*. In the mouse, nicotine-evoked dopamine overflow in the striatum was abolished in animals lacking the β2 subunit;[54] the locus of action of nicotine was not established in these studies. Clearly further work is needed to resolve the relative contributions of different nAChR subtypes, and of presynaptic vs. somatodendritic nAChR, to dopamine release *in vivo*.

2.5　nAChR LOCALIZATION

While the "functional pharmacology" approach provides strong evidence for the nicotinic modulation of transmitter release, the limited number of subtype-selective tools and a general lack of knowledge about the subunit composition of native nAChR have constrained the information generated about which nAChR subtypes are involved. Moreover, the precise localization of nAChR (for example in relation to synapses) cannot be addressed. Therefore techniques that map the anatomical localization of nAChR or nicotinic subunits can usefully complement functional studies. This section summarizes various techniques deployed for mapping studies (see Figure 2.3), with emphasis on their relevance to nAChR involved in modulating transmitter release. For a more extensive review of nAChR distribution, see Sargent.[55]

2.5.1　Receptor Ligand Binding

The distribution of binding sites for nAChR-selective radioligands in the CNS has been well documented using autoradiographic techniques. Thus [³H]ACh, [³H]nicotine and [³H]cytisine have been used to label predominantly α4β2 nAChR,[56,57] whereas [¹²⁵I]αBgt and [³H]MLA identify α7 nAChR.[56,58] [³H]epibatidine binding sites are more heterogeneous.[59,60] They can be resolved into a large "cytisine-sensitive" population (corresponding to sites labelled by [³H]nicotine and [³H]cytisine, hence predominantly α4β2), and a smaller "cytisine-resistant" population.[60] A proportion of the latter is blocked by αconotoxin MII, and these sites can be labelled directly by [¹²⁵I]αconotoxin MII.[36] They presumably correspond to nAChR containing α3/α6 and β2 subunits.[61] Loss of [¹²⁵I]αconotoxin MII binding in β3 null mutant mice[62] also implicates the β3 subunit in these nAChR. The distribution of [¹²⁵I]αconotoxin MII binding sites is limited to a few brain nuclei, including the striatum where they are correlated with the presynaptic, αconotoxin MII-sensitive nAChR on dopamine terminals.[31,32,36] Zoli et al.[59] compared the distributions of radioligand binding sites in wildtype mice and in mice lacking the gene for the β2 nAChR subunit, and deduced the presence of at least four classes of nAChR subtype.

Autoradiography defines brain regions enriched in particular binding sites but lacks the resolution needed to determine cellular or subcellular localization. Lesion

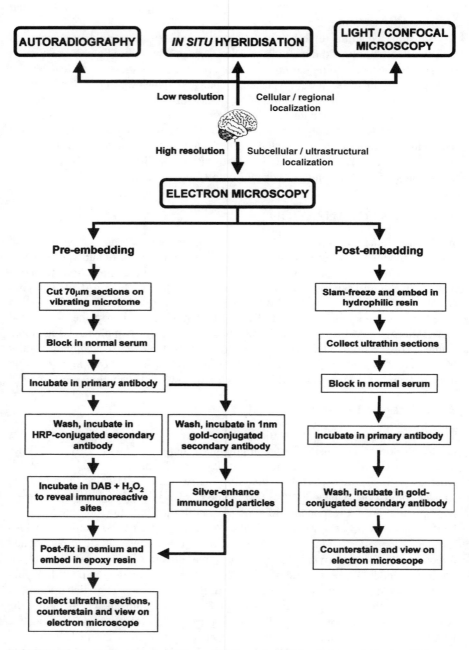

FIGURE 2.3 Techniques for localizing nAChRs and nAChR subunits within the CNS.

studies have been carried out in conjunction with autoradiography to define the existence of nAChR in particular pathways and to infer the occurrence of presynaptic nAChR. Lesion of catecholamine and serotonergic pathways (by intraventricular injection of 6-hydroxydopamine or 6-hydroxytryptamine, respectively) decreased [³H]ACh binding in certain projection areas, namely striatum and hypothalamus, consistent with presynaptic α4β2* nAChR on terminals in these regions.[40] More specific 6-hydroxdopamine lesions of nigrostriatal and mesolimbic pathways decreased [³H]nicotine labelling in the striatum and nucleus accumbens, as well as in the substantia nigra and VTA,[63] consistent with α4β2 nAChR on dopamine terminals (as well as on cell bodies) that degenerate after lesion. Recently, ibotenic acid lesions of prefrontal cortex were found to decrease the density of [¹²⁵I]αBgt binding sites in the VTA,[52] providing further evidence for the presence of α7 nAChR on glutamate terminals (see Sections 2.3 and 2.4).

Fluorescently labelled or enzyme-associated ligands offer an alternative strategy to autoradiography. The high affinity and slow dissociation of αBgt, coupled with its proteinaceous nature, have facilitated such approaches. Lentz and Chester[64] used a horseradish peroxidase (HRP)-αBgt conjugate to label nAChR in rat brain synaptosomes and tissue sections examined at the EM level. Labelling was associated with the postsynaptic densities and synaptic clefts of a small proportion of synapses; axon terminals at labelled synapses were devoid of dense cored vesicles (synonymous with biogenic amines) but contained small clear vesicles. FITC-αBgt was also used in the 1970s to compare the numbers of binding sites in the hippocampus and striatum.[65] The lower density in the latter area was interpreted as evidence against the hypothesis of a presynaptic nicotinic modulation of dopamine release: at this time the identity and heterogeneity of neuronal nAChR were not appreciated. The sensitivity of this approach can be amplified by using biotinylated αBgt, together with avidin-FITC to examine the distribution of α7-containing nAChR in chick ciliary ganglia neurons, using confocal microscopy.[66] This strategy also facilitates the use of avidin-gold, implemented in the same study for analysis at the ultrastructural level, using electron microscopy, of α7-nAChR on somatic spines.[66] The latter technique could be applied to label presynaptic α7-nAChR in the CNS.

2.5.2 nAChR Subunit Expression

In situ hybridization has been used to map the distribution of expression of particular nAChR subunits (e.g., Wada et al.[67] and Han et al.[68] reviewed in Sargent[55]). Hybridization of radiolabelled or enzyme-linked oligonucleotide probes is visualized by autoradiography or addition of a colored substrate, respectively. The major limitation of this approach with respect to presynaptic nAChR is that labelling is limited to perikarya, and does not reflect the destination of the protein. RT-PCR has been employed for qualitative analysis of nAChR subunit expression at the single cell level (e.g., Charpantier et al.[69] and Sheffield et al.[70]). However, levels of mRNA transcript may not reflect levels of translated protein or assembled nAChR pentamer. Nevertheless, this technique can define the range of subunits expressed by particular neurons and hence establishes those available for nAChR composition.

2.5.3 IMMUNOCYTOCHEMISTRY

2.5.3.1 General Principles

Immunocytochemistry utilizes the high affinity interaction of an antibody with a specific epitope on a target protein to localize that protein by visualization of the antibody, usually by using a tagged secondary or tertiary antibody. Monoclonal antibodies provide high specificity with less likelihood of cross reactivity, but the chances of specific labelling of the target protein under various fixation conditions may be diminished by having only one species of antibody present. The chances of success are increased by using polyclonal antibodies (comprising a variety of anti-bodies directed against the immunizing epitope or protein), but this is at the expense of potentially higher background labelling. Most subunit-selective antibodies used against nAChR subunits are directed against the extracellular N-terminal region or the cytoplasmic loop between membrane spanning regions three and four.[71] Using antibodies raised in different species allows labelling with more than one antibody to be undertaken simultaneously.

2.5.3.2 Immunocytochemistry at the Light/Confocal Microscope Level

At the light level, immunocytochemistry is relatively straightforward, typically car-ried out on aldehyde fixed tissue, and using vibrotome, cryostat, or wax sections. The second antibody is conjugated to an enzyme (typically HRP) or biotin (for the more sensitive ABC method); addition of substrate (typically diaminobenzidine, DAB) gives a colored reaction product that provides a permanent record of labelling. Alternatively, Swanson et al.[72] used iodinated monoclonal antibodies to map nAChR in rat and chicken brains. The mild fixation and high penetration conditions allow many antibodies to work successfully and serial sections are easy to collect. For example, this approach has been used to map the distributions of the nAChR $\beta 2$[72,73] and $\alpha 7$[74] subunits in the rat CNS. The limitations of this technique are the low resolution and the difficulty in labelling more than one protein for comparative purposes. However, Sorenson et al.[75] complemented the ABC-DAB method (which yields a brown product) with a peroxidase-antiperoxidase-benzidine dihydrochloride method (blue crystals) to demonstrate double labelling of cell bodies and dendrites in rat substantia nigra pars compacta by nAChR $\alpha 4$ subunit and tyrosine hydroxylase. This approach is not commonly applied, due to the very high toxicity of benzidine dihydrochloride, which can only be used under license in some countries.

Another way in which the colocalization of two proteins can be addressed is to immunolabel consecutive sections for the different proteins of interest. Comparative morphological analysis will indicate if the same structures are present and labelled in adjacent sections. This was done by Sorenson et al.[75] to demonstrate the colo-calization of $\alpha 4$ and $\beta 2$ nAChR subunits in cell bodies in the substantia nigra pars compacta. However it should be noted that demonstration that two nAChR subunits colocalize in the same cellular compartment does not necessarily mean that they are present in the same pentameric nAChR; biochemical methods would be needed to confirm this.

Double labelling is usually achieved by immunofluorescence microscopy. The procedure is similar to that outlined previously except that the second antibody either has a fluorescent tag or is biotinylated and reacted with a fluorescently labelled avidin group. Immunofluorescent labelling of rat substantia nigra with an antibody against the nAChR α3 subunit is shown in Figure 2.4A. Two or more protein targets can be labelled using different fluorophores. This method has demonstrated that tyrosine hydroxylase immunoreactivity colocalizes in cell bodies in the rat substantia nigra with nAChR α4[76] and nAChR α6[34] subunit immunoreactivities. If viewed in the confocal microscope, it is possible to construct 3-dimensional images. One disadvantage compared with conventional immunocytochemistry is the lack of permanent slides due to fluorescence bleaching. Both techniques are limited to relatively low resolution analysis, insufficient for reliably detecting axons and presynaptic elements in brain sections. Swanson et al.[72] inferred axonal and dendritic transport of [^{125}I]mAb 270, based on the anatomical regions labelled. For example, the labelling of substantia nigra pars compacta and ventral tegmental area was assumed to reflect labelling of dopaminergic cell bodies, and labelling of associated terminal fields (striatum and nucleus accumbens) was assumed to be "almost certainly due to axonal transport." Supporting evidence that the labelled subunit (now known to be β2) was transported via the optic tract to terminal fields was provided by the loss of contralateral labelling after unilateral enucleation, which eliminates the retinal ganglion cell bodies whose axons constitute the optic tract.[72] Hill et al.[73] observed "a fine and diffuse staining" of β2-like immunoreactivity in several regions of the rat brain, including the striatum and hippocampal molecular layer, which they interpreted as "staining of nerve terminal fields." Double immunofluorescent labelling of hippocampal neurons in culture was used to show the colocalization of the nAChR α7 subunit and the presynaptic marker synaptotagmin,[77] consistent with a presynaptic role for α7 nAChR. In contrast, nAChR β2 immunoreactivity was largely confined to soma and proximal processes, with little colocalization with synaptotagmin in these neurons.

Direct evidence for nAChR at presynaptic terminals has recently been provided by the successful application of confocal immunofluorescence microscopy to synaptosomes. Nayak et al.[78] reported that the nAChR α4 (but not α3/α5) subunit immunoreactivity and 5HT3 receptor immunoreactivity colocalize to the same striatal synaptosomes. Complementary functional studies and immunoprecipitation experiments indicated that the α4 and 5HT3 subunits contribute to separate receptor entities. Using comparable techniques, we have immunolocalized the nAChR β2 subunit in rat striatal synaptosomes (Figure 2.4B). In the same study, it was found that nAChR β2 subunit and tyrosine hydroxylase immunoreactivities are colocalized in a population of synaptophysin-positive structures, consistent with striatal dopaminergic terminals. Therefore, this approach, relatively straightforward to perform and analyze, offers potential for a qualitative analysis of the colocalization of different nAChR subunits, or an nAChR subunit with other neurochemical markers. Quantitative analysis is hampered by the tendency of synaptosomes to form clumps; hence, individual terminals cannot be reliably discerned.

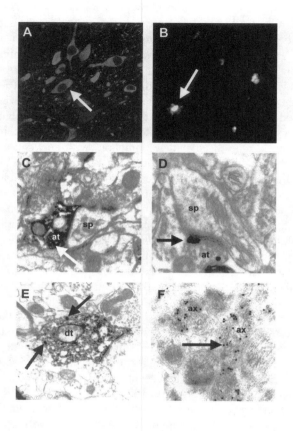

FIGURE 2.4 Immunocytochemical localization of nAChR subunits in rat brain. A. (confocal microscopy). nAChR α3 subunit immunoreactivity in the cell bodies (arrow) and processes of nigrostriatal dopaminergic neurons within the substantia nigra pars compacta; B. (confocal microscopy). nAChR β2 subunit immunoreactivity in Percoll-purified striatal synaptosomes (arrow); C. (pre-embedding EM). nAChR β2 subunit immunoreactivity (DAB reaction product, arrow) in an axon terminal (ax) forming a symmetric synapse with a dendritic spine (sp) within the dorsal striatum; D. (pre-embedding EM). nAChR β2 subunit immunoreactivity (silver-enhanced gold particles, arrow) in an axon terminal (ax) forming a symmetric synapse with a dendritic spine (sp) within the dorsal striatum; E. (double pre-embedding EM). Colocalization of the nAChR β2 subunit (silver-enhanced gold particles, arrows) and tyrosine hydroxylase (DAB reaction product) in a dendrite (dt) within the substantia nigra pars compacta; F. (double post-embedding EM). Colocalization of the nAChR β2 subunit (small gold particles) and tyrosine hydroxylase (large gold particles) in axons (ax) within the dorsal striatum. (Magnification: panels A and B, × 630; panels C–F, ×20,000.)

2.5.3.3 Electron Microscopy

Visualization of receptors at synapses requires the resolution of electron microscopy (EM). Stronger fixation conditions are necessary for preservation of morphology at the subcellular level and, as a consequence, antibodies that work successfully at the light microscopy level may not recognize protein epitopes under the harsher conditions used for EM. The most common EM technique is "pre-embedding" (Figure 2.3), in which tissue sections are immunolabelled prior to embedding in resin and sectioning. The ABC-DAB reaction is most often used; osmication makes the DAB reaction product more electron dense and hence easier to recognize. However, the product is diffusible, with a high affinity for membranes, so interpretation of the precise subcellular localization of the antigen under examination may be misleading.

Pre-embedding EM was used to investigate the localization of nAChR α4-like immunoreactivity in the rat cerebellum.[79] DAB immunoreaction product was observed in presynaptic axon terminals making synapses with Purkinje cell bodies and dendrites. While synaptic boutons were clearly labelled, no interpretation of the membrane localization of the nAChR α4 subunit could be inferred. Similarly, it has been shown that a proportion of axon terminals forming symmetric synapses onto dendritic spines within the dorsal striatum are immunopositive for the nAChR β2 subunit (Figure 2.4C).

The limitations of DAB diffusion within labelled structures can be avoided by using a gold-tagged second antibody, and subsequent silver enhancement to increase size and contrast. However, this also has disadvantages, as the method is less sensitive than the immunoperoxidase technique and penetration of the tissue section by the large tagged antibody can be poor. This method was used by Hill et al.[73] to refine the cytoplasmic localization of nAChR β2-like immunoreactivity observed using the DAB reaction. Silver grains were associated with plasma membranes of dendrites and synaptic boutons, supporting a membrane localization of the β2 subunit. However, a dearth of labelling of synaptic profiles was noted, suggesting a problem of accessibility or an extrasynaptic location of nAChR containing this subunit. It was possible to gold-label synaptosome membranes with an antibody directed against an extracellular epitope of the nAChR β2 subunit[22] (see Figure 2.1D), using the pre-embedding gold technique. When applied to sections of rat striatum, nAChR β2 subunit immunogold particles in axon terminals within the dorsal striatum (Figure 2.4D) were observed. By combining the two pre-embedding techniques shown in Figures 2.4C and D, it was possible to colocalize the nAChR β2 subunit and tyrosine hydroxylase within the same tissue section (Figure 2.4E).

For double labelling with different antibodies tagged with large and small gold particles, respectively, the post-embedding technique is used (Figure 2.3). Here the tissue is first embedded in hydrophilic resin; ultrathin sections are cut and then labelled sequentially with first and second antibodies. Good access to the tissue, and visualization of gold particles without the need for silver enhancement, allows quantitative analysis to be performed, as undertaken by Fujiyama et al. on the GABA$_A$ receptor subunit in the rat striatum.[80] Disadvantages are the poor preservation of tissue morphology, and loss of antigenicity such that many antibodies fail to work in this technique. The method has been used successfully by Arroyo-Jimenez et al. [76] to examine the ultrastructural localization of the nAChR α4 subunit in the rat substantia nigra pars compacta. Labelling of postsynaptic densities was clearly demonstrated; the presynaptic localization of nAChR was not addressed in this study. Using the same technique, nAChR β2 subunit immunoreactivity in axon boutons within the rat striatum has been shown[22] consistent with pre-embedding observations (Figure 2.4C and D). Double labelling experiments in which the nAChR β2 subunit and tyrosine hydroxylase were localized using small and large gold particles, respectively, have since been performed. Results (Figure 2.4F) show colocalization of the two proteins in axons within the dorsal striatum, indicative of the presence of the β2 subunit in dopaminergic terminals.

This brief overview of immunocytochemical methods, with examples pertinent to nAChR subunits, highlights the potential of this approach for evaluating the precise localization of presynaptic nAChR with respect to their neurochemical environment. In particular, it should be possible to correlate the anatomical relationships between nAChR- positive structures and cholinergic boutons and varicosities in order to identify the source of endogenous agonist for nAChR. Immunocytochemical approaches, such as those discussed previously, complement biochemical and functional studies; the dedicated application of these techniques to the study of presynaptic nAChR may be anticipated to confirm and extend knowledge of these receptors derived from other studies.

2.6   CONCLUSIONS

While the case for the presynaptic nicotinic modulation of transmitter release is very strong, knowledge of the subtypes of nAChR involved is currently limited (see Figure 2.2). There is a good case for involvement of the β2 subunit in nAChRs that facilitates the release of dopamine from striatal nerve terminals. Evidence (reviewed earlier) comes from the pharmacological analysis of nicotinic agonist-evoked dopamine release, studies in knock-out mice, lesions coupled with radioligand autoradiography, and immunocytochemistry. The pharmacological evidence supports at least two subtypes of β2-containing nAChR: α4β2* and α3/α6β2*. There is also compelling evidence (mainly from pharmacological analysis coupled with electrophysiological methods) for an association of the α7 subunit with glutamate terminals. The particular properties of α7 nAChR may favor this relationship, enabling nicotinic stimulation to participate in a presynaptic component of LTP-like phenomena. Other subtypes of nAChR are likely to promote the release of other neurotransmitters in various brain regions. There is much to learn before the significance of nAChR diversity with respect to their functional roles in modulating synaptic transmission can be fully appreciated.

ACKNOWLEDGMENTS

Work in the authors' laboratory is supported by grants from the Biological and Biotechnological Sciences Research Council and The Wellcome Trust. The localization of nAChR subunits by electron microscopy undertaken in our laboratory has been carried out in collaboration with Professor J. Paul Bolam, University of Oxford.

REFERENCES

1. Role, L.W. and Berg, D.K., Nicotinic receptors in the development and modulation of CNS synapses, *Neuron,* 16:1077–1085, 1996.
2. Wonnacott, S., Presynaptic nicotinic ACh receptors, *Trends Neurosci.,* 20:92–98, 1997.
3. MacDermott, A.B., Role, L.W., and Siegelbaum, S.A., Presynaptic ionotropic receptors and the control of transmitter release, *Ann. Rev. Neurosci.,* 22:443–485, 1999.
4. Vizi, E.S. and Lendvai, B., Modulatory role of presynaptic nicotinic receptors in synaptic and nonsynaptic chemical communication in the central nervous system, *Brain Res. Rev.,* 30:219–235, 1999.
5. Kaiser, S.A. and Wonnacott, S., Nicotinic Receptor Modulation of Neurotransmitter Release, in *Neuronal Nicotinic Receptors: Pharmacology and Therapeutic Opportunities,* S.P. Arneric and J.D. Brioni, Eds., Wiley-Liss, Inc., 1998, 141–159.
6. Gaddum, J.H., The technique of superfusion, *Br. J. Pharmacol.,* 8:82–87, 1953.
7. Turner, T.J., Pearce, L.B., and Goldin, S.M., A superfusion system designed to measure release of radiolabeled neurotransmitters on a subsecond time scale, *Anal. Biochem.,* 178:8–16, 1989.
8. Anderson, D.J., Puttfarcken, P.S., Jacobs, I., and Faltynek, C., Assessment of nicotinic acetylcholine receptor-mediated release of [^{3}H]-norepinephrine from rat brain slices using a new 96-well format assay, *Neuropharmacology,* 39:2663–2672, 2000.
9. Puttfarcken, P.S., Jacobs, I., and Faltynek, R., Characterization of nicotinic acetylcholine receptor-mediated [^{3}H]-dopamine release from rat cortex and striatum, *Neuropharmacology,* 39:2673–2680, 2000.
10. Cahill, P.S. and Wightman, R.M. Simultaneous amperometric measurement of ascorbate and catecholamine secretion from individual bovine adrenal medullary cells. *Anal. Chem.,* 67:2599–2605, 1995.
11. Sacaan, A.I., Dunlop, J.L., and Lloyd, G.K., Pharmacological characterization of neuronal acetylcholine gated ion channel receptor-mediated hippocampal norepinephrine and striatal dopamine release from rat brain slices, *J. Pharmacol. Exp. Ther.,* 274:224–230, 1995.
12. Clarke, P.B. and Reuben, M., Release of [^{3}H]-noradrenaline from rat hippocampal synaptosomes by nicotine: mediation by different nicotinic receptor subtypes from striatal [^{3}H]-dopamine release, *Br. J. Pharmacol.,* 117:595–606, 1996.
13. Lu, Y., Grady, S., Marks, M.J., Picciotto, M., Changeux, J.P., and Collins, A.C., Pharmacological characterization of nicotinic receptor-stimulated GABA release from mouse brain synaptosomes, *J. Pharmacol. Exp. Ther.,* 287:648–657, 1998.
14. Reuben, M. and Clarke, P.B., Nicotine-evoked [^{3}H]5-hydroxytryptamine release from rat striatal synaptosomes, *Neuropharmacology,* 39:290–299, 2000.
15. Wilkie, G.I., Hutson, P., Sullivan, J.P., and Wonnacott, S., Pharmacological characterization of a nicotinic autoreceptor in rat hippocampal synaptosomes, *Neurochem. Res.,* 21:1141–1148, 1996.
16. Grady, S.R., Meinerz, N.M., Cao, J.Z., Reynolds, A.M., Picciotto, M.R., Changeux, J.P., McIntosh, J.M., Marks, M.J., and Collins, A.C., Nicotinic agonists stimulate acetylcholine release from mouse interpeduncular nucleus: a function mediated by a different nAChR than dopamine release from striatum, *J. Neurochem.,* 258–268, 2001.
17. Giorguieff, M.F., Le Floch, M.L., Glowinski, J., and Besson, M.J., Involvement of cholinergic presynaptic receptors of nicotinic and muscarinic types in the control of the spontaneous release of dopamine from striatal dopaminergic terminals in the rat, *J. Pharmacol. Exp. Ther.,* 200:535–544, 1977.

18. Cheramy, A., Godeheu, G., L'Hirondel, M., and Glowinski, J., Cooperative contributions of cholinergic and NMDA receptors in the presynaptic control of dopamine release from synaptosomes of the rat striatum, *J. Pharmacol. Exp. Ther.* 276:616–625, 1996.

19. Kaiser, S. and Wonnacott, S., alpha-bungarotoxin-sensitive nicotinic receptors indirectly modulate [³H]dopamine release in rat striatal slices via glutamate release, *Mol. Pharmacol.*, 58:312–318, 2000.

20. Nicholls, D.G., Release of glutamate, aspartate, and gamma-aminobutyric acid from isolated nerve terminals, *J. Neurochem.*, 52:331–341, 1989.

21. Fosse, V.M., Kolstad, J., and Fonnum, F.A., Bioluminescence method for the measurement of L-glutamate: applications to the study of changes in the release of L-glutamate from lateral geniculate nucleus and superior colliculus after visual cortex ablation in rats, *J. Neurochem.*, 47:340–349, 1986.

22. Wonnacott, S., Kaiser, S., Mogg, A., Soliakov, L., and Jones, I.W., Presynaptic nicotinic receptors modulating dopamine release in the rat striatum, *Eur. J. Pharmacol.*, 393:51–58, 2000.

23. Schilström, B., Svensson, H.M., Svensson, T.H., and Nomikos, G.G., Nicotine and food induced dopamine release in the nucleus accumbens of the rat: putative role of alpha7 nicotinic receptors in the ventral tegmental area, *Neuroscience,* 85:1005–1009, 1998.

24. Schilström, B., Nomikos, G.G., Nisell, M., Hertel, P., and Svensson, T.H., N-methyl-D-aspartate receptor antagonism in the ventral tegmental area diminishes the systemic nicotine-induced dopamine release in the nucleus accumbens, *Neuroscience,* 82:781–789, 1998.

25. Dwoskin, L.P., Teng, L., Buxton, S.T., Ravard, A., Deo, N., and Crooks, P.A., Minor alkaloids of tobacco release [³H]dopamine from superfused rat striatal slices, *Eur. J. Pharmacol.*, 276:195–199, 1995.

26. Marshall, D., Soliakov, L., Redfern, P., and Wonnacott, S., Tetrodotoxin-sensitivity of nicotine-evoked dopamine release from rat striatum, *Neuropharmacology,* 35:1531–1536, 1996.

27. Sershen, H., Balla, A., Lajtha, A., and Vizi, E.S., Characterization of nicotinic receptors involved in the release of noradrenaline from the hippocampus, *Neuroscience,* 77:121–130, 1997.

28. Gray, E.G. and Whittaker, V.P., The isolation of nerve endings from brain: an electron microscopic study of cell fragments derived by homogenization and centrifugation, *J. Anat.*, 96:79–88, 1962.

29. Dunkley, P.R., Heath, J.W., Harrison, S.M., Jarvie, P.E., Glenfield, P.J., and Rostas, J.A., A rapid Percoll gradient procedure for isolation of synaptosomes directly from an S1 fraction: homogeneity and morphology of subcellular fractions, *Brain. Res.,* 441:59–71, 1988.

30. Rapier, C., Lunt, G.G., and Wonnacott, S., Nicotinic modulation of [³H]dopamine release from striatal synaptosomes: pharmacological characterisation, *J. Neurochem.*, 54:937–945, 1990.

31. Kulak, J.M., Nguyen, T.A., Olivera, B.M., and McIntosh, J.M., Alpha-conotoxin MII blocks nicotine-stimulated dopamine release in rat striatal synaptosomes, *J. Neurosci.*, 17:5263–5270, 1997.

32. Kaiser, S.A., Soliakov, L., Harvey, S.C., Luetje, C.W., and Wonnacott, S., Differential inhibition by alpha-conotoxin-MII of the nicotinic stimulation of [³H]dopamine release from rat striatal synaptosomes and slices, *J. Neurochem.*, 70:1069–1076, 1998.

33. Le Novère, N., Zoli, M., and Changeux, J.P., Neuronal nicotinic receptor alpha 6 subunit mRNA is selectively concentrated in catecholaminergic nuclei of the rat brain, *Eur. J. Neurosci.*, 8:2428–2439, 1996.

34. Goldner, F.M., Dineley, K.T., and Patrick, J.W., Immunohistochemical localization of the nicotinic acetylcholine receptor subunit alpha6 to dopaminergic neurons in the substantia nigra and ventral tegmental area, *Neuroreport*, 8:2739–2742, 1997.

35. Vailati, S., Hanke, W., Bejan, A., Barabino, B., Longhi, R., Balestra, B., Moretti, M., Clementi, F., and Gotti, C., Functional alpha6-containing nicotinic receptors are present in chick retina. *Mol. Pharmacol.*, 56:11–19, 1999.

36. Whiteaker, P., McIntosh, J.M., Luo, S., Collins, A.C., and Marks, M.J., ^{125}I-alpha-conotoxin MII identifies a novel nicotinic acetylcholine receptor population in mouse brain, *Mol. Pharmacol.*, 57:913–925, 2000.

37. Kuryatov, A., Olale, F., Cooper, J., Choi, C., and Lindstrom, J., Human alpha6 AChR subtypes: subunit composition, assembly, and pharmacological responses, *Neuropharmacology*, 39:2570–2590, 2000.

38. Sharples, C.G., Kaiser, S., Soliakov, L., Marks, M.J., Collins, A.C., Washburn, M., Wright, E., Spencer, J.A., Gallagher, T., Whiteaker, P., and Wonnacott, S., UB-165: A novel nicotinic agonist with subtype selectivity implicates the alpha4beta2* subtype in the modulation of dopamine release from rat striatal synaptosomes, *J. Neurosci.*, 20:2783–2791, 2000.

39. Luo, S., Kulak, J.M., Cartier, G.E., Jacobsen, R.B., Yoshikami, D., Olivera, B.M., and McIntosh, J.M., alpha-conotoxin AuIB selectively blocks alpha3beta4 nicotinic acetylcholine receptors and nicotine-evoked norepinephrine release. *J. Neurosci.*, 18:8571–8579, 1998.

40. Schwartz, R.D., Lehmann, J., and Kellar, K.J., Presynaptic nicotinic cholinergic receptors labeled by [3H]acetylcholine on catecholamine and serotonin axons in brain, *J. Neurochem.*, 42:1495–1498, 1984.

41. Coggan, J.S., Paysan, J., Conroy, W.G., and Berg, D.K., Direct recording of nicotinic responses in presynaptic nerve terminals, *J. Neurosci.*, 17:5798–5806, 1997.

42. Radcliffe, K.A., Fisher, J.L., Gray, R., and Dani, J.A., Nicotinic modulation of glutamate and GABA synaptic transmission of hippocampal neurons, *Ann. NY Acad. Sci.*, 868:591–610, 1999.

43. Pereira, E.F., Alkondon, M., Maelicke, A., and Albuquerque, E.X. Functional diversity of nicotinic acetylcholine receptors in the mammalian central nervous system: physiological relevance, in *Neuronal Nicotinic Receptors: Pharmacology and Experimental Therapeutics*, S.P. Arneric and J.D. Brioni, Eds., Wiley-Liss, New York, 1999, 161-186.

44. Léna, C., Changeux, J.P., and Mulle, C., Evidence for "preterminal" nicotinic receptors on GABAergic axons in the rat interpeduncular nucleus, *J. Neurosci.*, 13:2680–2688, 1993.

45. Gray, R., Rajan, A.S., Radcliffe, K.A., Yakehiro, M., and Dani, J.A., Hippocampal synaptic transmission enhanced by low concentrations of nicotine, *Nature*, 383:713–716, 1996.

46. Alkondon, M., Rocha, E.S., Maelicke, A., and Albuquerque, E.X., Diversity of nicotinic acetylcholine receptors in rat brain. V. alpha- bungarotoxin-sensitive nicotinic receptors in olfactory bulb neurons and presynaptic modulation of glutamate release, *J. Pharmacol. Exp. Ther.*, 278:1460–1471, 1996.

47. Mansvelder, H.D. and McGehee, D.S., Long-term potentiation of excitatory inputs to brain reward areas by nicotine, *Neuron*, 27:349–357, 2000.

48. Alkondon, M., Pereira, E.F., Cortes, W.S., Maelicke, A., and Albuquerque, E.X., Choline is a selective agonist of alpha7 nicotinic acetylcholine receptors in the rat brain neurons, *Eur. J. Neurosci.,* 9:2734–2742, 1997.

49. Radcliffe, K.A. and Dani, J.A., Nicotinic stimulation produces multiple forms of increased glutamatergic synaptic transmission, *J. Neurosci.,* 18:7075–7083, 1998.

50. Nisell, M., Marcus, M., Nomikos, G.G., and Svensson, T.H., Differential effects of acute and chronic nicotine on dopamine output in the core and shell of the rat nucleus accumbens, *J. Neural. Transm.* 104:1–10, 1997.

51. Nisell, M., Nomikos, G.G., and Svensson, T.H., Systemic nicotine-induced dopamine release in the rat nucleus accumbens is regulated by nicotinic receptors in the ventral tegmental area, *Synapse,* 16:36–44, 1994.

52. Schilström, B., Fagerquist, M.V., Zhang, X., Hertel, P., Panagis, G., Nomikos, G.G., and Svensson, T.H., Putative role of presynaptic alpha7* nicotinic receptors in nicotine stimulated increases of extracellular levels of glutamate and aspartate in the ventral tegmental area, *Synapse,* 38:375–383, 2000.

53. Marshall, D.L., Redfern, P.H., and Wonnacott, S., Presynaptic nicotinic modulation of dopamine release in the three ascending pathways studied by *in vivo* microdialysis: comparison of naive and chronic nicotine-treated rats, *J. Neurochem.,* 68:1511–1519, 1997.

54. Picciotto, M.R., Zoli, M., Rimondini, R., Lena, C., Marubio, L.M., Pich, E.M., Fuxe, K., and Changeux, J.P., Acetylcholine receptors containing the beta2 subunit are involved in the reinforcing properties of nicotine, *Nature,* 391:173–177, 1998.

55. Sargent, P.B., The distribution of neuronal nicotinic acetylcholine receptors, in *Handbook of Experimental Pharmacology Vol. 144: Neuronal Nicotinic Receptors,* F. Clementi, D. Fornasari, and C. Gotti, Eds., Springer-Verlag, New York, 2000, 163–184.

56. Clarke, P.B., Schwartz, R.D., Paul, S.M., Pert, C.B., and Pert, A. Nicotinic binding in rat brain: autoradiographic comparison of [^{3}H]acetylcholine, [^{3}H]nicotine, and [^{125}I]-alpha-bungarotoxin, *J. Neurosci.* 5:1307–1315, 1985.

57. Pabreza, L.A., Dhawan, S., and Kellar, K.J., [^{3}H]cytisine binding to nicotinic cholinergic receptors in brain, *Mol. Pharmacol.,* 39:9–12, 1991.

58. Whiteaker, P., Davies, A.R., Marks, M.J., Blagbrough, I.S., Potter, B.V., Wolstenholme, A.J., Collins, A.C., and Wonnacott, S., An autoradiographic study of the distribution of binding sites for the novel alpha7-selective nicotinic radioligand [^{3}H]-methyllycaconitine in the mouse brain, *Eur. J. Neurosci.,* 11:2689–2696, 1999.

59. Zoli, M., Lena, C., Picciotto, M.R., and Changeux, J.P., Identification of four classes of brain nicotinic receptors using beta2 mutant mice, *J. Neurosci.,* 18:4461–4472, 1998.

60. Whiteaker, P., Jimenez, M., McIntosh, J.M., Collins, A.C., and Marks, M.J., Identification of a novel nicotinic binding site in mouse brain using [^{125}I]-epibatidine, *Br. J. Pharmacol.,* 131:729–739, 2000.

61. Cartier, G.E., Yoshikami, D., Gray, W.R., Luo, S., Olivera, B.M., and McIntosh, J.M., A new alpha-conotoxin which targets alpha3beta2 nicotinic acetylcholine receptors, *J. Biol. Chem.,* 271:7522–7528, 1996.

62. Cordero-Erausquin, M., Marubio, L.M., Klink, R., and Changeux, J.P., Nicotinic receptor function: new perspectives from knockout mice, *Trends Pharmacol. Sci.,* 21:211–217, 2000.

63. Clarke, P.B. and Pert, A., Autoradiographic evidence for nicotine receptors on nigrostriatal and mesolimbic dopaminergic neurons, *Brain. Res.,* 348:355–358, 1985.

64. Lentz, T.L. and Chester, J., Localization of acetylcholine receptors in central synapses, *J. Cell. Biol.,* 75:258–267, 1977.

65. Amenta, F., Bernardi, G., Floris, V., and Marciani, M.G., Localization of alpha-bungarotoxin binding sites within the rat corpus striatum, *Neuropharmacology,* 18:319–322, 1979.
66. Shoop, R.D., Martone, M.E., Yamada, N., Ellisman, M.H., and Berg, D.K., Neuronal acetylcholine receptors with alpha7 subunits are concentrated on somatic spines for synaptic signaling in embryonic chick ciliary ganglia, *J. Neurosci.,* 19:692–704, 1999.
67. Wada, E., Wada, K., Boulter, J., Deneris, E., Heinemann, S., Patrick, J., and Swanson, L.W., Distribution of alpha2, alpha3, alpha4, and beta2 neuronal nicotinic receptor subunit mRNAs in the central nervous system: a hybridization histochemical study in the rat, *J. Comp. Neurol.,* 284:314–335, 1989.
68. Han, Z.Y., Le Novère, N., Zoli, M., Hill, J.A., Champtiaux, N., and Changeux, J.P., Localization of nAChR subunit mRNAs in the brain of macaca mulatta, *Eur. J. Neurosci.,* 12:3664–3674, 2000.
69. Charpantier, E., Barneoud, P., Moser, P., Besnard, F., and Sgard, F. Nicotinic acetylcholine subunit mRNA expression in dopaminergic neurons of the rat substantia nigra and ventral tegmental area, *Neuroreport,* 9:3097–3101, 1998.
70. Sheffield, E.B., Quick, M.W., and Lester, R.A., Nicotinic acetylcholine receptor subunit mRNA expression and channel function in medial habenula neurons, *Neuropharmacology,* 39:2591–2603, 2000.
71. Lindstrom, J., The structures of neuronal nicotinic receptors in *Handbook of Experimental Pharmacology Vol. 144: Neuronal Nicotinic Receptors,* F. Clementi, D. Fornasari and C. Gotti, Eds., Springer-Verlag, New York, 2000, 101–147.
72. Swanson, L.W., Simmons, D.M., Whiting, P.J., and Lindstrom, J., Immunohistochemical localization of neuronal nicotinic receptors in the rodent central nervous system, *J. Neurosci.,* 7:3334–3342, 1987.
73. Hill, J.A.J., Zoli, M., Bourgeois, J.P., and Changeux, J.P., Immunocytochemical localization of a neuronal nicotinic receptor: the beta2-subunit, *J. Neurosci.,* 13:1551–1568, 1993.
74. Del Torro, D., Juiz, J.M., Peng, X., Lindstrom, J., and Criado, M., Immunocytochemical localization of the alpha7 subunit of the nicotinic acetylcholine receptor in the rat central nervous system, *J. Comp. Neurol.,* 349:325–342, 1994.
75. Sorenson, E.M., Shiroyama, T., and Kitai, S.T., Postsynaptic nicotinic receptors on dopaminergic neurons in the substantia nigra pars compacta of the rat, *Neuroscience,* 87:659–673, 1998.
76. Arroyo-Jimènez, M.M., Bourgeois, J.P., Marubio, L.M., Le Sourd, A.M., Ottersen, O.P., Rinvik, E., Fair, and Changeux, J.P., Ultrastructural localization of the alpha4-subunit of the neuronal acetylcholine nicotinic receptor in the rat substantia nigra, *J. Neurosci.,* 19:6475–6487, 1999.
77. Zarei, M.M., Radcliffe, K.A., Chen, D., Patrick, J.W., and Dani, J.A., Distributions of nicotinic acetylcholine receptor alpha7 and beta2 subunits on cultured hippocampal neurons, *Neuroscience,* 88:755–764, 1999.
78. Nayak, S.V., Ronde, P., Spier, A.D., Lummis, S.C., and Nichols, R.A., Nicotinic receptors co-localize with 5-HT3 serotonin receptors on striatal nerve terminals, *Neuropharmacology,* 39:2681–2690, 2000.
79. Nakayama, H., Shioda, S., Nakajo, S., Ueno, S., Nakashima, T., and Nakai, Y., Immunocytochemical localization of nicotinic acetylcholine receptor in the rat cerebellar cortex, *Neurosci. Res.,* 29:233–239, 1997.
80. Fujiyama, F., Fritschy, J.M., Stephenson, F.A., and Bolam, J.P., Synaptic localization of GABA$_A$ receptor subunits in the striatum of the rat, *J. Comp. Neurol.,* 416:158–172, 2000.

3 Effects of Nicotine on Dopaminergic Neurotransmission

Peter P. Rowell

CONTENTS

3.1 INTRODUCTION

For the past 3 decades, much of the research directed at studying nicotine's inter-actions with neurotransmitter systems in the brain has focused on dopamine (DA). This is primarily a result of the discovery of dopamine's involvement in motor function and drug reinforcement. Before that time, nicotine's central actions were thought to be mediated primarily by norepinephrine and epinephrine (Vogt, 1959; Burn, 1960) since it had previously been demonstrated that nicotine stimulated the release of these catecholamines from the adrenal medulla. In contrast, although it had been know for many years that DA was an endogenous catecholamine (Holtz et al., 1942), it was thought to function simply as an intermediate in the synthesis of norepinephrine and epinephrine. This reasoning began to change with the devel-opment of more specific and sensitive chemical assays for DA which replaced the bioassays used for the determination of catecholamines. The importance of DA as a neurotransmitter rather than a precursor was evident when the experiments of

0-8493-2386-X/02/$0.00+$1.50
© 2002 by CRC Press LLC

Carlsson and coworkers demonstrated that rabbit brains contained levels of DA much higher than could be accounted for by a precursor role, and that the administration of DOPA led to large increases in DA without a concomitant increase in norepinephrine or epinephrine levels (Carlsson et al., 1958; Carlsson, 1959).

3.2 NIGROSTRIATAL AND MESOLIMBIC DOPAMINERGIC SYSTEMS

Dopamine's role as a major central neurotransmitter, and its involvement in motor activity, were confirmed when it was found that one area of the brain in particular, the caudate nucleus, contained the highest concentration of DA of any tissue in the body (Bertler and Rosengren, 1959; Sano et al., 1959). This brain area was already known to mediate the extrapyramidal side effects of several drugs, and the experiments of Vander Eecken et al. (1960) and Adams et al. (1964) showed that patients with movement disorders had loss of tissue in the caudate nucleus. It was also discovered from post-mortem samples that there were much lower levels of DA in the caudate/putamen (striatum) as well as the cell bodies of the substantia nigra (SN) in Parkinson's patients than in these same brain areas from normal individuals (Ehringer and Hornykiewicz, 1960). Within two years the efficacy of using L-DOPA to alleviate the symptoms of Parkinson's disease was reported by Birkmayer and Hornykiewicz (1961) and Barbeau et al. (1962). Interestingly, within a few years it was found that cigarette smokers had a reduced risk of developing Parkinson's disease (Nefzger et al. (1968). This negative correlation between cigarette smoking and Parkinson's disease has now been confirmed in a number of studies (reviewed by Baron, 1986; Morens et al., 1995).

The elucidation of dopamine's role in mesolimbic neurons has an equally interesting history. The neurochemical basis of reinforcement made a major advance with the discovery by Olds and Milner (1954) that there were specific areas of the brain where electrodes could be placed such that animals would work to obtain electrical stimulation by means of lever pressing. In these studies, response rates of thousands per hour were obtained, and animals would forego most other activities, including eating and drinking, to obtain stimulation (Olds and Olds, 1963). Pharmacological experiments indicated that increased catecholamine concentrations were responsible for the self-stimulation behavior since amphetamine (which was known to elevate catecholamine levels) enhanced self-stimulation whereas reserpine (which was found to deplete catecholamines in the brain) decreased this activity (Stein, 1962). The specific catecholamine involved was subsequently determined to be DA since selective DA receptor antagonists blocked the self-stimulation in animals induced by amphetamine and cocaine, whereas noradrenergic antagonists did not (Risner and Jones, 1976, 1980; deWit and Wise, 1977). Comparable experiments were conducted in human subjects (Gunne et al., 1972) showing that DA was similarly involved in the rewarding effects of amphetamine and cocaine. The dopaminergic pathway was later identified by the use of neurotoxins to originate in the ventral tegmental area (VTA) with projections to the nucleus accumbens (NAc), amygdala, olfactory tubercle and the frontal cortex (Lyness et al., 1979; Roberts and Koob, 1982; Bozarth and Wise, 1986).

3.3 NICOTINE-STIMULATED DOPAMINE RELEASE FROM NIGROSTRIATAL NEURONS

The findings that the concentration of DA in the striatum is about ten times higher than any other brain area, and the recognition that this area plays a major role in motor function, suggested that nicotine might act to release DA from striatal tissue. The technique of *in vitro* perfusion of specific brain areas developed in the late 1960s allowed a determination to be made of the concentration and efficacy of drug-induced effects on neurotransmitter systems in living brain tissue. Besson and coworkers (1969) studied the effect of cholinergic stimulation on DA release in rat striatal tissue and found that the application of acetylcholine produced a marked increase in newly synthesized [^{3}H]DA release. The first direct evidence of nicotine-stimulated DA release in striatum was reported by Westfall (1974), and this finding was soon confirmed by a number of other investigators (Goodman, 1974; Giorguieff et al. 1977; Arqueros et al., 1978).

These initial studies, which employed relatively high concentrations of nicotine, indicated that nicotine-stimulated DA release occurred by a calcium-independent process. It was suggested by Arqueros that the mechanism of nicotine's action was by displacement of DA from synaptic vesicles in a manner analogous to amphetamine (Arqueros et al., 1978). This mechanism had actually been proposed several years earlier by Long and Chiou to explain nicotine-stimulated acetylcholine release (Chiou et al., 1970). Later experiments using lower concentrations of nicotine demonstrated that the effect was indeed calcium dependent (Westfall et al., 1987) and this has been confirmed in most subsequent studies. More recent experiments with selective Ca^{++} channel antagonists have shown that nicotine-evoked DA release requires the activation primarily of N-type calcium channels (Harsing et al., 1992; Prince et al., 1996; Soliakov and Wonnacott, 1996). Studies using synaptosomes along with nicotinic receptor antagonists have confirmed that the nicotine-evoked release of DA in the striatum can be produced by the drug's interaction with presynaptic nicotinic heteroreceptors located on the nerve terminals (Sakurai et al., 1982; Rapier et al., 1988; Wonnacott et al., 1989).

3.4 NICOTINE-STIMULATED DOPAMINE RELEASE FROM MESOLIMBIC NEURONS

A number of studies beginning in the late 1970s demonstrated that increased levels of DA at the terminal fields of mesolimbic neurons, particularly in the NAc, appeared to be a common feature of many reinforcing drugs (Robertson and Mogenson, 1978; Lyness et al., 1979; Singer et al., 1982; Wise and Bozarth, 1985; Smith et al., 1985; Taylor and Robbins, 1986). These studies suggested that nicotine, like the other psychostimulant drugs, might also elevate synaptic levels of DA in this brain area. The first evidence was demonstrated by Imperato and coworkers (1986) using the technique of *in vivo* microdialysis. They showed that the systemic injection of 0.6 mg/kg nicotine caused a 100% increase in DA release from the NAc, and this activity was blocked by the centrally acting nicotinic antagonist, mecamylamine. It was also shown that the NAc was several times more sensitive to the effects of nicotine than was the striatum. *In vitro* superfusion studies with isolated nucleus accumbens tissue soon demonstrated that this effect could be explained by the ability of nicotine to directly stimulate DA release by acting on presynaptic terminals (Rowell et al., 1987; Fung, 1989). This mechanism was later confirmed with *in vivo* microdialysis studies in which nicotine was directly infused into the NAc (Mifsud et al., 1989; Damsma et al., 1989).

Other brain areas which receive dopaminergic input from terminal fields of mesocorticolimbic neurons are the amygdala and cerebral cortex. As in the striatum and nucleus accumbens, nicotine has been found to directly stimulate the release of DA from both the amygdala (Rowell, 1987) and cortex (Summers and Giacobini, 1995; Whiteaker et al., 1995).

Using the technique of *in vitro* superfusion, a comparison can be made of the activity and concentration–effect relationship of nicotine-evoked release in several brain areas. Male Sprague Dawley rats are anesthetized, decapitated, and their brains rapidly removed. The striatum and frontal cortex are removed by free-hand dissection with punches of NAc and amygdala taken from coronal slices. In order to prepare synaptosomes, tissue from the respective brain areas is homogenized in ice-cold 0.32 M sucrose (10% w/v) at 450 rpm with a glass-Teflon homogenizer for 15 seconds. The homogenate is centrifuged at 1000 x g for 10 min followed by centrifugation of the resulting supernatant at $16,000 \times g$ for 20 min. The resulting pellet is resuspended in Krebs–HEPES buffer and preincubated at 37°C for 10 min whereupon 2 to 4 µCi of [^{3}H]DA ($\approx$40 Ci/mmol) is added and incubated for an additional 10 min. Fifty to 100 µl of the synaptosomal suspension is then transferred onto 25 mm GF-A glass-fiber filters in open Swinnex filter holders and perfused with Krebs buffer at a rate of 1 ml/min. Following a 30 to 60 min equilibrium period, the response to agonists can be determined. Using this procedure, the nicotine-stimulated release of [^{3}H]DA from the four primary terminal areas of dopaminergic neurons in the brain results in the concentration–response relationships shown in Figure 3.1.

Nicotine produces a release of several times baseline in striatum with a more modest effect in the nucleus accumbens, frontal cortex, and amygdala. The potency of nicotine in accumbens and striatum is somewhat greater than in the cortex. Nicotine is the least potent as well as the least efficacious in the amygdala. Wonnacott and coworkers have also compared the nicotine-stimulated elevation of DA in the striatum, nucleus accumbens, and frontal cortex using *in vivo* microdialysis in both naive and nicotine-treated rats (Marshall et al., 1997). They found that nicotine is more efficacious at releasing DA from striatum and NAc than the cortex, and that prior nicotine treatment of the animals results in an elevated response.

3.5 DEDROSOMATIC AND TERMINAL SITES OF NICOTINE'S ACTION

The experiments considered above provide evidence that nicotine can act directly at the terminals of mesocorticolimbic and nigrostriatal neurons to release nicotine, ostensibly by acting upon nicotinic receptors on the presynaptic nerve terminal.

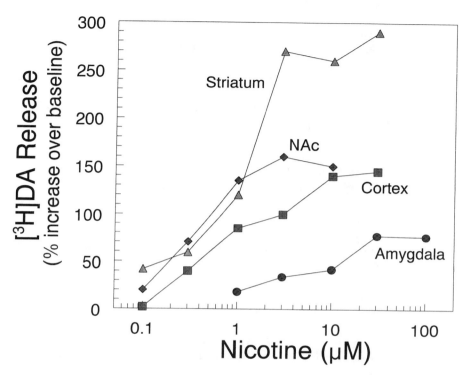

FIGURE 3.1 Concentration-response relationship of nicotine in synaptosomes from striatum, nucleus accumbens, frontal cortex, and amygdala. Tissue, prepared and perfused as described, was challenged with 30 second pulses of L-nicotine at the concentrations indicated. The resulting increase in the baseline release (extrapolated from the period preceding the pulse) is presented as the percent increase over baseline at each concentration.

Audioradiographic studies of Clarke using [³H]nicotine demonstrated that high affinity nicotinic receptors are present on dopaminergic terminals in these brain areas (Clarke et al., 1985; Clarke and Pert, 1985). These studies also showed that nicotinic receptors were on the soma of these neurons in the VTA and the SN. The experiments of Lichtensteiger and coworkers (1976, 1982) found that the local application of nicotine to the SN produced an increase in firing of dopaminergic neurons and an increase in DA metabolite levels in the striatum. This provided evidence that the elevation in DA levels in the striatum might be the result of its action on somato-dendritic receptors rather than those at the terminal. Similarly, microdialysis studies by Yoshida et al. (1993) have demonstrated that the local administration of nicotine into the VTA results in an increase in DA levels in the NAc. The finding that the application of tetrodotoxin abolishes the increase in nicotine-stimulated DA release in intact preparations of the NAc (Benwell et al., 1993) and striatum (Marshall et al., 1996) indicates that neuroconduction is required, thus providing further evidence that nicotine's interaction with receptors at the somatodendritic area of mesolimbic and nigrostriatal neurons is responsible for the effects of systemically administered nicotine.

The most convincing evidence for a dendrosomatic site of action for nicotine-stimulated DA release has come from the experiments of Nisell et al. (1994a, 1994b) in which the nicotinic antagonist, mecamylamine, was infused into either the NAc or VTA. These studies showed that the nAChRs in the VTA are primarily responsible for the stimulation of DA release in the NAc. This was further demonstrated by the experiments of Corrigall et al. (1994) showing that nicotine self-administration is blocked by the administration of nicotinic antagonists in the VTA but not in the NAc. Further microdialysis studies by Nisell and coworkers have demonstrated that the release of DA occurs primarily in the shell of the NAc compared to the core (Nisell et al., 1997). Interestingly, it has recently been shown that sensitization to nicotine's locomotor effect following chronic administration is accompanied by a decreased release of DA in the shell of the Nac, but an increase in DA release in the core (Cadoni and DiChiara, 2000).

From these studies it could be concluded that nAChR located on the dendrites and cell bodies of dopaminergic neurons in the VTA and SN are the receptors which preferentially give rise to nicotine-stimulated increase in DA levels at the nerve endings of the NAc and striatum, and that these dendrosomatic nAChRs are primarily responsible for the rewarding effects of nicotine. The nAChR in the SN appear to receive their major cholinergic input from neurons projecting from the pedunculopontine tegmental nucleus (Clarke et al., 1987; Beninato and Spencer, 1987; Blaha and Winn, 1993), whereas the cholinergic input to the VTA emanates primarily from the laterodorsal tegmental nucleus (Blaha et al., 1996) — an area important in nicotine self-administration (Lanca et al., 2000). While there are clearly presynaptic nAChRs located on the terminals of the nigrostriatal and mesocorticolimbic neurons which can mediate the release of DA in response to local nicotine administration, as discussed previously, the functional significance of these receptors with regards to the systemic- or self-administration of nicotine is unknown, nor is whether these receptors receive cholinergic input by which they might serve a modulatory role on dopaminergic neurotransmission at the synapse (Vizi and Lendvai, 1999).

3.6 DYNAMICS OF DA RELEASE

Although the studies cited above suggest that the nicotine-evoked release of DA from terminals of mesocorticolimbic and nigrostriatal neurons is primarily a direct result of the stimulation of postsynaptic receptors in the VTA and NA, respectively, the evidence indicates that nicotine's effect at the nerve terminals is modulated by the firing pattern of dopaminergic neurons which, in turn, is affected by amino acid neurotransmitter actions. In the early 1980s, electrophysiological measurements of the activity of dopaminergic neurons in the SN found that, in addition to spontaneous single depolarizations, characteristic bursting patterns of multiple spikes were also observed (Grace and Bunney, 1983, 1984). It was soon discovered that neurons in the VTA had higher burst firing activity than those in the SN (Grenhoff et al., 1986, 1988; Clark and Chiodo, 1988) and the administration of nicotine to animals dramatically increased the amount of burst firing in both of these dopaminergic pathways (Grenhoff et al., 1986).

Evidence that burst firing is controlled by excitatory amino acid neurotransmitters was demonstrated by the finding that the administration of NMDA agonists produced an increase in neuronal firing whereas the administration of NMDA antagonists dramatically reduced this activity (Overton and Clark, 1991, 1992; Chergui et al., 1993, Christoffersen and Meltzer, 1995). Further experiments have determined that the origin of the glutamatergic neurons responsible for input to the SN and VTA is primarily from the prefrontal cortex, although several other minor excitatory amino acid pathways to nigrostriatal and mesolimbic neurons have been described (Meltzer et al., 1997; Overton and Clark, 1997). In contrast to the facilitatory effect of glutamate on burst firing, it has been shown that GABAergic input can depress burst firing in nigrostriatal and mesolimbic neurons (Suaud-Chagny et al., 1992; Engberg et al., 1993; Erhardt et al., 1998; Paladini et al., 1999; Celada et al., 1999; Yin and French, 2000). Therefore, it appears that the firing of dopaminergic neurons is modulated *in vivo* by both excitatory and inhibitory amino acid neurotransmitters.

More direct measurements of glutamatergic modulation of nigrostriatal and mesolimbic neurotransmission come from *in vivo* microdialysis experiments in which the levels of DA in the striatum and NAc have been determined. It has been shown that stimulation of NMDA receptors at the terminals of the NAc and striatum results in an enhancement of DA release (Overton and Clark, 1992; Asencio et al., 1991). The local application of nicotine in the striatum also increases DA release, as expected; however, nicotine also produces a large increase in glutamate release, and nicotine-stimulated DA release is partially blocked by NMDA receptor antagonists (Toth et al., 1992; Kaiser and Wonnacott, 2000). Thus, nicotine-stimulated DA release in the striatum can occur by a direct and an indirect action. The direct action would be the result of nicotine's interaction with presynaptic nAChRs located on striatal nerve endings, whereas the indirect action would be produced by the stimulation of glutamate release which, in turn, would evoke DA release in the tissue (Vizi and Lendvai, 1999; Wonnacott et al., 2000; Kaiser and Wonnacott, 2000).

It is likely that this same dual action occurs at mesolimbic dopaminergic terminals in the NAc and in the cell bodies of the VTA as well. In the VTA it has been shown that NMDA receptor stimulation leads to an enhanced release of DA from the NAc (Suaud-Chagny et al., 1992; Gonon and Sundstrom, 1996; Westerink et al., 1996; Schilström et al., 1998a; Kretschmer, 1999). Also, the elevation of DA release produced by nicotine can be attenuated by the coadministration of NMDA antagonists into the VTA (Schilström et al., 1998a), suggesting that cholinergic and glutamatergic neurons act in concert within the VTA to stimulate accumbens DA levels. It has also been shown that the sensitization of both locomotor activity and accumbens DA release, which occurs as a result of subchronic nicotine administration, is decreased by NMDA antagonists (Balfour et al., 1996, 1998; Shoaib et al., 1997b; Svensson et al., 1998). Interestingly, it appears that this decrease in nicotine's effect may be partially due to blockade of nAChR upregulation in the presence of NMDA antagonism (Shoaib et al., 1997b), although the specificity of the MK-801 on NMDA receptors in this latter study is open to question.

The nAChR subtypes responsible for these direct and indirect actions of nicotine have been discussed in detail in preceding chapters. Nevertheless, a few comments are in order as they relate to nicotine-stimulated DA release. The nAChRs which are present on dopaminergic neurons appear to consist primarily of the $\alpha_3\beta_2$ and $\alpha_4\beta_2$ subtypes. At striatal terminals, the application of the relatively selective $\alpha_3\beta_2$ antagonists, neuronal bungarotoxin (Grady et al., 1992) or α conotoxin MII (Kulak et al., 1997; Kaiser et al., 1998) produces a substantial blockade of release, although a significant component of the release appears to be mediated by $\alpha_4\beta_2$ nAChRs as well (Kaiser et al., 1998; Sharples et al., 2000). In any case, the direct actions of nicotine appear to involve nAChRs containing the β_2 subunit (Grady et al., 1992; Wonnacott et al., 2000), although a small proportion of receptors at the terminals may contain the β_4 subunit based on the activity of cytisine (Reuben et al., 2000).

At the cell bodies of these nigrostriatal neurons, it has been shown that mRNA for the α_4 subunit (Sorenson et al., 1998; Arroyo-Jim nez et al., 1999) and α_6 subunit (Le Novère et al., 1996; Goldner et al., 1997) predominate. A prominent localization of the mRNA for the α_6 subunit along with β_3 has also been found in the VTA (Le Novère et al., 1996). These studies suggest that a mixture of nAChR subtypes exist at the cell bodies of nigrostriatal and mesocorticolimbic dopaminergic neurons resulting in complex dynamics for nicotinic stimulation.

In addition to these nAChRs on dopaminergic neurons, it appears that the predominant receptors on the terminals of neighboring neurons which indirectly mediate the release of DA are of the α_7 subtype. A discussion of the localization and pharmacology of the nAChRs in the striatum is found in Chapter 2 and has been characterized in recent studies by Wonnacott and coworkers (Wonnocott et al., 2000; Kaiser and Wonnacott, 2000). In the VTA, α_7 nAChRs also appear to mediate DA release from the NAc (Schilström et al., 1998b). Within the NAc, while the local administration of mecamylamine almost completely blocks nicotine-stimulated DA release (Marshall et al., 1997; Fu et al., 1999), the α_7 antagonists, methylcacaconite and α-bungarotoxin, are able to decrease part of the response to nicotine if the mecamylamine blockade is submaximal (Fu et al., 1999), indicating that, as in the striatum, α_7 receptors at mesolimbic nerve terminals are involved in modulating the release of DA as well. The α_7 receptors have also been implicated in the activation of glutamatergic neurons in the VTA which indirectly stimulate DA release by an action on NMDA and/or AMPA receptors (Desce et al., 1991; Pidoplichko et al., 1997).

From this information it is apparent that the effect of nicotine on dopaminergic neurotransmission is a complex process. At the dopaminergic terminal regions, nAChRs appear to be present on glutamatergic nerve endings in the NAc and CP such that DA levels are enhanced via nicotine-stimulated glutamate release. Nicotinic receptors are clearly present on the dopaminergic terminal as well so nicotine can directly modulate DA release at the synapse. Nicotine's primary activity following systemic administration appears to result from the stimulation of presynaptic nAChR on glutamatergic nerve terminals in the VTA and SN which causes an increase in burst firing of dopaminergic neurons projecting to the NAc and CP. The ability of nicotine to produce a more direct depolarization of dopaminergic neurons in the

VTA and SN via postsynaptic nAChRs on the dendrites and soma of these nerves may contribute to this effect.

3.7 FUNCTIONAL ACTIVITY OF UPREGULATED nAChRs ON DOPAMINERGIC NEURONS

As indicated above, by the late 1970s it was clear that nicotine could stimulate the release of DA by acting on nAChRs in the brain. At this time, the ligand most often used to characterize nAChR binding sites was [^{125}I]alpha-bungarotoxin (αBTX), and it had been demonstrated in dozens of studies that αBTX binding sites were present in a number of brain areas (see reviews by Morley et al., 1979; Oswald and Freeman, 1981). Although the chronic treatment of nicotine had been shown to result in tolerance to a number of the behavioral and neurochemical effects of nicotine, no studies at that time indicated that chronic nicotine treatment produced changes in [^{125}I]αBTX binding sites in the brain.

In 1983, Schwartz and Kellar (using [^{3}H]acetylcholine) and Marks and coworkers (using [^{3}H]nicotine) independently discovered that the chronic treatment of animals with nicotine produced an increase in binding sites in the rat cortex (Schwartz and Kellar, 1983) and several mouse brain areas (Marks et al., 1983). The ability of chronic nicotine administration to produce nAChR upregulation has been confirmed in dozens of studies; it has also been shown that the post-mortem brains of cigarette smokers have an increased density of nAChR (Benwell et al., 1988; Breese et al., 1997). These results were initially somewhat unexpected, since the chronic administration of nicotine, a receptor agonist, would have been expected to cause a decrease in receptor density in the brain (Creese and Sibley, 1981). An explanation for this nicotine-induced nAChR upregulation was that the chronic drug treatment produced a long-term nAChR desensitization and thus a functional antagonism of nicotinic receptors in the brain (Schwartz and Kellar, 1983; Marks et al., 1983).

Although the chronic administration of nicotine to animals has been shown to produce an increase in nAChR density in the brain, it is not apparent what effect this treatment might have on the activity of these upregulated nicotinic receptors. The first study to examine the effect of chronic nicotine treatment on DA release was that of Connelly and Littleton (1983) in which nicotine was administered for 2 weeks by inhalation to achieve blood levels of 90 to 100 ng/ml. They found no difference in the amount of [^{3}H]DA released from whole brain synaptosomes prepared from treated rats compared to controls. Similarly, Harsing et al. (1992) found no difference in nicotine-stimulated release of [^{3}H]DA in mouse striatal slices between 2-week nicotine-treated animals and controls. Damsma and coworkers (1989) assessed the endogenous levels of DA in NAc by *in vivo* microdialysis and also found no difference between animals treated for 2 weeks with nicotine compared to controls.

In contrast to these results, a number of investigators have reported that chronic nicotine treatment results in a loss of nAChR-mediated neurotransmitter release assayed *in vitro*. Westfall and Perry (1986) first found that 2-week administration of the nicotinic agonist dimethylphenylpiperazinium (DMPP) resulted in a decrease in the ability of DMPP to release DA from rat striatal slices, although the dose employed was rather high and DMPP, a quaternary amine, does not easily gain access to central sites. Later, Lapchak, and coworkers (1989) found that twice daily administration of nicotine (0.6 mg/kg) produced an upregulation of nAChRs, but a decrease in the levels of acetylcholine, as well as the methylcarbamylcholine- and K^+-stimulated release of acetylcholine from striatal (but not hippocampal or cortical) slices. Lapin et al. (1989) found that chronic nicotine treatment results in a decrease in DA turnover and release in the NAc, and Collins' group has found that chronic nicotine administration results in a decrease in nicotine-stimulated [^{3}H]DA release from mouse striatal synaptosomes (Marks et al., 1993b; Grady et al., 1997).

Results from other labs using *in vitro* techniques, however, have shown that chronic nicotine treatment results in an enhanced response to nicotine-evoked neurotransmitter release. Fung (1989) found that the administration of 1.5 mg/kg/day nicotine to rats for 2 weeks resulted in an increase in endogenous levels of DA in the NAc as well as a 55% increase in the ability of NAc slices to release newly formed [^{3}H]DA. It has also been shown that treatment of rats for 1 week with the nicotinic agonist, (+)anatoxin-a, results in both an upregulation of striatal nAChRs and a 43% increase in nicotine-stimulated [^{3}H]DA release from striatal synaptosomes (Rowell and Wonnacott, 1990). In an attempt to clarify whether increased or decreased nicotine-stimulated release occurred, Yu and Wecker (1994) investigated the release of four neurotransmitters in rat striatal slices following chronic nicotine treatment. They found that chronic nicotine treatment increased the nicotine-evoked release of DA and 5-HT while decreasing the release of acetylcholine, with no change in norepinephrine release.

Investigations of the effects of chronic nicotine treatment on nAChR desensitization *in vivo* and resulting effects on DA levels have been assessed by microdialysis. A number of studies have found that chronic nicotine treatment of animals results in an enhanced ability of a nicotine challenge to elevate DA levels in the NAc (Benwell and Balfour, 1992; Marshall et al., 1997; Miyata et al., 1996; Reid et al., 1998) as well as the striatum (Marshall et al., 1997) and prefrontal cortex (Nisell et al., 1996). In the latter study, however, a nicotine challenge in chronically treated animals did not elevate DA levels in the NAc beyond that found in control animals. It has also been reported that chronic nicotine treatment differentially affects the elevation in DA levels produced by stress, chronic nicotine generally producing an enhanced effect in the striatum but an attenuation of the effect of stress in the NAc, hippocampus, and prefrontal cortex (George et al., 1998; Takahashi et al., 1998; Pawlak et al., 2000).

3.8 NICOTINE-INDUCED DESENSITIZATION OF DA RELEASE

As discussed above, the most tenable explanation to account for the ability of chronic nicotine treatment to produce an upregulation of nAChRs in the brain is that nicotine acts as a functional antagonist by producing long-term nAChR desensitization (Marks et al., 1983; Schwartz and Kellar, 1985; Wonnacott, 1990; Ochoa and

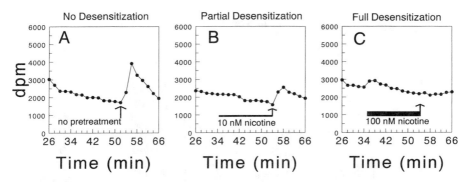

FIGURE 3.2 Nicotine-induced desensitization of [³H]DA release from rat striatal synaptosomes. Striatal synaptosomes, prepared as described in Figure 3.1 were exposed to 0, 10 or 100-nM nicotine for 20 min followed by a 2 min challenge with 5 μM nicotine. Samples of the superfusate were collected every 2 min; the resulting profile of [³H]DA release during this time is presented.

McNamee, 1990) or inactivation (Aoshima, 1984; Simasko et al., 1986; Egan and North, 1986; Lukas, 1991; Rowell and Duggan, 1998). The ability of nicotine to produce desensitization of the nAChRs responsible for nicotine-evoked release of DA can readily be demonstrated to occur at very low concentrations of the drug. The use of *in vitro* superfusion of synaptosomes offers the advantage that drug access to and removal from the tissue can occur rapidly so that drug concentration, time-course, temperature, and other variables can be carefully controlled while still assessing the effects of treatment on endogenous receptors. Studies on striatal synaptosomes have shown that nicotine rapidly (within minutes of exposure) produces almost complete nAChR desensitization at very low concentration, with an ED50 of about 10 nM (Grady et al., 1994; Rowell and Hillebrand, 1994; Marks et al., 1994; Lippiello et al., 1995). This can be seen in Figure 3.2, which shows the effect of pretreatment of rat striatal synaptosomes with nanomolar concentrations of nicotine. It can be seen that a challenge with 5 μM nicotine (arrow) produces a robust release of DA in nontreated tissue (panel A), but has virtually no effect after a 20 min exposure to 100 nM nicotine (panel C), although this concentration of nicotine produces some initial activation.

Since the studies cited indicate that nAChRs desensitize at nanomolar concentrations of nicotine, and desensitization is responsible for nAChR upregulation, it is important to consider the brain levels of nicotine achieved during administration of the drug. In human cigarette smokers, as might be expected, there is considerable individual variability in the nicotine blood levels achieved. Blood levels can range from 30 to 500 nM (5 to 90 ng/ml), depending upon the "smoking profile" of the individual (Russell, 1990); however, for most smokers, the blood levels of nicotine vary throughout the day from about 125 to 275 nM (Haines et al., 1974; Russell and Feyerabend, 1978; Bridges et al., 1990; Russell, 1990; Benowitz et al., 1997). The plasma half-life of nicotine in humans is 1 to 2 hours (Rosenberg et al., 1980; Kyerematen et al., 1982, 1990; Shiffman et al., 1992). Fortuitously, the plasma

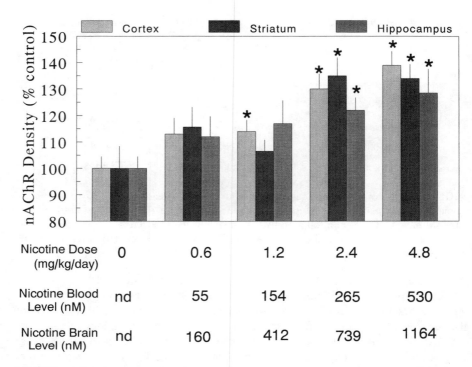

FIGURE 3.3 Effect of chronic treatment on brain nicotine and nAChR levels. Male Sprague Dawley rats were administered nicotine, 0.6–4.8 mg/kg/day at a rate of 20 μl/hr, subcutaneously for 10 days via indwelling catheters connected through a Harvard swivel to an infusion pump. Plasma and whole brain (minus cerebellum and the areas dissected for nAChR analysis) nicotine levels were determined by GC with NP detection and nAChR density was determined by specific saturated binding with [³H]cytisine.

half-life of nicotine in rats is almost the same as in humans: about one hour (Adir et al., 1976; Miller et al., 1977; Kyerematen et al., 1988; Benowitz et al., 1990). Therefore, nicotine blood levels comparable to those found in human smokers can be achieved by frequent administration or, more commonly, the constant infusion of 1.5 to 4 mg/kg/day nicotine via osmotic minipumps.

The relationship between nicotine brain levels and nAChR upregulation in animals is readily assessed by the chronic administration of nicotine and subsequent measurement of brain nAChR density and brain nicotine levels. Animals can be administered nicotine at any dose or frequency desired by means of an indwelling catheter connected to a nicotine-filled syringe controlled by a programmable pump. With this procedure nicotine has been administered at intervals from hourly to twice per day as well as constantly infused. Figure 3.3 shows brain and blood levels of nicotine achieved by the administration of 0.6 to 4.8 mg/kg/day via constant infusion along with the resulting changes in nAChR density in three brain areas.

Note that when rats are chronically treated with various doses of nicotine, a dose-dependent increase in nAChR takes place in the three brain areas examined. An assay of the blood and brain nicotine levels indicates that concentrations of nicotine much higher than the ED_{50} for nicotine-induced desensitization of DA release are achieved. At the highest doses of nicotine administered, brain levels of nicotine which will significantly stimulate DA release from NAc, striatal, and cortical tissue (≥ 500 nM) are achieved.

When one compares the concentrations of nicotine sufficient to achieve full desensitization of central nAChR *in vitro* (50 to 100 nM) with the blood levels of nicotine (125 to 275 nM) found in studies on experimental animals as well as human smokers that give rise to nAChR upregulation, one might assume that central nAChRs should be fully desensitized at all times *in vivo*. This is even more likely considering findings that the levels of nicotine in the brain following nicotine administration are 1.5 to 3 times higher than in the blood (Benowitz, 1990; Plowchalk et al., 1992; Henningfield et al., 1993; Sastry et al., 1995; Rowell and Li, 1997). This begs the question of why animals (and human smokers) self-administer nicotine if their receptors are in a continuous state of desensitization.

Behavioral or microdialysis experiments in living animals should be able to shed light on the dopaminergic responses to a nicotine exposure in chronically treated animals. Unfortunately, most studies of chronic nicotine treatment that study such things as nicotine self-administration, locomotion, place preference, or DA levels via *in vivo* microdialysis have administered nicotine to rats at infrequent intervals (once or twice per day) so there would be no continuous exposure to the drug between sessions. With a half-life of only about 1 hour, the concentration of nicotine in the brain between experiments could fall to very low levels so any nAChR desensitization might have recovered before the next session. For example, it can be calculated that the nicotine blood level from a 1 mg/kg injection would fall below that required to maintain significant nAChR desensitization (≈ 5 nM) after about 8 hours. Thus, it is not surprising that daily injections of nicotine could lead to an increase in nicotine-evoked DA release from NAc (Benwell and Balfour, 1992; Marshall et al., 1997; Balfour et al., 1998).

This typical nicotine delivery regimen in animal studies is clearly different from that with most cigarette smokers where frequent smoking, with consequent high nicotine levels, should result in continuous nAChR desensitization. Only before the first cigarette of the day might nicotine levels fall below that producing desensitization. Even then, nicotine blood levels do not typically fall below 30 nM (≈ 5 ng/ml) due to the somewhat longer half-life of nicotine in humans, as well as more recent nicotine exposure just before retiring (Benowitz et al., 1990). Therefore, unlike most studies in animals, the brain nAChRs in humans should still be desensitized before each exposure to nicotine.

A few studies have investigated nicotine's behavioral and neurochemical effects *in vivo* during delivery regimens that would presumably result in nicotine levels sufficient to maintain receptor desensitization. Hakan and Ksir (1991) studied locomotor activity of rats injected with nicotine every 20 min. They observed a progressive decline in activity with subsequent injections of nicotine up to a maximal effect at 60 min. Since locomotor activity, like positive reinforcement, is mediated in large part through DA release from the NAc (Clarke et al., 1988), this would be an indirect indication of nicotine-stimulated DA release in the brain. Sharp and coworkers have measured nicotine-stimulated norepinephrine release via microdialysis in rats injected with nicotine at various time intervals (Sharp and Matta, 1993; Fu et al., 1998). They found that a nicotine injection rapidly desensitized the response to subsequent injections. Benwell et al. (1995) found that constant infusion of 4 mg/kg/day nicotine for 14 days abolished the sensitizing effect of systemic nicotine injections to increase DA release in the VTA, indicating that the nAChRs are desensitized *in vivo* by this treatment.

The animal studies cited indicate that frequent or constant administration of nicotine does indeed produce an *in vivo* desensitization of nicotine-evoked responses in the brain. Why, then, would animals continue to self-administer nicotine for long periods of time and cigarette smokers smoke throughout the day if nAChRs were in a state of desensitization? One suggestion is that nicotine is self-administered to keep a population of receptors in a continual state of desensitization (Balfour, 1994; Pidoplichko et al., 1997), implying that a return of these receptors to the active state leads to dysphoria and/or withdrawal. While this is a possibility, it seems unlikely since functional MRI studies of nicotine administration in smokers indicate that it, like many other drugs of dependence with positive reinforcing properties, increases rather than decreases neuronal activity in the NAc (Stein et al., 1998). Unless nicotine is modulating the activity of an inhibitory neurotransmitter or pathway, it appears that nAChR activation, rather than desensitization, results from nicotine administration.

A more likely possibility is that, even during desensitization, nAChRs may be able to produce an increased release of DA above the basal (non-nicotine) condition. Collins and coworkers have shown that continued exposure of brain tissue to nicotine results in a prolonged elevation of DA release (about 20% of the initial stimulation response) as long as the drug is present (Grady et al., 1997). They determined that this persistent release is mediated by the same nAChRs responsible for initial stimulation. In a similar fashion, experiments in our lab have shown that, although very low concentrations of nicotine do not produce an initial transient neurotransmitter spike (see Figure 3.2), prolonged exposure does lead to a gradual elevation in release above baseline (Rowell, 1995). These results suggest that, although the blood levels of nicotine in cigarette smokers and nicotine-treated animals may be above that sufficient to achieve nAChR desensitization, the desensitized receptors are, nonetheless, able to produce an elevation in DA release above that produced in the absence of the drug. Of course, these experiments were performed with isolated nerve endings of the striatum. It is quite probable that nAChRs in the VTA provide an even greater response because of increased glutamate release resulting in enhanced burst firing of the neuron with sustained elevated DA levels in the NAc.

Experiments directly comparing the effects of nicotine on nAChRs of the VTA with those on the terminals of the NAc have, in fact, suggested that there are differences in the activation and/or desensitization properties of nAChRs in these two tissues. Differences in the extent of desensitization between terminal and dendrosomatic receptors are indicated by the experiments of Nisell et al. (1994a) in which the continued application of nicotine in the NAc increased DA levels for only

20 min while the response from nicotine application to the VTA was maintained throughout an 80 min application. While electrophysiological studies of Dani and coworkers have shown that nAChRs in the VTA are desensitized by nanomolar concentrations of nicotine (Pidoplichko et al., 1997; Fisher et al., 1998), the desensitization characteristics are variable, and receptors, even within the same brain area, can behave quite differently based on the influence of neighboring neurons (Dani et al., 2000). In fact, Yin and French (2000) have demonstrated that, while nicotine stimulates both dopaminergic and nondopaminergic neurons in the VTA, the nAChRs on dopaminergic neurons are comparatively resistant to desensitization.

Clarke and coworkers have recently conducted *in vitro* superfusion studies in which they have characterized and compared the efficacy of nicotinic agonists to stimulated the release of [^{3}H]DA from striatal terminals with their activity on dendrosomes prepared from the SN (Reuben et al., 2000). These investigators took advantage of the finding that the dopamine transporter (DAT) is located in relatively high density on the dendrites and dendritic spines of neurons of the SN (Nirenberg et al., 1996) and that dendrosomes prepared from the SN incorporate and release [^{3}H]DA in response to K^+ depolarization (Hefti and Lichtensteiger, 1978; Silbergeld and Walters, 1979; Marchi et al., 1991). It was found that, at the synaptosomes of the terminals, nicotine is somewhat less efficacious than epibatidine in evoking [^{3}H]DA release, whereas at the dendrosomes of the SN, nicotine has equal or greater activity than epibatidine (Reuben et al., 2000). Moreover, whereas the epibatidine evoked [^{3}H]DA release from striatal synaptosomes was almost completely blocked by mecamylamine, at large portion of the release from dendrosomes was mecamylamine insensitive. These experiments indicate differences in the nAChR activation properties and/or processes for nicotine-evoked release of DA between nerve terminals compared to the dendrites of nigrostriatal dopaminergic neurons.

Similar *in vitro* superfusion experiments are underway comparing nicotine's effects at the terminals and dendrites of mesolimbic neurons. In these experiments, the nicotine-evoked efflux of $^{86}Rb^+$ is compared in tissue slices of the NAc and VTA. The brain is cooled and cross-sectional cuts are made just anterior to the NAc (at 3.3 mm rostral to bregma), just posterior to the VTA (at 6.6 mm caudal to bregma) and then midway between these two sections to yield two approximately 5-mm-thick coronal brain sections. These are placed on a cold block with the area of interest facing upwards and 800 mm punches are removed from each side at the appropriate regions. These cylinders are placed on a McIlwine tissue slicer and 4 to 5 400 μm slices of the NAc or VTA are taken and placed in artificial CSF buffer. The entire dissection and slicing procedure is performed in a cold room at approximately 5°C. The slices are warmed to 32°C and "loaded" with $^{86}Rb^+$ for 30 min (Marks et al., 1993a) at which time individual slices are placed on 22-mm-diameter GF/C filters in an open chamber two-pump superfusion system (Grady et al., 1992) and perfused with artificial CSF buffer at 1 ml/min. The results of a 1 min challenge with nicotine in these two brain areas is shown in Figure 3.4.

These results suggest that nicotine is slightly more potent as an agonist in the VTA compared to the NAc. The efficacy of nicotine in the VTA plateaus at about a 35% increase in basal release in the VTA, while the activity in the NAc continues to increase with increasing concentrations of nicotine. The slopes of the concentration-response

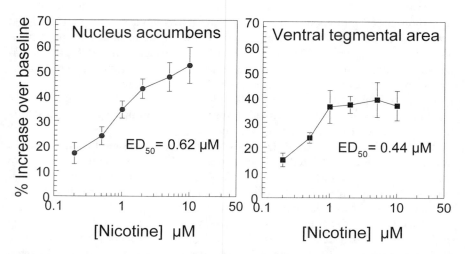

FIGURE 3.4 Nicotine-stimulated $^{86}Rb^+$ efflux from NAc and VTA slices. Tissue slices, prepared as described earlier, were superfused for 30 min in oxygentated aCSF, at which point they were exposed to a 1-min pulse of nicotine at the indicated concentrations. Results are means ± S.E.M. (n = 10–16).

curves between 0.2 and 1 μM nicotine are similar in both tissues. Since these experiments are conducted in relatively intact tissue slices, it is possible that part of the $^{86}Rb^+$ efflux measured is coming from the terminals of neighboring nondopaminergic neurons which also possess nAChRs, as evidenced by the finding that part of the nicotine-stimulated $^{86}Rb^+$ efflux is calcium dependent.

3.9 EFFECTS OF NICOTINE ON OTHER PROTEINS OF DOPAMINERGIC NEURONS

The experiments considered thus far have centered on the ability of nicotine to interact with nAChRs on the membranes of dopaminergic neurons and directly or indirectly stimulate the release of DA. In addition to this activity, there is some evidence to indicate that nicotine also interacts with other proteins of dopaminergic neurons to produce an enhancement of DA levels in the synapse. In the case of DA synthesis, nicotine does not appear to have any direct effect on the synthetic enzymes. In particular, the activity of tyrosine hydroxylase, the rate-limiting enzyme in the synthesis of DA, is not directly affected by nicotine treatment *in vivo* or *in vitro* (Naquira et al., 1978; Carr et al. 1989), although nicotine can increase tyrosine hydroxylase mRNA indirectly (Gueorguiev et al., 1999). The increased turnover of DA that can be measured as a result of nicotine treatment is apparently an indirect response which accompanies the stimulation of neurotransmitter release. No experimental evidence exists that nicotine directly affects the vesicular uptake pump or storage proteins. Interestingly, however, the nicotinic alkaloid, lobeline, has been shown to inhibit both the uptake of DA into synaptic vesicles and the DA transporter

(DAT) (Teng et al., 1997, 1998). Nicotine also does not appear to act directly on dopaminergic receptors, at least those of the D2 subtypes, as determined with *in vitro* experiments (Fung and Lau, 1996; Carr et al., 1989). Although chronic nicotine administration does not appear to produce a widespread alteration in dopaminergic receptor affinity or density (Kirch et al., 1992; Fung and Lau, 1992; Levin et al., 1997; Court et al., 1998), it may selectively alter DA autoreceptor function (Reilly et al., 1987; Janson et al., 1992; Harsing et al., 1992; Takaki, 1995).

Although nicotine does not directly affect the DAT, Izenwasser et al. (1991) found that very low concentrations of nicotine inhibit DA uptake in an intact striatal tissue preparation. The maximal inhibition was about 50%, and the IC_{50} was found to be 0.005 nM. The ability of nicotine to act at low picomolar concentrations would make this the most potent of nicotine's activities reported so far. Nicotine's inhibition of DA uptake was stereospecific, with the (-) isomer about 20 times more potent than the (+); other nicotinic agonists (carbachol and DMPP) were also effective, although at much higher concentrations. Because nicotine does not directly inhibit the DAT in synaptosomes (Carr et al., 1989; Izenwasser et al., 1991) it was suggested that nicotine acts indirectly to release other neurotransmitters or neuromodulators which in turn inhibit the DAT. Experiments have been unable to show an inhibition of DA uptake in intact striatal mince or slices (Rowell and Hill, 1993). Interestingly, it has been found that non-nicotine components of cigarette smoke or nicotine metabolites, which inhibit the activity of the DAT (Carr et al., 1991, 1992; Dwoskin et al., 1992) might exist and could act synergistically with nicotine, along with an as yet-unidentified monoamine oxidase (MAO) inhibitor present in cigarette smoke, to enhance the activity of nicotine *in vivo*.

With respect to DA metabolism, while nicotine does not inhibit MAO directly (Oreland et al., 1981; Carr and Basham, 1991), a component of tobacco appears to be a potent MAO inhibitor. Oreland et al. (1981) and Yong and Perry (1986) found that the platelet activity of MAO was lower in smokers than nonsmokers, and Yu and Bolton (1987) determined that aqueous solutions of cigarette smoke could irreversibly inhibit MAO *in vitro*. Studies in our lab were the first to find that some component of cigarette smoke produces a decrease in the activity of MAO in the brain (Carr and Rowell, 1990). It was found that, in animals which had been chronically exposed to cigarette smoker for two to three weeks, a significant decrease in the activity of cortical MAO-B took place. In humans, the studies of Fowler and coworkers (Fowler et al., 1996a, 1996b) found that the activities of both MAO-A and MAO-B are decreased in the brains of cigarette smokers. The component of tobacco responsible for this effect has not been definitively identified; it is possibly a quinoline or quinone derivative (Mendez-Alvarez et al., 1997; Khalil et al., 2000). A decreased metabolism of DA brought about by MAO inhibition from tobacco component(s) may possibly act synergistically with nicotine-stimulated DA release to potentiate the reinforcement and dependence of cigarette smoking (Fowler et al., 1998).

Finally, in addition to nicotine, alkaloids present in tobacco, pyrolysis products from cigarette smoke, and/or metabolic products of tobacco alkaloids are able to stimulate the release of DA from brain tissue. These include cotinine, nornicotine, anabaseine, and anabasine (Dwoskin et al., 1993, 1995, 1999; Teng et al., 1997).

Although somewhat less potent than nicotine, it is possible that these compounds could also potentiate the effects of nicotine to increase synaptic DA levels (Crooks and Dwoskin, 1997).

3.10 SUMMARY

In conclusion, nicotine's effect on dopaminergic neurotransmission is obviously a very complex process. The presence of various nAChR subtypes or combinations of subtypes at different locations on dopaminergic neurons, or on neighboring neurons, provides the brain with a wide range of possible processes by which to modulate DA release. Furthermore, the concentration and duration of exposure to nicotine influences what combination of nAChR activation, desensitization, inactivation, and/or upregulation will result. Nicotine, along with other tobacco smoke components, may also interact with DA storage, reuptake, and/or metabolism to increase the synaptic levels of DA in selected brain areas. In addition, although not considered here, nicotine can modulate the synaptic levels of many other biogenic amine and polypeptide neurotransmitters or neuromodulators which, along with DA, can affect a wide variety of responses and behaviors. It is no wonder then that, after more than 100 years of active investigation, we are only just beginning to unravel the complex neurochemical and behavioral consequences resulting from the administration of this most interesting pharmacological agent.

REFERENCES

Adir J., Miller R.P., and Rotenberg K.S. (1976), Disposition of nicotine in the rat after intravenous administration, *Res. Commun. Chem. Pathol. Pharmacol.*, 13:173–183.

Aoshima, H. (1984) A second, slower inactivation process in acetylcholine receptor-rich membrane vesicles prepared from *Electrophorus electricus, Arch. Biochem. Biophys.*, 235:312–318.

Arqueros L., Naquira D., and Zunino E. (1978), Nicotine-induced release of catecholamines from rat hippocampus and striatum, *Biochem. Pharmacol.*, 27:2667–2674.

Arroyo-Jim nez, M.M., Bourgeois, J.P., Marubio, L.M., Le Sourd, A.M., Ottersen, O.P., Rinvik, E., Fairen, A., and Changeux, J.P. (1999), Ultrastructural localization of the alpha4-subunit of the neuronal acetylcholine nicotinic receptor in the rat substantia nigra, *J. Neurosci.*, 19:6475–6487.

Asencio, H., Bustos, G., Gysling, K., and Labarca, R. (1991), N-Methyl-D-aspartate receptors and release of newly-synthesized [^{3}H]dopamine in nucleus accumbens slices and its relationship with neocortical afferences, *Prog. Neuropsychopharmacol. Biol. Psych.*, 15:663–676.

Balfour, D.J. (1994). Neural mechanisms underlying nicotine dependence, *Addiction,* 89:1419–1423

Balfour, D.J., Birrell, C.E., Moran, R.J., and Benwell, M.E. (1996), Effects of acute D-CPPene on mesoaccumbens dopamine responses to nicotine in the rat, *Eur. J. Pharmacol.*, 316:153–156.

Balfour, D.J., Benwell, M.E., Birrell, C.E., Kelly, R.J., and Al-Aloul M. (1998), Sensitization of the mesoaccumbens dopamine response to nicotine, *Pharmacol. Biochem. Behav.*, 59:1021–1030.

Barbeau, A., Sourkes, T.L., and Murphy, G.F. (1962), Les catecolamines dans la maladie de Parkinson, in *Monoamines et Systeme Nerveux Centrale*, Ajuriaguerra, J., Ed. Georg, Geneva, 247–262.

Baron, J.A. (1986), Cigarette smoking and Parkinson's disease, *Neurology*, 36:1490–1496.

Beninato, M. and Spencer, R.F. (1987), A cholinergic projection to the rat substantia nigra from the pedunculopontine tegmental nucleus, *Brain Res.*, 412:169–174.

Benowitz, N.L., Zevin, S., and Jacob, P., III. (1997), Sources of variability in nicotine and cotinine levels with use of nicotine nasal spray, transdermal nicotine, and cigarette smoking, *Br. J. Clin. Pharmacol.*, 43:259–267.

Benowitz, N.L. (1990), Pharmacokinetic considerations in understanding nicotine dependence, in *The Biology of Nicotine Dependence*, Bock, G. and Marsh, J., Eds., John Wiley & Sons, New York, *Ciba Found. Symp.*, 152, 186–200.

Benowitz, N.L., Porchet, H., and Jacob, P. (1990), Pharmacokinetics, metabolism and pharmacodynamics of nicotine, in *Nicotine Psychopharmacology: Molecular, Cellular and Behavioral Aspects*, Wonnacott, S., Russell, M.A.H., and Stolerman, I.P., Eds., Oxford University Press, New York, 112–157.

Benwell, M.E. and Balfour, D.J. (1992), The effects of acute and repeated nicotine treatment on nucleus accumbens dopamine and locomotor activity, *Br. J. Pharmacol.*, 105:849–856, 1992.

Benwell, M.E., Balfour, D.J.K., and Anderson, J.M. (1988), Evidence that tobacco smoking increases the density of (-)[^{3}H]nicotine binding sites in human brain, *J. Neurochem.*, 50:1243–1247.

Benwell, M.E., Balfour, D.J., and Lucchi, H.M. (1993), Influence of tetrodotoxin and calcium on changes in extracellular dopamine levels evoked by systemic nicotine, *Psychopharmacology*, 112:467–474.

Benwell, M.E., Balfour, D.J., and Birrell, C.E. (1995), Desensitization of the nicotine-induced mesolimbic dopamine responses during constant infusion with nicotine, *Br. J. Pharmacol.*, 114:454–460.

Bertler, A. and Rosengren, E. (1959), Occurrence and distribution of dopamine in brain and other tissue, *Experiment*, 15:10.

Birkmayer, W. and Hornykiewicz, O. (1961), Der L-3,4-dioxyphenylalanin (DOPA), Effekt bei der Parkinson-akinese, *Wien. Klin. Wschr.*, 73:787–788.

Blaha, C.D. and Winn, P. (1993), Modulation of dopamine efflux in the striatum following cholinergic stimulation of the substantia nigra in intact and pedunculopontine tegmental nucleus-lesioned rats, *J. Neurosci.*, 13:1035–1044.

Blaha, C.D., Allen, L.F., Das, S., Inglis, W.L., Latimer, M.P., Vincent, S.R., and Winn, P. (1996), Modulation of dopamine efflux in the nucleus accumbens after cholinergic stimulation of the ventral tegmental area in intact, pedunculopontine tegmental nucleus-lesioned, and laterodorsal tegmental nucleus-lesioned rats, *J. Neurosci.*, 16:714–722.

Bozarth, M.A. and Wise, R.A. (1986), Involvement of the ventral tegmental dopamine system in opioid and psychomotor stimulant reinforcement, in *Problems of Drug Dependence*, Harris, L.S., Ed., U.S.G.P.O., Washington, DC, 190–196.

Breese, C.R., Marks, M.J., Logel, J., Adams, C.E., Sullivan, B., Collins, A.C., and Leonard, S. (1997), Effect of smoking history on [^{3}H]nicotine binding in human post-mortem brain, *J. Pharmacol. Exp. Ther.*, 282:7–13.

Bridges, R.B., Combs, J.G., Humble, J.W., Turbek, J.A., Rehm, S.R., and Haley, N.J. (1990), Population characteristics and cigarette yield as determinants of smoke exposure, *Pharmacol. Biochem. Behav.*, 37:17–28.

Burn, J.H. (1960), The action of nicotine on the peripheral circulation, *Ann. N.Y. Acad. Sci.*, 90:81–84.

Cadoni, C. and Di Chiara, G. (2000), Differential changes in accumbens shell and core dopamine in behavioral sensitization to nicotine, *Eur. J. Pharmacol.*, 387:R23–25.

Carlsson, A., Lindqvist, M., Magnusson, T., and Waldeck, B. (1958), On the presence of 3-hydroxytyramine in brain, *Science,* 127:471.

Carlsson, A. (1959), The occurrence, distribution and physiological role of catecholamines in the nervous system, *Pharmacol. Rev.*, 11:490–493.

Carr, L.A. and Basham, J.K. (1991), Effects of tobacco smoke constituents on MPTP-induced toxicity and monoamine oxidase activity in the mouse brain, *Life Sci.*, 48:1173–1177.

Carr, L.A. and Rowell., P.P. (1990), Attenuation of 1-methyl-4-phenyl-1,2,3,6-tetrahydropy-ridine-induced neurotoxicity by tobacco smoke, *Neuropharmacology*, 29:311–314.

Carr, L.A., Rowell, P.P., and Pierce, W.M. Jr. (1989), Effects of subchronic nicotine administration on central dopaminergic mechanisms in the rat, *Neurochem. Res., 14*:511–515.

Carr, L.A., Basham, J.K., and Rowell, P.P. (1991), Inhibition of dopamine uptake in striatal synaptosomes by tobacco smoke fractions, *Int. Symp. Dopamine*, 1:99–101.

Carr, L.A., Basham, K., York, B.K. and Rowell, P.P. (1992), Inhibition of uptake of 1-methyl-4-phenylpyridinium ion (MPP+) and dopamine in striatal synaptosomes by tobacco smoke components, *Eur. J. Pharmacol.*, 215:285–287.

Celada, P., Paladini, C.A., and Tepper, J.M. (1999), GABAergic control of rat substantia nigra dopaminergic neurons: role of globus pallidus and substantia nigra pars reticulata, *Neuroscience,* 89:813–825.

Chergui, K., Charlety, P.J., Akaoka, H., Saunier, C.F., Brunet, J.L., Buda, M., and Svensson, T.H., and Chovet G. (1993), Tonic activation of NMDA receptors causes spontaneous burst discharge of rat midbrain dopamine neurons *in vivo*, *Eur. J. Neurosci.*, 5:137–144.

Chiou, C.Y., Long, J.P., Potrepka, R., and Spratt, J.L. (1970), The ability of various nicotinic agents to release acetylcholine from synaptic vesicles, *Arch. Int. Pharmacodyn. Ther.*, 187:88–96.

Christoffersen, C.L. and Meltzer, L.T. (1995), Evidence for N-methyl-D-aspartate and AMPA subtypes of the glutamate receptor on substantia nigra dopamine neurons: possible preferential role for N-methyl-D-aspartate receptors, *Neuroscience,* 67:373–381.

Clark, D. and Chiodo, L.A. (1988), Electrophysiological and pharmacological characterization of identified nigrostriatal and mesoaccumbens dopamine neurons in the rat, *Synapse,* 2:474–485.

Clarke, P.B., Hommer, D.W., Pert, A., and Skirboll, L.R. (1987), Innervation of substantia nigra neurons by cholinergic afferents from pedunculopontine nucleus in the rat: neuroanatomical and electrophysiological evidence, *Neuroscience,* 23:1011–1019.

Clarke, P.B., Fu, D.S., Jakubovic, A., and Fibiger, H.C. (1988), Evidence that mesolimbic dopaminergic activation underlies the locomotor stimulant action of nicotine in rats, *J. Pharmacol. Exp. Ther.*, 246:701–708.

Clarke, P.B. and Pert, A. (1985), Autoradiographic evidence for nicotine receptors on nigrostriatal and mesolimbic dopaminergic neurons, *Brain Res.*, 348:355–358.

Clarke, P.B., Schwartz, R.D., and Paul, S.M., Pert, C.B., and Pert, A. (1985), Nicotinic binding in rat brain: autoradiographic comparison of [^{3}H]acetylcholine, [^{3}H]nicotine, and [^{125}I]-α-bungarotoxin, *J. Neurosci.*, 5:1307-1315.

Connelly, M.S. and Littleton, J.M. (1983), Lack of stereoselectivity in ability of nicotine to release dopamine from rat synaptosomal preparations, *J Neurochem.*, 41:1297-1302.

Corrigall, W.A., Coen, K.M., and Adamson, K.L. (1994), Self-administered nicotine activates the mesolimbic dopamine system through the ventral tegmental area, *Brain Res.*, 653:278–284.

Court, J.A., Lloyd, S., Thomas, N., Piggott, M.A., Marshall, E.F., Morris, C.M., Lamb, H., Perry, R.H., Johnson, M., and Perry, E.K. (1998), Dopamine and nicotinic receptor binding and the levels of dopamine and homovanillic acid in human brain related to tobacco use, *Neuroscience,* 87:63–78.

Creese, I. and Sibley, D.R. (1980), Receptor adaptation to centrally acting drugs, *Ann. Rev. Pharmacol. Toxicol.,* 21:357–391.

Crooks, P.A. and Dwoskin, L.P. (1997), Contribution of CNS nicotine metabolites to the neuropharmacological effects of nicotine and tobacco smoking, *Biochem. Pharmacol.,* 54:743–753.

Damsma, G., Day, J., and Fibiger, H.C. (1989), Lack of tolerance to nicotine-induced dopamine release in the nucleus accumbens, *Eur. J. Pharmacol.,* 168:363–368.

Dani, J.A., Radcliffe, K.A., and Pidoplichko, V.I. (2000), Variations in desensitization of nicotinic acetylcholine receptors from hippocampus and midbrain dopamine areas, *Eur. J. Pharmacol.,* 393:31–38.

Desce, J.M., Godeheu, G., Galli, T., Artaud, F., Cheramy, A., and Glowinski, J. (1991), Presynaptic facilitation of dopamine release through D,L-alpha-amino-3-hydroxy-5-methyl-4-isoxazole propionate receptors on synaptosomes from the rat striatum, *J. Pharmacol. Exp. Ther.,* 259:692–698.

deWit, H. and Wise, R.A. (1977), Blockade of cocaine reinforcement in rats with the dopamine receptor blocker pimozide but not with the noradrenergic blockers phentolamine or phenoxybenzamine, *Can. J. Psychol.,* 31:195–203.

Dwoskin, L.P., Leibee, L.L., Jewell, A.L., Fang, Z.X., and Crooks, P.A. (1992), Inhibition of [3H]dopamine uptake into rat striatal slices by quaternary N-methylated nicotine metabolites, *Life Sci.,* 50:233–237.

Dwoskin, L.P., Buxton, S.T., Jewell, A.L., and Crooks, P.A. (1993), S(-)nornicotine increases dopamine release in a calcium-dependent manner from superfused rat striatal slices, *J. Neurochem.,* 60:2167–2174.

Dwoskin, L.P., Teng, L.H., Buxton, S.T., Ravard, A., Niranjan, D., and Crooks, P.A. (1995), Minor alkaloids of tobacco release [3H]dopamine from superfused rat striatal slices, *Eur. J. Pharmacol.,* 276:195–199.

Dwoskin, L.P., Teng, L., Buxton, S.T., and Crooks, P.A. (1999), (S)-(-)Cotinine, the major brain metabolite of nicotine, stimulates nicotinic receptors to evoke [3H]dopamine release from rat striatal slices in a calcium-dependent manner, *J. Pharmacol. Exp. Ther.,* 288:905–911.

Egan, T.M. and North, R.A. (1986), Actions of acetylcholine and nicotine on rat locus coeruleus neurons *in vitro, Neuroscience,* 19:565–571.

Ehringer, H. and Hornykiewicz, O. (1960), Verteilung von noradrenalin und dopamin (3-hydroxytyramin) im gehirn des menschen und ihr verhalten bei erkrankungen des extrapyramidalen systems, *Klin Wchnschr,* 38: 1236–1239.

Engberg, G., Kling-Petersen, T., and Nissbrandt, H. (1993), GABAB-receptor activation alters the firing pattern of dopamine neurons in the rat substantia nigra, *Synapse,* 15:229–238.

Erhardt, S., Andersson, B., Nissbrandt, H., and Engberg, G. (1998), Inhibition of firing rate and changes in the firing pattern of nigral dopamine neurons by gamma-hydroxybutyric acid (GHBA) are specifically induced by activation of GABA(B) receptors, *Naunyn Schmiedeberg's Arch Pharmacol.,* 357:611–619.

Fisher, J.L., Pidoplichko, V.I., and Dani, J.A. (1998), Nicotine modifies the activity of ventral tegmental area dopaminergic neurons and hippocampal GABAergic neurons, *J. Physiol. Paris,* 92:209–213.

Fowler, J.S., Volkow, N.D., Wang, G.J., Pappas, N., Logan, J., MacGregor, R., Alexoff, D., Shea, C., Schlyer, D., Wolf, A.P., Warner, D., Zezulkova, I., and Cilento, R. (1996a), Inhibition of monoamine oxidase B in the brains of smokers, *Nature,* 379:733–736.

Fowler, J.S., Volkow, N.D., Wang, G.J., Pappas, N., Logan, J., Shea, C., Alexoff, D., MacGregor, R.R., Schlyer, D.J., Zezulkova, I., and Wolf, A.P. (1996b), Brain monoamine oxidase A inhibition in cigarette smokers, *Proc. Natl. Acad. Sci. U.S.A.,* 93:14065–14069.

Fowler, J.S., Volkow, N.D., Wang, G.J., Pappas, N., Logan, J., MacGregor, R., Alexoff, D., Wolf, A.P., Warner, D., Cilento, R., and Zezulkova, I. (1998), Neuropharmacological actions of cigarette smoke: brain monoamine oxidase B (MAO B) inhibition, *J. Addict. Dis.,* 17:23–34.

Fu, Y., Matta, S.G., Valentine, J.D., and Sharp, B.M. (1998), Desensitization and resensitization of norepinephrine release in the rat hippocampus with repeated nicotine administration, *Neurosci. Lett.,* 241:147–150.

Fu, Y., Matta, S.G., and Sharp, B.M. (1999), Local alpha-bungarotoxin sensitive nicotinic receptors modulate nicotine-induced dopamine release in the nucleus accumbens, *Soc. Neurosci. Abst.,* 25:187.

Fung, Y.K. (1989), Effects of chronic nicotine pretreatment on (+)amphetamine and nicotine-induced synthesis and release of [^{3}H]dopamine from [^{3}H]tyrosine in rat nucleus accumbens, *J. Pharm. Pharmacol.,* 41:66–68.

Fung, Y.K. and Lau, Y.S. (1992), Chronic effects of nicotine on mesolimbic dopaminergic system in rats, *Pharmacol. Biochem. Behav.,* 41:57–63.

Fung, Y.K., Schmid, M.J., Anderson, T.M., and Lau, Y.S. (1996), Effects of nicotine withdrawal on central dopaminergic systems, *Pharmacol. Biochem. Behav.,* 53:635–640.

George, T.P., Verrico, C.D., and Roth, R.H. (1998), Effects of repeated nicotine pre-treatment on mesoprefrontal dopaminergic and behavioral responses to acute footshock stress, *Brain Res.,* 801:36–49.

Giorguieff, M.F., Le Floc'h, M.L., Glowinski, J., and Besson, M.J. (1977), Involvement of cholinergic presynaptic receptors of nicotinic and muscarinic types in the control of the spontaneous release of dopamine from striatal dopaminergic terminals in the rat, *J. Pharmacol. Exp. Ther.,* 200:535–544.

Goldner, F.M., Dineley, K.T., and Patrick, J.W. (1997), Immunohistochemical localization of the nicotinic acetylcholine receptor subunit alpha6 to dopaminergic neurons in the substantia nigra and ventral tegmental area, *Neuroreport,* 8:2739–2742.

Gonon, F. and Sundstrom, L. (1996), Excitatory effects of dopamine released by impulse flow in the rat nucleus accumbens *in vivo, Neuroscience,* 75:13–18.

Goodman, F.R. (1974), Effects of nicotine on distribution and release of 14C-norepinephrine and 14C-dopamine in rat brain striatum and hypothalamus slices, *Neuropharmacology,* 13:1025–1032.

Grace, A.A. and Bunney, B.S. (1983), Intracellular and extracellular electrophysiology of nigral dopaminergic neurons--3. Evidence for electrotonic coupling, *Neuroscience,* 10:333–48.

Grace, A.A. and Bunney, B.S. (1984), The control of firing pattern in nigral dopamine neurons: burst firing, *J. Neurosci.,* 4:2877-2890.

Grady, S.R., Marks, M.J., Wonnacott, S., and Collins, A.C. (1992), Characterization of nicotinic receptor-mediated [^{3}H]dopamine release from synaptosomes prepared from mouse striatum, *J. Neurochem.,* 59: 848–856.

Grady, S.R., Marks, M.J., and Collins, A.C. (1994), Desensitization of nicotine-stimulated [³H]dopamine release from mouse striatal synaptosomes, *J. Neurochem.*, 62:1390–1398, 1994.

Grady, S.R., Grun, E.U., Marks, M.J., and Collins, A.C. (1997), Pharmacological comparison of transient and persistent [³H]dopamine release from mouse striatal synaptosomes and response to chronic L-nicotine treatment, *J. Pharmacol. Exp. Ther.*, 282:32–43.

Grenhoff, J., Aston-Jones, G., and Svensson, T.H. (1986), Nicotinic effects on the firing pattern of midbrain dopamine neurons, *Acta Physiol. Scand.*, 128:351–358.

Grenhoff, J., Ugedo, L., and Svensson, T.H. (1988), Firing patterns of midbrain dopamine neurons: differences between A9 and A10 cells, *Acta Physiol. Scand.*, 134:127–132.

Gueorguiev, V.D., Zeman, R.J., Hiremagalur, B., Menezes A., and Sabban E.L. (1999), Differing temporal roles of Ca²⁺ and cAMP in nicotine-elicited elevation of tyrosine hydroxylase mRNA, *Am. J. Physiol.*, 276:C54–65.

Haines, C.F. Jr., Mahajan, D.K., Miljkovc, D., Miljkovic, M., and Vesell, E.S. (1974), Radio-immunoassay of plasma nicotine in habituated and naive smokers, *Clin. Pharmacol. Ther.*, 16:1083–1089.

Hakan, R.L. and Ksir, C. (1991), Acute tolerance to the locomotor stimulant effects of nicotine in the rat, *Psychopharmacology,* (Berl) 104:386–390.

Harsing, L.G., Jr, Sershen, H., and Lajtha, A. (1992), Dopamine efflux from striatum after chronic nicotine: evidence for autoreceptor desensitization, *J. Neurochem.*, 59:48–54.

Hefti, F. and Lichtensteiger, W. (1978), Dendritic dopamine: studies on the release of endogenous dopamine from subcellular particles derived from dendrites of rat nigro-striatal neurons, *Neurosci. Lett,* 10:65–70.

Henningfield, J.E., Stapleton, J.M., Benowitz, N.L., Grayson, R.F., and London, E.D. (1993), Higher levels of nicotine in arterial than in venous blood after cigarette smoking, *Drug Alcohol Depend.*, 33:23–29.

Holtz, P, Credner, K., and Koepp, W. (1942), Die enzymatische entstehung von Oxytyramin im organismus und die ohysiologische bedeutung der dopadecarboxylase, *Naunyn Schmiedeberg's Arch. Exp. Path. Pharmak.*, 200:356–388.

Izenwasser, S., Jacocks, H.M., Rosenberger, J.G., and Cox, B.M. (1991), Nicotine indirectly inhibits [³H]dopamine uptake at concentrations that do not directly promote [³H]dopamine release in rat striatum, *J. Neurochem.*, 56:603–610.

Janson, A.M., Hedlund, P.B., Hillefors, M., and von Euler, G. (1992), Chronic nicotine treatment decreases dopamine D2 agonist binding in the rat basal ganglia, *Neurore-port,* 3:1117–1120.

Kaiser, S.A. and Wonnacott, S. (2000), α-Bungarotoxin-sensitive nicotinic receptors indirectly modulate [³H]dopamine release in rat striatal slices via glutamate release, *Mol. Pharmacol.*, (in press).

Kaiser, S.A., Soliakov, L., Harvey, S.C, Luetje C.W., and Wonnacott S. (1998), Differential inhibition by α∠conotoxin-MII of the nicotinic stimulation of [³H]dopamine release from rat striatal synaptosomes and slices, *J. Neurochem.*, 70:1069–1076.

Khalil, A.A., Steyn, S., Castagnoli, N. Jr. (2000), Isolation and characterization of a monoamine oxidase inhibitor from tobacco leaves, *Chem. Res. Toxicol.*, 13:31–35.

Kirch, D.G., Taylor, T.R., Creese, I., Xu, S.X., and Wyatt, R.J. (1992), Effect of chronic nicotine treatment and withdrawal on rat striatal D1 and D2 dopamine receptors, *J. Pharm. Pharmacol.*, 44:89–92.

Kretschmer, B.D. (1999), Modulation of the mesolimbic dopamine system by glutamate: role of NMDA receptors, *J. Neurochem.*, 73:839–848.

Kulak, J.M., Nguyen, T.A., Olivera, B.M, and McIntosh, J.M. (1997), Alpha-conotoxin MII blocks nicotine-stimulated dopamine release in rat striatal synaptosomes, *J. Neurosci.*, 17:5263–5270.

Kyerematen, G.A., Damiano, M.D., Dvorchik, B.H, and Vesell, E.S. (1982), Smoking-induced changes in nicotine disposition: application of a new HPLC assay for nicotine and its metabolites, *Clin. Pharmacol. Ther.*, 32:769–780.

Kyerematen, G.A., Taylor, L.H., deBethizy, J.D., and Vesell, E.S. (1988), Pharmacokinetics of nicotine and 12 metabolites in the rat. Application of a new radiometric high performance liquid chromatography assay, *Drug Metab. Dispos.*, 16:125–129.

Kyerematen, G.A., Morgan, M.L., Chattopadhyay, B., deBethizy, J.D., and Vesell, E.S. (1990), Disposition of nicotine and eight metabolites in smokers and nonsmokers: identification in smokers of two metabolites that are longer lived than cotinine, *Clin. Pharmacol. Ther.*, 48:641–651.

Lanca, A.J., Adamson, K.L., Coen, K.M., Chow, B.L., and Corrigall, W.A. (2000), The pedunculopontine tegmental nucleus and the role of cholinergic neurons in nicotine self-administration in the rat: a correlative neuroanatomical and behavioral study, *Neuroscience,* 96:735–742.

Lapin, E.P., Maker, H.S., Sershen, H., and Lajtha, A. (1989), Action of nicotine on accumbens dopamine and attenuation with repeated administration, *Eur. J. Pharmacol.*, 160:53–59.

Le Novère, N., Zoli, M., and Changeux, J.P. (1996), Neuronal nicotinic receptor alpha6 subunit mRNA is selectively concentrated in catecholaminergic nuclei of the rat brain, *Eur. J. Neurosci.*, 8:2428–2439.

Levin, E.D., Torry, D., Christopher, N.C,. Yu X., Einstein, G., and Schwartz-Bloom, R.D. (1997), Is binding to nicotinic acetylcholine and dopamine receptors related to working memory in rats? *Brain Res. Bull.*, 43:295–304.

Lichtensteiger, W., Felix, D., Lienhart, R., and Hefti, F. (1976), A quantitative correlation between single unit activity and fluorescence intensity of dopamine neurones in zona compacta of substantia nigra, as demonstrated under the influence of nicotine and physostigmine, *Brain Res.*, 117:85–103.

Lichtensteiger, W., Hefti, F., Felix, D., Huwyler, T., Melamed, E., and Schlumpf, M. (1982), Stimulation of nigrostriatal dopamine neurones by nicotine, *Neuropharmacology,* 21:963–968.

Lippiello, P.M., Bencherif, M., Prince, R.J. (1995), The role of desensitization in CNS nicotinic receptor function, in *Effects of Nicotine in Biological Systems II*, Clarke, P.B.S., Quik, M., Adlkofer, F., and Thurau, K., Eds., *Birkhauser Verlag,* Basel, 79–85.

Lukas, R.J. (1991), Effects of chronic nicotinic ligand exposure on functional activity of nicotinic acetylcholine receptors expressed by cells of the PC12 rat pheochromocytoma or the TE671/RD human clonal line, *J. Neurochem.*, 56:1134–1145.

Lyness, W.H., Friedle, N.M., and Moore, K.E. (1979), Destruction of dopaminergic nerve terminals in nucleus accumbens: effect on d-amphetamine self-administration, *Pharmacol. Biochem. Behav.*, 11:553–556.

Marchi, M., Augliera, A., Codignola, A., Lunardi, G., Fedele, E., Fontana, G., and Raiteri, M. (1991), Cholinergic modulation of [³H]dopamine release from dendrosomes of rat substantia nigra, *Naunyn Schmiedeberg's Arch. of Pharmacol.*, 344:275–280.

Marks, M.J., Burch, J.B., and Collins, A.C. (1983), Effects of chronic nicotine infusion on tolerance development and nicotinic receptors, *J. Pharmacol. Exp Ther.*, 226:817–825.

Marks, M.J., Farnham, D.A., Grady, S.R., and Collins, A.C. (1993a), Nicotinic receptor function determined by stimulation of rubidium efflux from mouse brain synaptosomes, *J. Pharmacol. Exp. Ther.,* 264:542–552.

Marks, M.J., Grady, S.R., and Collins, A.C. (1993b), Downregulation of nicotinic receptor function after chronic nicotine infusion, *J. Pharmacol. Exp. Ther.,* 266:1268–1276.

Marks, M.J., Grady, S.R., Yang, J.M., Lippiello, P.M., and Collins, A.C. (1994), Desensitization of nicotine-stimulated ^{86}Rb$^+$ efflux from mouse brain synaptosomes, *J. Neurochem.,* 63:2125–2135.

Marshall, D., Soliakov, L., Redfern, P., and Wonnacott, S. (1996), Tetrodotoxin-sensitivity of nicotine-evoked dopamine release from rat striatum, *Neuropharmacology,* 35:1531–1536.

Marshall, D.L., Redfern, P.H., and Wonnacott, S. (1997), Presynaptic nicotinic modulation of dopamine release in the three ascending pathways studied by in vivo microdialysis: comparison of naïve and chronic nicotine-treated rats, *J. Neurochem.,* 68:1511–1519.

Meltzer, L.T., Christoffersen, C.L., and Serpa, K.A. (1997), Modulation of dopamine neuronal activity by glutamate receptor subtypes, *Neurosci. Biobehav. Rev.,* 21:511–518.

Mendez-Alvarez, E., Soto-Otero, R., Sanchez-Sellero, I., and Lopez-Rivadulla Lamas, M. (1997), Inhibition of brain monoamine oxidase by adducts of 1,2,3,4-tetrahydroisoquinoline with components of cigarette smoke, *Life Sci.,* 60:1719–1727.

Mifsud, J.C., Hernandez, L., and Hoebel, B.G. (1989), Nicotine infused into the nucleus accumbens increases synaptic dopamineas measured by *in vivo* microdialysis, *Brain Res.,* 478:365–367.

Miller, R.P., Rotenberg, K.S., and Adir, J. (1977), Effect of dose on the pharmacokinetics of intravenous nicotine in the rat, *Drug Metab. Dispos.,* 5:436–443.

Miyata, H., Ando, K., and Yanagita, T. (1996), Comparison of the effects of nicotine and methamphetamine on extracellular dopamine in the nucleus accumbens of behaviorally sensitized rats, *Nihon Shinkei Seishin Yakurigaku Zasshi,* 16:41–47.

Morens, D.M., Grandinetti, A., Reed, D., White, L.R., and Ross, G.W. (1995), Cigarette smoking and protection from Parkinson's disease: false association or etiologic clue? *Neurology,* 45:1041–1051.

Morley, B.J., Kemp, G.E., and Salvaterra, P. (1979), α-bungarotoxin binding sites in the CNS, *Life Sci.,* 24:859–872.

Naquira, D., Zunino, E., Arqueros, L., and Viveros, O.H. (1978), Chronic effects of nicotine on catecholamine synthesizing enzymes in rats, *Eur. J. Pharmacol.,* 47:227–229.

Nefzger, M.D., Quadfasel, F.A., and Karl, V.C. (1968), A retrospective study of smoking in Parkinson's disease, *Am. J. Epidemiol.,* 88:149–158.

Nirenberg, M.J., Vaughan, R.A., Uhl, G.R., Kuhar, M.J., and Pickel, V.M. (1996), The dopamine transporter is localized to dendritic and axonal plasma membranes of nigrostriatal dopaminergic neurons, *J. Neurosci.,* 16:436–447.

Nisell, M., Nomikos, G.G., and Svensson, T.H. (1994a), Infusion of nicotine in the ventral tegmental area or the nucleus accumbens of the rat differentially affects accumbal dopamine release, *Pharmacol. Toxicol.,* 75:348–352.

Nisell, M., Nomikos, G.G., and Svensson, T.H. (1994b), Systemic nicotine-induced dopamine release in the rat nucleus accumbens is regulated by nicotinic receptors in the ventral tegmental area, *Synapse,* 16:36–44.

Nisell, M., Nomikos, G.G., Hertel, P., Panagis, G., and Svensson, T.H. (1996), Condition-independent sensitization of locomotor stimulation and mesocortical dopamine release following chronic nicotine treatment in the rat, *Synapse,* 22:369–381.

Nisell, M., Marcus, M., Nomikos, G.G., and Svensson, T.H. (1997), Differential effects of acute and chronic nicotine on dopamine output in the core and shell of the rat nucleus accumbens, *J. Neural Transm.*, 104:1–10.

Ochoa, E.L., Li, L., and McNamee, M.G. (1990), Desensitization of central cholinergic mechanisms and neuroadaptation to nicotine, *Mol. Neurobiol.*, 4:251–287.

Olds, J. and Milner, P. (1954), Positive reinforcement produced by electrical stimulation of septal areas and other regions of rat brain, *J. Comp. Physiol. Psychol.*, 47:419–427.

Olds, J. and Olds, M.E. (1963), Approach-avoidance analysis of rat diencephalon, *J. Comp. Neurol.*, 120: 259–295.

Oreland, L., Fowler, C.J., and Schalling, D. (1981), Low platelet monoamine oxidase activity in cigarette smokers, *Life Sci.*, 29:2511–2518.

Oswald, R.E. and Freeman, J.A. (1981), Alpha-bungarotoxin binding and central nervous system nicotinic acetylcholine receptors, *Neuroscience*, 6:1–14.

Overton, P. and Clark, D. (1991), N-methyl-D-aspartate increases the excitability of nigrostriatal dopamine terminals, *Eur. J. Pharmacol.*, 201:117–120.

Overton, P. and Clark, D. (1992), Electrophysiological evidence that intrastriatally administered N-methyl-D-aspartate augments striatal dopamine tone in the rat, *J. Neural Transm. Park Dis. Dement.*, 4:1–14.

Overton, P.G. and Clark, D. (1997), Burst firing in midbrain dopaminergic neurons, *Brain Res. Brain Res. Rev.*, 25:312–334.

Paladini, C.A., Iribe, Y., and Tepper, J.M. (1999), GABA-A receptor stimulation blocks NMDA-induced bursting of dopaminergic neurons *in vitro* by decreasing input resistance, *Brain Res.*, 832:145–151.

Pawlak, R., Takada, Y., Takahashi, H., Urano, T., Ihara, H., Nagai, N., and Takada, A. (2000), Differential effects of nicotine against stress-induced changes in dopaminergic system in rat striatum and hippocampus, *Eur. J. Pharmacol.*, 387:171–177.

Pidoplichko, V.I., DeBiasi, M., Williams, J.T., and Dani, J.A. (1997), Nicotine activates and desensitizes midbrain dopamine neurons, *Nature*, 390:401–404.

Plowchalk, D.R., Andersen, M.E., and deBethizy, J.D. (1992), A physiologically based pharmacokinetic model for nicotine disposition in the Sprague-Dawley rat, *Toxicol. Appl. Pharmacol.*, 116:177–188, 1992.

Prince, R.J., Fernandes, K.G., Gregory, J.C., Martyn, I.D., and Lippiello, P.M. (1996), Modulation of nicotine-evoked [³H]dopamine release from rat striatal synaptosomes by voltage-sensitive calcium channel ligands, *Biochem. Pharmacol.*, 52:613–618

Rapier, C., Lunt ,G.G., and Wonnacott, S. (1988), Stereoselective nicotine-induced release of dopamine from striatal synaptosomes: concentration dependence and repetitive stimulation, *J. Neurochem.*, 50:1123–1130.

Reid, M.S., Ho, L.B., and Berger, S.P. (1998), Behavioral and neurochemical components of nicotine sensitization following 15-day pretreatment: studies on contextual conditioning, *Behav. Pharmacol.*, 9:137–148.

Reilly, M.A., Lapin, E.P., Maker, H.S., and Lajtha, A. (1987), Chronic nicotine administration increases binding of [³H]domperidone in rat nucleus accumbens, *J. Neurosci. Res.*, 18:621–625.

Reuben, M., Boye, S., and Clarke, P.B. (2000), Nicotinic receptors modulating somatodendritic and terminal dopamine release differ pharmacologically, *Eur. J. Pharmacol.*, 393:39–49.

Risner, M. and Jones, B.E. (1976), Role of noradrenergic and dopaminergic processes in amphetamine self-administration, *Pharmacol. Biochem. Behav.*, 5:477–482

Risner, M.E. and Jones, B.E. (1980), Intravenous self-administration of cocaine and norcaine by dose, *Psychopharmacology*, 71:83–89.

Roberts, D.C.S. and Koob, G.F. (1982), Disruption of cocaine self-adminstration following 6-hydroxydopamine lesions of the ventral tegmental area in rats, *Pharmacol. Biochem. Behav.*, 17:901–904.

Robertson, A. and Mogenson, G.J. (1978), Evidence for a role for dopamine in self-stimulation of the nucleus accumbens of the rat, *Can. J. Psychol.*, 32:67–76.

Rosenberg, J., Benowitz, N.L., Jacob, P., and Wilson, K.M. (1980), Disposition kinetics and effects of intravenous nicotine, *Clin. Pharmacol. Ther.*, 28:517–522.

Rowell, P.P. (1987), Current concepts on the effects of nicotine on neurotransmitter release in the central nervous system, *Adv. Behav. Biol.*, 31:191–208.

Rowell, P.P. (1995), Nanomolar concentrations of nicotine increase the release of [3H]dopamine from rat striatal synaptosomes, *Neurosci. Lett.*, 189:171–175.

Rowell, P.P., Carr, L.A., and Garner, A.C. (1987), Stimulation of [3H]dopamine release by nicotine in rat nucleus accumbens, *J. Neurochem.*, 49:1449–1454.

Rowell, P.P. and Wonnacott, S. (1990), Evidence for functional activity of upregulated nicotine binding sites in rat striatal synaptosomes, *J. Neurochem.*, 55, 2105–2110.

Rowell, P.P. and Hill, A.S. (1993), Apparent inability of nicotine to inhibit dopamine uptake into rat striatal tissue *in vitro*, *Pharmacologist*, 35:134.

Rowell, P.P. and Hillebrand, J.A. (1994), Desensitization of nicotine-stimulated dopamine release from rat striatal synaptosomes, *J. Neurochem.*, 63:561–569.

Rowell, P.P. and Li, M. (1997), Dose-response relationship for nicotine-induced up-regulation of rat brain nicotinic receptors, *J. Neurochem.*, 68:1982–1989.

Rowell, P.P. and Duggan, D.S. (1998), Long-lasting inactivation of nicotinic receptor function in vitro by treatment with high concentrations of nicotine, *Neuropharmacology,* 37:103–111.

Russell, M.A.H. (1990), Nicotine intake and its control over smoking, in *Nicotine Psychopharmacology: Molecular, Cellular and Behavioral Aspects*, Wonnacott, S., Russell, M.A.H. and Stolerman, I.P., Eds., Oxford University Press, New York, 374–418.

Russell, M.A. and Feyerabend, C. (1978), Cigarette smoking: a dependence on high-nicotine boli, *Drug Metab. Rev.*, 8:29–57.

Sakurai, Y., Takano, Y., Kohjimoto, Y., Honda, K., and Kamiya, H.O. (1982), Enhancement of [3H]dopamine release and its [3H]metabolites in rat striatum by nicotinic drugs, *Brain Res.*, 242:99–106.

Sano, I., Gamo, T., Kakimoto, Y., Taniguchi, K. Takesada, M., and Michinuma, K. (1959), Distribution of catechol compounds in human grain, *Biochemica Biochysica Acta,* 32:586–587.

Sastry, B.V., Chance, M.B., Singh, G., Horn, J.L., and Janson, V.E. (1995), Distribution and retention of nicotine and its metabolite, cotinine, in the rat as a function of time, *Pharmacology,* 50:128–136.

Schilström, B., Nomikos, G.G., Nisell, M., Hertel, P., and Svensson, T.H. (1998a), N-methyl-D-aspartate receptor antagonism in the ventral tegmental area diminishes the systemic nicotine–induced dopamine release in the nucleus accumbens, *Neuroscience,* 82:781–789.

Schilström, B., Svensson, H.M., Svensson, T.H., and Nomikos, G.G. (1998b), Nicotine and food induced dopamine release in the nucleus accumbens of the rat: putative role of alpha7 nicotinic receptors in the ventral tegmental area, *Neuroscience,* 85:1005–1009.

Schwartz, R.D. and Kellar, K.J. (1983), Nicotinic cholinergic receptor binding sites in the brain: regulation *in vivo*, *Science,* 220:214–216.

Schwartz, R.D. and Kellar, J. (1985), *In vivo* regulation of [3H]acetylcholine recognition sites in brain by nicotinic cholinergic drugs, *J. Neurochem.*, 45:427–433.

Sharp, B.M. and Matta, S.G. (1993), Detection by *in vivo* microdialysis of nicotine-induced norepinephrine secretion from the hypothalamic paraventricular nucleus of freely moving rats: dose-dependency and desensitization, *Endocrinology,* 133:11–19.

Sharples, C.G., Kaiser, S., Soliakov, L., Marks, M.J., Collins, A.C., Washburn, M., Wright, E., Spencer, J.A., Gallagher, T., Whiteaker, P., and Wonnacott, S. (2000), A novel nicotinic agonist with subtype selectivity implicates the alpha4beta2* subtype in the modulation of dopamine release from rat striatal synaptosomes, *J Neurosci,* 20:2783–2791.

Shiffman, S., Zettler-Segal, M., Kassel, J., Paty, J., Benowitz, N.L., and O'Brien, G. (1992), Nicotine elimination and tolerance in non-dependent cigarette smokers, *Psychopharmacology* (Berl), 109:449–456.

Shoaib, M., Schindler, C.W., Goldberg, S.R., and Pauly, J.R. (1997), Behavioral and biochemical adaptations to nicotine in rats: influence of MK801, an NMDA receptor antagonist, *Psychopharmacology* (Berl), 134:121–130.

Silbergeld, E.K. and Walters, J.R. (1979), Synaptosomal uptake and release of dopamine in substantia nigra: effects of gamma-aminobutyric acid and substance, *P. Neurosci. Lett.,* 12:119–126.

Simasko, S.M., Soares, J.R., and Weiland, G.A. (1986), Two components of carbamylcholine-induced loss of nicotinic acetylcholine receptor function in the neuronal cell line PC12, *Mol. Pharmacol.,* 30:6–12.

Singer, G., Wallace, M., and Hall, R. (1982), Effects of dopaminergic nucleus accumbens lesions on the acquisition of schedule induced self injection of nicotine in the rat, *Pharmacol. Biochem. Behav.,* 17:579–581.

Smith, J.E., Guerin, G.F., Co, C., Barr, T.S., and Lane, J.D. (1985), Effects of 6-OHDA lesions of the central medial nucleus accumbens on rat intravenous morphine self-administration, *Pharmacol. Biochem. Behav.,* 23:843–849.

Soliakov, L. and Wonnacott, S. (1996), Voltage-sensitive Ca^{2+} channels involved in nicotinic receptor-mediated [^{3}H]dopamine release from rat striatal synaptosomes, *J. Neurochem.,* 67:163–170.

Sorenson, E.M., Shiroyama, T., and Kitai, S.T. (1998), Postsynaptic nicotinic receptors on dopaminergic neurons in the substantia nigra pars compacta of the rat, *Neurosci.,* 87:659–673.

Stein, E.A., Pankiewicz, J., Harsch, H.H., Cho, J.K., Fuller, S.A., Hoffmann, R.G., Hawkins, M., Rao, S.M., Bandettini, P.A., and Bloom, A.S. (1998), Nicotine-induced limbic cortical activation in the human brain: a functional MRI study, *Am. J. Psychiatry,* 155:1009–1015.

Stein, L. (1962), Effects and interactions of imipramine, chlorpromazine, reserpine, and amphetamine on self-stimulation: possible neurophysiological basis of depression, in *Recent Advances in Biological Psychiatry,* Plenum Press, New York, 288–308.

Suaud-Chagny, M.F., Chergui, K., Chouvet, G., and Gonon, F. (1992), Relationship between dopamine release in the rat nucleus accumbens and the discharge activity of dopaminergic neurons during local in vivo application of amino acids in the ventral tegmental area, *Neuroscience,* 49:63–72.

Summers, K.L. and Giacobini, E. (1995), Effects of local and repeated systemic administration of (-) nicotine on extracellular levels of acetylcholine, norepinephrine, dopamine, and serotonin in rat cortex, *Neurochem. Res.,* 20:753–759.

Svensson, T.H., Mathe, J.M., Nomikos, G.G., and Schilström, B. (1998), Role of excitatory amino acids in the ventral tegmental area for central actions of non-competitive NMDA-receptor antagonists and nicotine, *Amino Acids,* 14:51–56.

Takahashi, H., Takada, Y., Nagai, N., Urano, T., and Takada, A. (1998), Effects of nicotine and footshock stress on dopamine release in the striatum and nucleus accumbens, *Brain Res. Bull.*, 452:157–162.

Takaki, T. (1995), Chronic treatment with nicotine enhances the sensitivity of dopamine autoreceptors that modulate dopamine release from the rat striatum, *Nihon Shinkei Seishin Yakurigaku Zasshi*, 15:335–1344.

Taylor, J.R. and Robbins, T.W. (1986), 6-Hydroxydopamine lesions of the nucleus accumbens, but not of the caudate nucleus, attenuate enhanced responding with reward-related stimuli produced by intra-accumbens d-amphetamine, *Psychopharmacology*, (Berl) 90:390–397.

Teng, L., Crooks, P.A., Sonsalla, P.K., and Dwoskin, L.P. (1997), Lobeline and nicotine evoke [^{3}H]overflow from rat striatal slices preloaded with [^{3}H]dopamine: differential inhibition of synaptosomal andvesicular [^{3}H]dopamine uptake, *J. Pharmacol. Exp. Ther.*, 280:1432–1444.

Teng, L., Crooks, P.A., and Dwoskin, L.P. (1998), Lobeline displaces [^{3}H]dihydrotetrabenazine binding and releases [^{3}H]dopamine from rat striatal synaptic vesicles: comparison with d-amphetamine, *J. Neurochem.*, 71:258–265.

Toth, E., Sershen, H., Hashim, A., Vizi, E.S., and Lajtha, A. (1992), Effect of nicotine on extracellular levels of neurotransmitters assessed by microdialysis in various brain regions: role of glutamic acid, *Neurochem. Res.*, 17:265–271.

Vander Eecken, H.V., Adams, R.D., and Van Bogaert, L. (1960), Striopallidal-nigral degeneration; an hitherto undescribed lesion in paralysis agitans, *J. Neuropathol. Exp. Neurol.*, 19:159–161.

Vizi, E.S. and Lendvai, B. (1999), Modulatory role of presynaptic nicotinic receptors in synaptic and non-synaptic chemical communication in the central nervous system, *Brain Res. Brain Res. Rev.*, 30:219–35.

Vogt, M. (1959), *Proc. 4th Int. Cong. Biochem*, Vol. 3, Brucker, F., Ed., Pergamon, London, 279.

Westerink, B.H., Kwint, H.F., and deVries, J.B. (1996), The pharmacology of mesolimbic dopamine neurons: a dual-probe microdialysis study in the ventral tegmental area and nucleus accumbens of the rat brain, *J. Neurosci.*, 16:2605–2611.

Westfall, T.C. (1974), Effect of nicotine and other drugs on the release of [^{3}H]norepinephrine and [^{3}H]dopamine from rat brain slices, *Neuropharmacology*, 13:693–700.

Westfall, T.C., Perry H., and Vickery, L. (1987), Mechanisms of nicotine regulation of dopamine release in neostriatum, in *Tobacco Smoking and Nicotine*, Martin, W.R., Van Loon, G.R., Iwamoto, E.T., and Davis, L., Eds., Plenum Press, New York, 209–223.

Whiteaker, P., Garcha, H.S., Wonnacott, S., and Stolerman, I.P. (1995), Locomotor activation and dopamine release produced by nicotine and isoarecolone in rat, *Br. J. Pharmacol.*, 116:2097–105.

Wise, R.A. and Bozarth, M.A. (1985), Brain mechanisms of drug reward and euphoria, *Psychiatr. Med.*, 3:445–460.

Wonnacott, S. (1990), The paradox of nicotinic acetylcholine receptor upregulation by nicotine, *Trends Pharmacol. Sci.*, 11:216–219.

Wonnacott, S., Irons, J., Rapier, C., Thorne, B., and Lunt, G.G. (1989), Presynaptic modulation of transmitter release by nicotinic receptors, *Prog. Brain Res.*, 79:157–163.

Wonnacott, S., Kaiser, S., Mogg, A., Soliakov, L., and Jones, I.W. (2000), Presynaptic nicotinic receptors modulating dopamine release in the rat striatum, *Eur. J. Pharmacol.*, 393:51–58.

Yin, R. and French, E.D. (2000), A comparison of the effects of nicotine on dopamine and non-dopamine neurons in the rat ventral tegmental area: an *in vitro* electrophysiological study, *Brain Res. Bull.*, 51:507–514.

Yong, V.W. and Perry, T.L. (1986), Monoamine oxidase B, smoking, and Parkinson's disease, *J. Neurol. Sci.*, 72:265–272.

Yoshida, M., Yokoo, H., Tanaka, T., Mizoguchi, K., Emoto, H., Ishii, H., and Tanaka, M. (1993), Facilitatory modulation of mesolimbic dopamine neuronal activity by a mu-opioid agonist and nicotine as examined with in vivo microdialysis, *Brain Res.*, 624:277–280.

Yu, P.H. and Boulton, A.A. (1987), Irreversible inhibition of monoamine oxidase by some components of cigarette smoke, *Life Sci.*, 41:675–682.

Yu, Z.J. and Wecker, L. (1994), Chronic nicotine administration differentially affects neurotransmitter release from rat striatal slices, *J. Neurochem.*, 63:186–194.

4 Nicotinic Receptors in the Periphery

Susan Jones, Hao Lo, and Sidney A. Simon

CONTENTS

0-8493-2386-X/02/$0.00+$1.50

4.1 OVERVIEW OF PERIPHERAL NEURONAL nAChRS

4.1.1 STRUCTURE, COMPOSITION, AND DISTRIBUTION

Nicotinic acetylcholine receptors (nAChRs) belong to the ligand-gated ion channel receptor superfamily: nAChRs are composed of 5 subunits arranged to form an integral ion channel that can open upon binding the neurotransmitter, ACh, or exogenous ligands, such as nicotine. Muscle type nAChRs are found at the neuromuscular junction and in the electric organs of fish. The muscle nAChRs are very well-characterized and will be noted but not discussed in this chapter. More recently, neuronal type nAChRs were identified in neurons in the central and peripheral nervous systems and also in epithelia. The nAChR family is found throughout the central nervous system (CNS), in the peripheral autonomic nervous system, in adrenal chromaffin cells, and also in skin, the cornea, cochlear, and bronchial epithelial cells (Nguyen et al., 2000b). Their functions, which are just beginning to be elucidated, are a subject of intense investigation (Cordero-Erausquin et al., 2000; Wessler et al., 1999).

The naming of nAChRs (in the CNS and PNS) followed that of the muscle nAChR. Muscle nAChRs are composed of 5 subunits: 2 $\alpha 1$, $\beta 1$, γ/ϵ, and δ. In neurons, 8 further α subunits have been identified ($\alpha 2$–$\alpha 9$), and 3 further subunits that have been named $\beta 2$, $\beta 3$, and $\beta 4$ (or, in some cases, non-α subunits) (Arias, 2000). Within the same species, α subunits in neurons show about 50% homology with α subunits in muscle. Both muscle and neuronal nAChR subunit genes encode peptides that have 4 hydrophobic domains that are putative transmembrane regions: here, the homology approaches 100%. Both α and β subunits appear to contribute to the physiological and pharmacological properties of nAChRs (Arias, 2000). Among the different subunits, only $\alpha 7$, $\alpha 8$, and $\alpha 9$ can form functional receptors alone; $\alpha 2$–$\alpha 4$ require β subunits to make functional receptors. Recent *in vitro* studies, however, have indicated that this classification may be too simplistic in that $\alpha 7$ subtypes may combine with other subunits, thus markedly increasing the possibilities of the functional nAChRs subtypes (Crabtree et al., 1997; Gotti et al., 1994). Studies have shown that $\alpha 5$ subunits will not form functional receptors in combination with β subunits alone, but can contribute to functional receptors with β subunits and other α subunits. Also $\beta 3$ subunits may combine with $\alpha 6$ and $\beta 2$ subunits to form functional channels (Cordero-Erausquin et al., 2000). In general, however, the precise combinations of subunits that comprise functional nAChRs in neurons, glia, or epithelial cells are not known.

In this chapter, we will review the current status of peripheral nAChRs in neurons and in epithelia. Although nAChRs are found in many epithelia, we will focus on their roles in skin since this has been the most extensively studied peripheral system and also because some of their functions have been elucidated (Grando, 1997; Grando and Horton, 1997). The emphasis of this chapter will be on how different methodological approaches have contributed to present understanding of peripheral nAChR structure and function.

4.1.2 Function

Like the muscle AChR, neuronal AChRs are cation-selective ion channels that depolarize cells when they are gated open by endogenous ligands. Certain nAChR subtypes have a relatively high permeability to Ca^{2+} ions (see Figure 4.7), which can subsequently act as a signaling molecule (Séguéla et al., 1993; Vernino et al., 1992). In the PNS, nAChRs are found at synapses, as well as extrasynaptically, where they can modulate transmitter release (Sargent and Garrett, 1995; Schechter and Rosecrans, 1971). In sensory neurons in the PNS they may signal the onset of pain and inflammation and in non-neuronal cells they serve many functions, including modulating cell adhesion (Grando and Horton, 1997).

4.1.2.1 Autonomic Nervous System

The autonomic nervous system exerts involuntary control over all peripheral systems (except for skeletal muscle), and thus controls physiological processes such as circulation, respiration, digestion, metabolism, excretion, and homeostasis (Goodman and Gilman, 1990). The autonomic nervous system includes sympathetic, parasympathetic, and enteric outflow divisions. All preganglionic neurons, all postganglionic parasympathetic neurons, a few postganglionic sympathetic neurons, and most enteric neurons are cholinergic, meaning they release ACh. At synapses formed by cholinergic neurons, fast chemical synaptic transmission is mediated by ACh binding to neuronal AChRs on the target cell and opening the integral cation channel, leading to the generation of excitatory postsynaptic potentials, which may summate to cause general excitation of the target cell. The physiological role of nAChRs in the peripheral nervous system appears very simple: fast excitatory synaptic transmission in autonomic ganglia (Zhang et al., 1996). However, the complexity of the subunit expression patterns in autonomic ganglia belies such simplicity.

Studies of peripheral neuronal AChRs have relied heavily on two preparations in particular. The chick ciliary ganglion (CG), containing parasympathetic neurons that innervate the eye (Dryer, 1994), has largely been used to characterize avian nAChR subunits, although some studies have also been carried out using the chick sympathetic ganglion neurons. For mammalian studies, the rat superior cervical ganglion (SCG) has proved useful; the SCG contains sympathetic neurons that innervate the head and neck. Both CG and SCG preparations offer the advantage of being useful for *in vitro* as well as *in vivo* techniques, with the neurons easily dissociated and amenable to cell culture.

4.1.2.2 Sensory systems

For many years it has been known that ACh, nicotine, and other cholinergic compounds evoke sensations of pain when they interact with dry (skin) or wet (tongue, cornea) epithelia (Dressier et al., 1998; Keele, 1962). The burning sensation was taken as evidence that AChRs are present in nociceptors in dorsal root (DRG), trigeminal (TG — Figure 4.1) and nodose (ND) ganglion neurons (Baccaglini and Cooper, 1982; Liu et al., 1993; Morita and Katayama, 1989; Sucher et al., 1990).

A

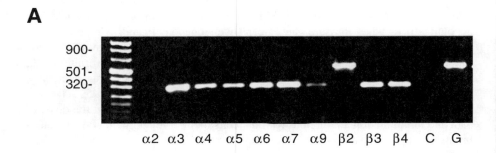

B

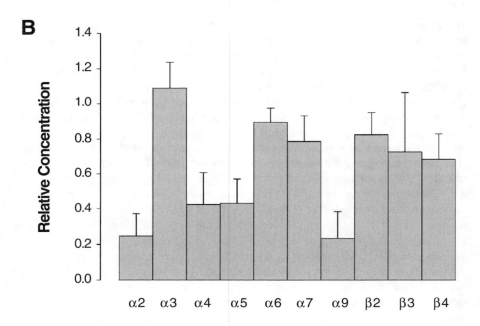

FIGURE 4.1 Detection of nAChR transcripts in the rat trigeminal ganglion. The figure shows an erthidium –bromide stained agarose gel revealing the presence of α2–α7, α9 and β2–β4 subunits. Shown are the MW markers (on the left panel), the various subunits, a control (C) where no primer was added, and the housekeeping gene GAPDH (G). The bottom panel shows the relative (to GAPDH) amount of each subunit in TG ganglia (modified from Liu et al., 1998).

Indeed nAChRs are found in nociceptive capsaicin-sensitive neurons (Liu et al., 1996) so that when TG neurons are activated by nicotine, a cascade of physiological responses is evoked that dilutes the stimuli (salivation, tearing) and prevents it from being absorbed into the body (apnea, coughing, sneezing) (Duner-Engstrom et al., 1986; Izumi and Karita, 1993). Pretreatment with the nAChR antagonist, mecamylamine, inhibits the burning sensation (Dressier et al., 1998). Despite their ubiquitous presence, the roles of peripheral nAChRs in nociceptors are not well understood.

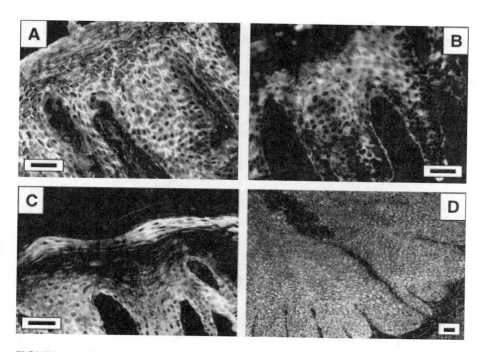

FIGURE 4.2 Nicotinic acetylcholine receptors in gingival (A and B) or esophageal (C and D) epithelia as revealed by immunocytochemistry. Panel A was stained with α5–LI, panel F with β2, and panels C and D with α7. The secondary antibody was an FITC-labeled donkey antibody. The scale bar is 50 μm (modified from Nguyen et al., 2000 b).

AChRs that are not in neurons have been shown to be involved in many cellular functions. These include cytoskeletal organization, release of cytokines, cell division, and adhesion and migration (Grando and Horton, 1997; Wessler et al., 1998, 1999). Many of these processes are regulated by nAChRs via increases in calcium influx that may arise from calcium diffusion through the nAChRs and/or via the opening of voltage gated calcium channels. Keratinocytes and other epithelial cells (e.g., corneocytes, airway epithelia) that may have muscarinic and/or nicotinic acetylcholine receptors can synthesize, degrade, store, and release ACh, indicating that it may mediate both autocrine and paracrine receptors (Grando, 1997) (Figure 4.2). The nAChRs in bronchial epithelial cells (BECs) have been found to be similar to those in ganglionic neurons in regard to their channel properties, subunit compositions, and ability to increase intracellular calcium (Maus et al., 1998).

4.2 METHODOLOGICAL APPROACHES IN NEURONAL nAChR RESEARCH

Present understanding of nAChRs in the periphery results from the endeavors of several decades of research, utilizing almost every conceivable neurobiological technique, ranging from molecular studies of nAChR genes and their distribution to functional studies of their role in peripheral tissues (Cordero-Erausquin et al., 2000; Grando and Horton, 1997; Wessler et al., 1998, 1999). The purpose of this chapter is to illustrate how a range of techniques has contributed to this field of research. Brief explanations of the techniques utilized with a consideration of the problems and pitfalls will be given; however, for an exhaustive account of methods readers are referred to specialized technical references for this purpose.

4.2.1 CLASSICAL PHARMACOLOGY STUDIES OF PERIPHERAL NEURONAL nAChRs

The earliest studies hinting at the non-uniformity of peripheral neuronal nAChRs and suggestive of specialized nAChR subtypes for performing specific functions came from the use of pharmacological tools, namely agonists, antagonists, and radiolabeled ligands for nAChR binding sites.

4.2.1.1 Autonomic Nervous System

Nicotinic receptors were initially characterized on the basis of their responses to selective agonists and antagonists (see Figures 4.4–4.7; McGehee and Role, 1995). In nAChR research, the use of neuronal toxins as antagonists has highlighted the existence of multiple nAChR subtypes, as well as emphasizing the difficulties in the pharmacological approach to delineating these subtypes. A number of snake-derived toxins occupy one or both of the agonist binding sites on nAChRs and prevent agonist-induced opening of channels (Arias, 2000). A major dilemma for nAChR pharmacology throughout the 1970s involved the snake toxin, α-bungarotoxin (α-BnTx), shown to be a potent antagonist of nAChRs in skeletal muscle. Numerous studies demonstrated that radiolabeled α-BnTx could bind with high affinity to neuronal preparations of both sympathetic and parasympathetic autonomic ganglia; however, a functional block of cholinergic synaptic transmission in autonomic ganglia using the toxin was not convincingly shown, except in cases where toxin purity was questionable. This work has been reviewed in detail by Chiapinelli (1985), and will not be revisited here. The obvious conclusion from these studies was that radioligand binding studies had revealed the presence of α-BnTx binding sites, while functional pharmacological studies, mainly using physiological assays, revealed that this α-BnTx-binding site was not responsible for mediating synaptic transmission in autonomic ganglia.

Subsequently, a second snake toxin, previously called toxin F and now κ-BnTx, was found to block synaptic transmission or responses to applied nAChR agonists in autonomic ganglia (Ravdin and Berg, 1979; Chiapinelli, 1983; Chiapinelli and Dryer, 1984; Loring et al., 1984; Loring and Zigmond, 1988). Localization studies using the two toxins suggest that, as expected, κ-BnTx labels synaptic sites in the ganglia (Loring and Zigmond, 1987), while α-BnTx labels extrasynaptic sites (Jacob and Berg, 1983; Fumagalli and De Renzis, 1984; Loring et al., 1985). It is now known that α-BnTx labels homomeric $\alpha7$ and $\alpha9$ nAChRs, whereas κ-BnTx labels $\alpha2\beta2$, $\alpha3\beta2$, $\alpha4\beta2$ and $\alpha3\beta4$ type nAChRs (McGhee and Role, 1995). It should be

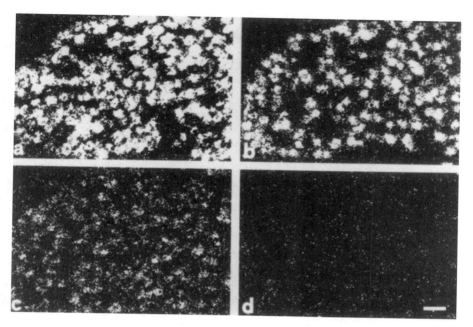

FIGURE 4.3 Expression of individual mRNAs encoding subunits of nAChRs in superior cervical ganglion neurons (SCG). Panels show dark field micrographs following emulsion autoradiography with (a) α3, (b) β4, (c) α4-1 antisense probes: (d) shows control with β4 sense probe (from Rust et al., 1994, with permission).

noted that, despite its name, the neurotoxin κ-BnTx does not block all neuronal nAChRs and, at high concentrations, can block muscle AChRs (Sargent, 1993) (Figure 4.3).

4.2.1.2 Sensory Systems

In sensory systems, the snake toxins α-BnTx and κ-BnTx have also been used to determine the types of receptors present in peripheral tissues. In extracellular recordings from rabbit corneal neurons, responses to ACh were inhibited by κ-BnTx, suggesting the presence of α7 and/or α9 subunits (MacIver and Tanelian, 1993). Rat TG neurons have binding sites for α-and κ-BnTx, as well as currents that are inhibited by α-BnTx (Simon, unpublished observations; Liu et al., 1993). In bronchial epithelial cell (BEC) lines, as well as in keratinocytes, κ-BnTx inhibited currents evoked by nicotine. BECs were also specifically labeled by α-BnTx, suggesting that these cells have multiple types of nAChRs (Maus et al., 1998). The finding that α7 and α9 bindings sites are in peripheral sensory neurons and in epithelial cells begs the question as to their function, especially given their relatively high Ca^{2+} permeabilities. In keratinocytes, the α9 subunit seems to be involved in cell adhesion (Nguyen et al., 2000a).

All subunit types are present in sensory ganglia (Figure 4.1). RT-PCR experiments on mammalian sensory ganglia reveal that they have α2–α7 and α9 as well as β2–β4 subunits (Keiger and Walker, 2000; Liu et al., 1998; Puttfarcken et al., 1997). The burning sensation of nicotine on the human tongue can be blocked by mecamylamine, suggesting the involvement of peripheral α3β4 and α4β2 receptors (Dressier et al., 1998). In F11 cells that originate from DRGs measurements of ^{86}Rb efflux were activated in the order ± epibatidine > DMPP> nicotine > ABT-418> cytisine (Puttfarcken et al., 1997). All these fluxes were inhibited by 100 μM mecamylamine (IC_{50} = 0.16 μM) and were weakly inhibited by DHβE, supporting the presence of functional α3β4 subunits (Puttfarcken et al., 1997). In this study it was also noted that α4β2 receptors may also be present. Whole nerve studies of ethmoid nerve revealed that i.p. injections of mecamylamine or DHβE inhibit the activation of the TG neurons by nicotine, but not those evoked by cyclohexanone (Alimohammadi and Silver, 1999). Since DHβE is an antagonist known to inhibit α4β2 receptor subtypes, and mecamylamine is known to bind to α3β4 and α4β2 subtypes, it is possible that these subtypes are present in the terminals of peripheral trigeminal neurons.

4.2.2 CLONING REVEALS MULTIPLE SUBUNITS OF NEURONAL nAChRs

Information on nAChR heterogeneity exploded with the advent of molecular cloning techniques. Progressing from partial purification of protein sequences, cloning allowed the determination of the nucleic acid sequences encoding proteins that combined to form functional nAChRs as determined in classical pharmacological studies. Studies in the post-cloning era draw heavily on this genetic information, and therefore some consideration will be given to experiments that have led to present knowledge of neuronal nAChR subunits. (For reviews, see Deneris et al., 1991; Sargent et al., 1993; McGehee and Role, 1995; Lindstrom, 1996).

Genes encoding the subunits of the muscle nAChR had been identified previously (see Lindstrom, 1996, for review), and the cloning of the first neuronal nAChR subunit mRNA took advantage of the known sequence of the muscle subunit mRNA. Using a radioactive probe prepared from a complementary DNA (cDNA) clone encoding a mouse muscle α subunit, the RNA of PC12 cells (a rat-derived sympathetic neuronal cell line) was screened at low stringency for hybridizing sequences. A hybridizing sequence was routinely found under these conditions, but not under high stringency conditions, suggesting that it was homologous, but not identical, to the mouse muscle α subunit (Boulter et al., 1986). A cDNA library was prepared from PC12 cell RNA, and the positive clone was identified and sequenced; a probe made from this sequence was found to hybridize with RNA prepared from brain tissue and from the adrenal medulla (Boulter et al., 1986). This subunit was subsequently named α3. A second, homologous rat α subunit was isolated and sequenced by Goldman et al. (1987) and named α4.

Then followed the isolation and sequencing of a β subunit by low stringency screening of a cDNA library prepared from rat neuronal RNA, using a probe prepared from the α3 subunit clone; the resulting hybridizing partial clone was isolated and used for high stringency screening of a PC12 cell-derived cDNA library (Boulter et al., 1987). The subunit was named β2 (the muscle AChR contains β1) because it did not have the cysteine residues required for ACh binding to the α subunits, and it could substitute for the β subunit in muscle nAChRs. The discovery of β2 allowed the confirmation that the α3, α4, and β2 clones' encoded functional nAChR subunits: pair-wise injection of either α3 and β2 or α4 and β2 into *Xenopus* oocytes resulted in physiological responses being recorded when ACh was applied to the oocytes (Boulter et al., 1987). These responses were blocked by κ-BnTx, but not by α-BnTx, suggesting that they resemble the neuronal nAChRs found at sympathetic ganglion cholinergic synapses.

Genomic and cDNA encoding an α2 subunit (Wada et al., 1988) and a β3 subunit (Deneris et al., 1989) were cloned. A cDNA clone encoding a fourth rat β subunit (β4) was isolated from PC12 cells (Duvoisin et al., 1989). When β4 mRNA was injected into *Xenopus* oocytes along with either α2, α3, or α4 it produced functional nAChRs. A cDNA clone for α5 was isolated from brain and PC12 cells (Boulter et al., 1990).

The isolation and cloning of α2, α3, α4, and nonα subunits from chick were described by Nef et al. (1988). Couturier et al. (1990a) reported three genes, closely linked in the avian genome, that encode α3, α5, and nonα3 (β4) proteins. In expression studies, nonα3 could form functional nAChRs with other subunits, while α5 could not.

Proteins that bind α-BnTx had been purified from both chick and rat brains and the N-terminus of the α-BnTx binding protein had been partially protein sequenced (reviewed by Clarke, 1992). Using an oligonucleotide from this N terminus protein, two clones were isolated from a chick brain cDNA library that encoded α-BnTx binding proteins (Schoepfer et al., 1990); these were later known as α7 and α8. These clones were homologous to sequences for ligand-gated ion channels, with four putative membrane spanning segments; they also had two cysteine residues that might contribute to an ACh binding site. No physiological or pharmacological studies were performed to determine their identity. In the same year, Couturier et al. (1990b) reported the isolation and sequencing of a clone encoding a novel α subunit, which they named α7, from chick brain. This sequence had limited homology with other neuronal nAChR subunits, but when α7 mRNA was injected alone into *Xenopus* oocytes to form a receptor composed of only α7 subunits, the receptor was found to respond to ACh and nicotine. Furthermore, α-BnTx blocked the effect of ACh on the expressed α7 receptor. Thus, the cloning of the α7 subunit revealed that an α-BnTx binding protein could bind ACh and also produce a functional response.

A rat α7 clone was subsequently isolated from a rat brain cDNA library (Séguéla et al., 1993). By injecting mRNA for the rat α7 subunit into *Xenopus* oocytes, it was found that, not only was this subunit sensitive to α-BnTx, but it was also very permeable to Ca^{2+} ions, and therefore a source of ACh-induced Ca^{2+} entry into the cell.

The most recently discovered subunit (α9) was also isolated and cloned from a cDNA library (Elgoyhen et al., 1994). It was found that α9 subunits form homomeric ion channels that are also blocked by α-BnTx. These subunits are found associated with hair cells in the cochlea, trigeminal ganglion neurons (see Figure 4.1), the olfactory bulb (Keiger and Walker, 2000), and also keratinocytes (Nguyen et al., 2000a).

Thus, between 1986 and 1994, 8 α (α2–α9) and 3 β (β2–β4) nAChR subunits were identified. From this information, several lines of investigation, in particular, have contributed a vast amount of information to understanding of nAChR function in the periphery.

4.2.3 DETECTION OF NUCLEIC ACIDS ENCODING NEURONAL nAChR SUBUNITS

With the sequences of eight α and three β neuronal subunits known, the next pressing questions were (1) where different subunit combinations were expressed, and (2) which of these subunits, in combination or alone, would form functional neuronal nAChRs. To begin to address these questions, initial experiments that utilized nucleic acid detection protocols, such as Northern blotting, in situ hybridization in combination with autoradiography, and the polymerase chain reaction (PCR) to amplify nucleic acid signals were conducted.

In Northern blotting, a radioactive probe of either genomic or complimentary DNA (cDNA) encoding proteins of interest is used to detect hybridizing sequences in samples of RNA from a particular cell population. In this procedure, RNA fragments from the cellular preparation are initially separated by gel electrophoresis. Then the fragments are transferred en masse to nitrocellulose paper and hybridized with the radiolabeled DNA probes for a specific subunit sequence. The presence (or absence) of the subunit RNAs is determined using autoradiography.

In situ hybridization, as the name implies, allows the detection of hybridizing sequences in sections of tissue (see Figure 4.4). Hybridization is defined by the hydrogen binding of two complementary single strands of nucleic acids. For in situ hybridization, as well as for Northern blotting, the three types of possible probes are cDNA, RNA, and oligonucleotide. The former two are the result of subunit cloning and the latter must be synthesized. DNA is first denatured by heat and then incubated with radiolabeled probes specific for particular subunits. Radioactive DNA-RNA or DNA-DNA hybrids are detected using autoradiography (Tecott et al., 1987; Wisden et al., 1991).

In the RT-PCR method (see Figure 4.1), a sequence of steps ensures the successful amplification of DNA using only a small amount of RNA encoding proteins of interest. First, mRNA from the tissue of interest is reverse transcribed (RT). Second, the two strands of the resulting double-stranded cDNA are separated. Next, short preselected primers (either commercially bought or newly synthesized) are hybridized to the DNA strands. Finally, a special heat-sensitive DNA polymerase synthesizes and elongates new strands of complementary DNA. The last three steps can be repeated many times, amplifying the total amount of DNA produced. It is advantageous to use RT-PCR because of its sensitivity and specificity. RT-PCR is sensitive enough that mRNA sequences of interest can be detected in the cytoplasm from a single cell (see Figure 4.5), it is highly specific because hybridization is exceedingly stringent. Also, a perfect match between primer and complementary sequence is not required, and the actual nucleotide sequence of the genomic region of interest (between two primers) need not be identified.

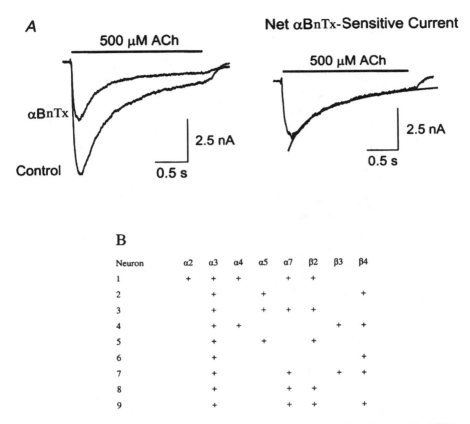

FIGURE 4.4 Individual neurons may contain several types and subunits of neuronal nAChRs. A. Inward current activated by 500 μM ACh applied to a SCG neuron in the absence and presence of 50 nM α-BnTx. The side panel shows the net α-BnTx-sensitive current. The holding potential was –60 mV. B. Results of single cell PCR experiments representing the subunit mRNAs among intracardiac parasympathetic neurons. Note that all the neurons have several subunits. Figure adapted from Cuevas et al., 2000 (A) and Poth et al., 1997 (B).

4.2.3.1 Autonomic Nervous System

In peripheral neuronal tissue, mRNA encoding neuronal nAChR subunits was first detected by Boyd et al. (1988); Northern blot analysis and *in situ* hybridization identified α3 mRNA in chick CG, while α2 or α4 were not detected. In chick CG and SCG, α3 along with nα3 (β4) transcripts were detected by Northern blots (Couturier et al., 1990a). Subsequently, Corriveau and Berg (1993) extended these findings by reporting that the chick CG expressed 5 nAChR subunit genes: α7 in highest abundance, α3, and small amounts of α5 along with β2 and β4. Quantification of the amount of each subunit was achieved by RNase protection experiments. They also reported very few transcripts of α2, α4, α8, or β3 subunits.

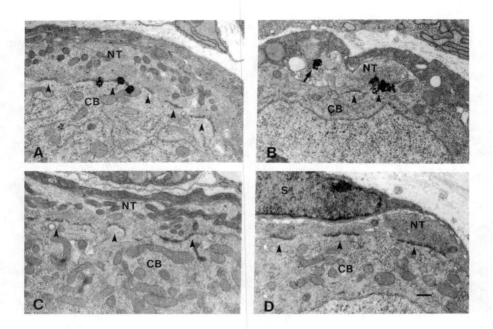

FIGURE 4.5 Localization of synaptic nAChRs. EM autoradiographs of ^{125}I-toxin F (κ-BnTx) binding to ciliary neurons in the presence of 2 µM α-BnTx (A and B). Occasionally, grains were found in clusters (B). Dispersed grains were also observed near other membrane areas (arrow in B). Few autoradiographic grains were observed near synaptic membranes in the presence of 2µM toxin F(c) or 100µMDHβE and 2µMγ–BnTx(D). NT-presynaptic nerve terminal, S-support cell. Scale bar is 0.5 µm (from Loring and Zigmond, 1988, with permission).

In the rat, both sympathetic and parasympathetic ganglia were found to have detectable levels of α3, α7, β4, α4, and β2 (Rust et al., 1994; Mandelzys et al., 1994). In SCG neurons, α2, α5, α6 and β3 were not detected in either sympathetic or parasympathetic ganglia by Rust et al. (1994), although Mandelzys et al. (1994) detected mRNA encoding α5.

The potential for diversity of subunit expression by populations of neurons was highlighted by Poth et al. (1997), who used single cell RT-PCR in combination with physiological studies to show that individual neurons isolated from the rat parasympathetic cardiac ganglion could express a varied repertoire of subunits. Of nine neurons tested, all expressed α3 mRNA with either β2 or β4, five expressed α7, only two expressed α4, three had α5, and two had β3. The variability in the physiological responses of individual neurons supported the expression studies. Electrophysiological studies also revealed that individual SCG neurons could also contain functional α-BnTx sensitive and insensitive currents (Figure 4.4).

4.2.3.2 Sensory Systems

Using *in situ* colocalization methods, Flores et al. (1996) suggested that rat TG contain $\alpha4\beta2$ and $\alpha3\beta4$ nAChRs. TG neurons also contain high affinity ^{3}H-nicotine binding sites, suggesting the presence of $\alpha4\beta2$ receptors (Walker et al., 1996) as well as epibatidine binding sites (Flores et al., 1996).

In non-neuronal bronchial epithelial cells (BECs), RT-PCR revealed the presence of $\alpha3$, $\alpha4$, $\alpha5$, $\alpha7$, $\beta2$, and $\beta4$ subunits (Maus et al., 1998). The presence of functional receptors containing these subunits was consistent with electrophysiological and pharmacological measurements. Human keratinocytes were found to contain $\alpha3$, $\alpha9$, $\beta2$, and $\beta4$ subunits (Grando et al., 1995; Nguyen et al., 2000b). RT-PCR experiments obtained from olfactory epithelium revealed the presence of all known mammalian subunits (Keiger and Walker, 2000).

In summary, Northern blot and *in situ* hybridization analysis have revealed that autonomic ganglia predominantly express mRNA for $\alpha3$, $\alpha5$, $\alpha7$, $\beta2$, and $\beta4$ subunits, but possibly not for $\alpha2$, $\alpha6$, or $\beta3$; sensory neurons express a full complement of α and β subunit mRNA, and non-neuronal tissues also express mRNA for multiple subunits.

4.2.4 Detection of Neuronal nAChR Subunit Proteins

Antibodies used in the study of nAChR subunit composition and function at the neuromuscular junction subsequently found application in the study of neuronal nAChRs. Monoclonal antibodies (mAb), specific for the subunits of the muscle and electric organ nAChR, included mAb 35, raised against purified nAChRs from eel electric organ. This mAb has been particularly useful in defining the synaptic nAChR in ganglia, and is used to distinguish the synaptic receptor from the extrasynaptic α-BnTx binding site. The mAb 35 has been widely employed in immunological studies of the subunit composition of peripheral neuronal nAChRs.

Antibodies used to investigate neuronal nAChR composition and functions have been applied in immunolabeling experiments. In Western blotting, a native protein mixture is separated by gel electrophoresis; the resultant separated proteins are transferred to nitrocellulose paper, the protein sheet is incubated with specific radiolabeled or enzyme-linked antibodies, and, finally, autoradiography or enzyme-substrate reactions are performed to visualize the presence of the receptor (Stott, 1994). This process allows detection of the receptor in crude peripheral nervous tissue. In immunohistochemistry, similar techniques are used to identify the presence and location of nAChRs in processed tissue sections (Cuello, 1993; see Figure 4.2).

Immunolabeling can also be used to immunoprecipitate proteins (to identify specific mAbs and for SDS-PAGE analysis of protein size) to determine if other proteins coassemble, and to functionally block responses mediated by that protein. In immunoprecipitation experiments, protein–antibody complexes are collected spontaneously or by using antibody-attached synthetic beads followed by centrifugation. To assess the amount of receptor protein present in a given sample, the tissue mixture of interest is incubated with radioactive agonist/antagonist (for example, epibatidine) prior to immunoprecipitation and liquid scintillation spectrometry (Flores et al., 1996).

Antibodies can also be used to affinity-purify specific proteins from tissue or, in immunodepletion studies, to prevent extraction of specific proteins from tissue, and therefore determine which proteins usually co-assemble with the depleted protein. The properties of the huge library of neuronal nAChR mAbs are summarized in Lindstrom (1996).

4.2.4.1 Autonomic Nervous System

Antibodies raised against the eel electric organ nAChR were first tested against neuronal nAChRs by Patrick and Stallcup (1977). This antibody caused a functional block of agonist-induced Na^+ influx in PC12 cells, but did not recognize the α-BnTx binding site and confirmed pharmacological evidence that the functional neuronal nAChR and the α-BnTx sites are distinct. More importantly, the study indicated that neuronal nAChRs have antigenic similarities to muscle nAChRs.

This similarity was further examined by Jacob et al. (1984); using a mAb to the "main immunogenic region" of an AChR purified from eel electric organ (mAb 35), they tested for cross reactivity with the neuronal nAChR in chick CG neurons (dissociated and cultured) by immunolabeling with HRP-conjugated mAb. The mAb 35 labeled synaptic regions and differed from binding of HRP-conjugated α-BnTx. Subsequently, it was found that the antigen recognized by mAb 35 is found in sympathetic ganglia as well as in CG neurons. In CG, the target protein sediments as a 10S species, consistent with a pentameric receptor (Smith et al., 1985). In fact, the mAb 35 site is the same site targeted by neuronal bungarotoxin (Halvorsen and Berg, 1987). Confirmation that mAbs to nAChRs could recognize extrasynaptic, as well as synaptic, regions in ganglion neurons also came from Sargent and Pang (1989).

Using the specific mAbs developed by Lindstrom and colleagues, Berg and colleagues have begun to elucidate the subunits contributing to native neuronal nAChRs in the chick CG. Beginning with immunopurification of nAChRs from CG and using mAb 35 sepharose, adsorbed proteins were then radiolabeled for autoradiography and immunoprecipitated to determine protein identity and homology with other known nAChR subunits (Halvorsen and Berg, 1990). Three proteins were obtained: α3 plus two unknowns. Following this, Conroy et al. (1992) identified one of the proteins found in chick CG as the α5 subunit by in vitro translation of α5 gene and immunoprecipitation of the synthesized protein to identify α5-specific mAbs. Native proteins, immunopurified on mAb 35 Actigel, could then be immunoprecipitated to determine whether other subunits co-assemble with α5 in native receptors. Sequential immunoprecipitation and immunodepletion experiments were used to confirm co-assembly of different α subunits: immunodepletion of α3 or β4 causes less α5 to be recovered.

Subunit composition in the CG was further elucidated by Vernallis et al. (1993). Following immunopurification of nAChRs from CG with mAb 35, immunoprecipitation with specific mAbs identified α3, α5, and β4; mAb 35 binding was measured after immunoprecipitation with specific mAbs to determine the proportions of subunit-containing receptors in CG. Immunodepletion on specific mAb Actigel followed by immunoblot analysis with other subunit mAbs was used to determine which subunits co-assemble. The study indicated that synaptic receptors contain at least three types of subunits ($\alpha3$, $\alpha5$, $\beta4$), while extrasynaptic receptors contain $\alpha7$, but lack $\alpha3$, $\alpha5$, and $\beta4$ type subunits. These findings were extended by the discovery that CG neurons also contain $\beta2$ subunits (Conroy and Berg, 1995). Using a similar approach, $\beta2$ was immunopurified and, using mAb immunohistochemistry, demonstrated that $\beta2$ mAb labels most neuronal soma, comparable to mAb 35 labeling. Immunoblot and immunodepletion analysis showed that $\beta2$-containing receptors are a subpopulation of $\alpha3\beta4$-containing receptors. It was concluded that all synaptic neuronal nAChRs that have $\alpha3$ also have $\beta4$, and vice versa; most $\alpha5$ is associated with $\alpha3$ and $\beta4$, and 20% of these receptors contain $\beta2$. Thus, chick CG neurons express at least 3, and in some cases, 4 different synaptic subunits. Thus, in the chick CG, $\alpha3$, $\alpha5$, $\alpha7$, $\beta2$, and $\beta4$ subunit proteins have been detected, which corresponds well with the detection of subunit mRNA.

Laser scanning confocal microscopy of mAb 35 and Cy3-conjugated α-BnTx in chick CG revealed that both receptor types were found in clusters widely distributed in embryonic and adult CG neurons in whole-mounted ganglia, while mAb 35 receptors alone showed a punctate distribution (Wilson-Horch and Sargent 1995). There was no evidence for colocalization of α-BnTx with synaptic vesicle proteins, although the clusters were close to synaptic sites, indicating that α-BnTx sites are perisynaptic. Clusters of mAb 35 binding were not synaptic, but punctate areas of mAb 35 binding showed overlap with the synaptic vesicle protein, while nonsynaptic mAb 35 protein overlapped with α-BnTx protein.

4.2.4.2 Sensory Systems

In a limited study of TG neurons, mAbs revealed the presence of $\alpha7$ and $\beta2$ subunits. Moreover it was noted that both subunits are likely to be present on the same neuron (Liu et al., 1998). In keratinocytes, mAb staining revealed that they contain both $\alpha3$ and $\alpha7$ subunits and that in the presence of long term exposure to nicotine, $\alpha7$ upregulated and $\alpha3$ downregulated. In respiratory epithelial cells, the presence of $\alpha3$, $\alpha4$, $\alpha5$, and $\alpha7$ subunits was revealed in immunocytochemical studies (Maus et al., 1998). Interestingly $\alpha4$ subunits were found on the epithelial cells lining the alveoli and $\alpha7$ subunits were found in the submucosal gland.

4.2.5 DETECTION OF NEURONAL nAChR SUBUNIT FUNCTION

A popular approach to illustrate functional roles for ligand-gated ion channel receptors has been to use electrophysiological studies of ionic conductance changes in response to neuronal nAChR activation with agonists. Standard two-electrode, voltage-clamp methods (one to record voltage, one to inject current) have been largely applied in expression studies using *Xenopus* oocytes. The advent of the "patch clamp" technique enabled high resolution recordings of single ion channel conductances in various membrane patch configurations, or the whole cell membrane conductance of small cells (Hamill et al., 1981). This technique produced an explosion of knowledge of the properties of many receptors and ion channels, including the neuronal nAChRs.

4.2.5.1 Cloned and Expressed Neuronal nAChRs

Having demonstrated the presence of nAChR subunit mRNA and protein in peripheral neuronal tissues, the next challenge was to demonstrate that these proteins were functional. The most convenient way of achieving this goal was to artificially express the protein products of the cloned subunits and measure the physiological responses to nAChR agonists. The *Xenopus* oocyte expression system has been widely used; for mammalian genes, mRNA was transcribed in vitro and injected into the oocyte, and for the avian genes, cDNA expression vectors with a promoter were used (Deneris et al., 1991, for review). The properties of cloned receptors have been reviewed by Sargent (1993), McGehee and Role (1995), and Lindstrom (1996). Growing concerns about potential differences between native receptors and those transiently expressed in oocytes led to the development of stable expression systems such as cell lines.

Demonstrations that cloned subunits could form functional receptors have been provided for most of the subunits isolated, except $\alpha 6$ and $\beta 3$. Mammalian $\alpha 3\beta 2$, $\alpha 4\beta 2$ (Boulter et al., 1987), mammalian $\alpha 2\beta 2$ (Wada et al., 1988), mammalian $\alpha 2\beta 4$, $\alpha 3\beta 4$, $\alpha 4\beta 4$ (Duvoisin et al., 1989), avian $\alpha 4n\alpha 4$ (Ballivet et al., 1988), avian $\alpha 3n\alpha 4$ (Couturier et al., 1990a), avian $\alpha 7$ (Couturier et al., 1990b), mammalian $\alpha 7$ (Séguéla et al., 1993), $\alpha 5$ in conjunction with other α and β subunits (Wang et al., 1996), and avian $\alpha 8$ (Gerzanich et al., 1994) form functional neuronal nAChRs in *Xenopus* oocytes. The most recent of the mammalian functional nicotinic receptors is composed entirely of homomeric $\alpha 9$ subunits (Elgoyhen et al., 1994).

Studies of the single channel properties of cloned receptors revealed differences in the conductance and kinetics when different subunits were expressed. For example, Papke et al. (1989) found that receptors assembled from $\alpha 2\beta 2$, $\alpha 3\beta 2$, or $\alpha 4\beta 2$ subunit pairs injected into *Xenopus* oocytes showed differing conductance states; Papke and Heinemann (1991) reported that $\alpha 3\beta 4$ expression in oocytes produced channels with increased probability of opening and increased bursting activity compared with channels formed from $\alpha 3\beta 2$ subunits. Thus, particular α and β subunits can modulate or alter the physiological properties of neuronal nAChR channels. Indeed, Ramirez-Latorre et al. (1996) found that coexpression of $\alpha 5$ subunits with $\alpha 4\beta 2$ subunits in oocytes markedly increased the peak currents through nAChRs; Sivilotti et al. (1997) also found a change in conductance when $\alpha 5$ was coexpressed with $\alpha 3\beta 4$ in oocytes. Individual subunits can also contribute to the desensitization properties of the neuronal nAChR: comparison of $\alpha 4n\alpha 1$ receptors with $\alpha 3n\alpha 1$ receptors revealed that the latter showed stronger desensitization, indicating that α subunits can contribute to the desensitization properties of the nAChR (Gross et al., 1991), as can β subunits because desensitization was more rapid in $\alpha 3\beta 2$ receptors than in $\alpha 3\beta 4$ receptors (Cachelin and Jaggi, 1991). Co-expression of $\alpha 5$ with $\alpha 3\beta 2$ or with $\alpha 3\beta 4$ increased the desensitization rates of both receptor types (Wang et al., 1996). Receptors assembled from $\alpha 3\beta 4$ (Verino et al., 1992), $\alpha 7$ (Couturier et al.,

1990b; Séguéla et al., 1993), α8 (Gerzanich et al., 1994), and α9 (Elgoyhen et al., 1994) have all been reported to be permeable to Ca^{2+} ions.

Pharmacological studies with cloned receptors indicate that both α and β subunits contribute to agonist potency. Gross et al. (1991) found that α3nα1 receptors were less sensitive to ACh than α4nα1 receptors; differences in agonist sensitivity of α3 receptors were also found depending on whether a β2 or β4 subunit was coexpressed (Cachelin and Jaggi, 1991). Using different combinations of α2, α3, α4, β2, and β4, Luetje and Patrick (1991) found that each subunit combination gave distinct patterns of agonist sensitivity. Covernton et al. (1994) reported differing orders of agonist potency depending on whether α3 was expressed with β2 or β4. The α5 subunit was found to increase the potency of ACh and nicotine at receptors containing α3β2 subunits (Wang et al., 1996) (Figure 4.6).

Thus, the diversity of subunits available to form functional neuronal nAChRs in peripheral neurons can potentially contribute to the diversity of the physiological and pharmacological properties of the receptors. But do native nAChRs exhibit the properties of the cloned receptors, as described in these studies? Colquhoun and colleagues have begun to compare the properties of cloned receptors with those of native receptors expressed by rat SCG neurons. An initial comparison of agonist potency indicated that nAChRs in rat SCG showed a similar, but not identical, agonist potency profile to α3β4 receptors in oocytes (Covernton et al., 1994). Analysis of single-channel properties did not identify a channel of identical conductance to rat SCG nAChRs among α3β4, α4β4, or α3β4α5 subunit combinations in oocytes (Sivilotti et al., 1997). Lewis et al. (1997), looking at single-channel conductance as well as agonist potency, found closer similarities between cloned (α3β4) receptors expressed in a stable cell line and rat SCG nAChRs, than between α3β4 receptors expressed in oocytes. The difficulties in comparing expressed receptors with native receptors might in part be explained by the fact that nAChR subunits in cells may combine with other subunits in the same superfamily but form different types of channels (Elsele et al., 1993), or they may be associated with other proteins that may modify their selectivity to agonists. Also, nAChR subunits may be at various states of phosphorylation, resulting in changes in many of their properties.

4.2.5.2 Native Receptors

The evaluation of the physiological and pharmacological properties of cloned receptors has naturally inspired a reevaluation of the properties of native receptors and brings us full circle back to classical studies. Have the studies on cloned receptors facilitated the characterization of native receptors? Lindstrom, in his 1996 review, commented that more is known about the properties of cloned neuronal nAChRs than about native nAChRs in neurons. In this regard nothing has changed.

Early patch clamp studies of the single-channel and whole-cell physiological properties of nAChRs in rat sympathetic (Mathie et al., 1987; Derkach et al., 1987; Mathie et al., 1991) and chick parasympathetic (Ogden et al., 1984) neurons suggested a largely uniform family of ligand-gated channels with brief open times (around 1 ms), a single channel conductance of between 20 to 50 pS, and bursts of

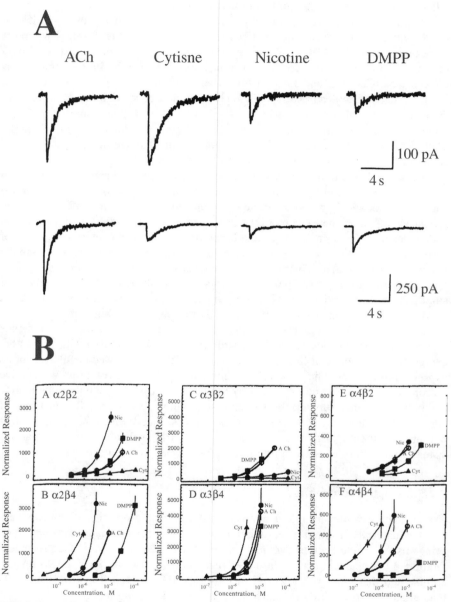

FIGURE 4.6 Whole cell currents in rat parasympathetic neurons evoked by four prototype nicotinic receptor agonists. A) The responses represent the inward currents evoked by 10 second applications of 100 μM ACh, cytisine, nicotine, and DMPP. The holding potential was –90 mV. B) The relative currents evoked by the four agonists in *Xenopus* oocytes that expressed in (A) α2β2, (B) α2β4, (C) α3β2, (D) α3β4, (E) α4β2, (F) α4β4. A) adapted from Poth et al. (1997) and B) adapted from Luetje and Patrick (1991) with permission.

openings in response to ACh. However, a hint of multiple conductance states was observed in rat sympathetic neurons (Mathie et al., 1991), indicating that heterogeneity of nAChRs may exist. In embryonic chick sympathetic neurons, three distinct conductance states were identified. Interestingly, the distribution of conductance states changed following synaptogenesis (Moss et al., 1989).

An early controversy that does seem to have been assisted by the cloning of nAChR subunits is the question of the function of α-BnTx binding sites in peripheral neurons. A reevaluation of their role began in the 1990s, following the isolation of genes encoding α7 subunits. The first evidence that peripheral neuronal α-BnTx sites were in fact functional came from Vijayaraghavan et al. (1992), who used calcium imaging techniques to demonstrate that, in CG, pulses of nicotine could induce increases in intracellular Ca^{2+} that were sensitive to blockade by α-BnTx (Figure 4.7). However, despite the evidence that the addition of nicotine could promote the transfer of Ca^{2+} across the membrane, electrophysiological recordings by these same authors still failed to reveal detectable membrane currents sensitive to α-BnTx. This dilemma was resolved by the finding that very rapid step applications of nicotine (within milliseconds — more than 10 times faster than the puffer pipette application method previously used) did in fact induce a fast activating, rapidly desensitizing membrane current in CG neurons that could be blocked by α-BnTx (Zhang et al., 1994). Furthermore, α-BnTx-sensitive receptors were subsequently shown to contribute to synaptic response in the CG (Ullian et al., 1997), completely revising the traditional idea that α-BnTx sites were purely extrasynaptic and probably nonfunctional in the ganglion.

In the rat, functional α-BnTx receptors were equally difficult to isolate. While nAChRs in rat SCG were shown to be permeable to Ca^{2+} ions, this Ca^{2+}-permeable conductance was not blocked by α-BnTx (Trouslard et al., 1993). In a pharmacological study, Mandelzys et al. (1995) found no evidence for α-BnTx-sensitive responses, even when using a rapid (ms) agonist application method; indeed, their study supported earlier ideas that rat SCG neurons express a uniform population of neuronal nAChRs.

A breakthrough came with the discovery of a novel type of α-BnTx-sensitive response in the parasympathetic neurons of rat intracardiac ganglia (Cuevas and Berg, 1998). Whole-cell patch clamp recordings from dissociated cells, using rapid agonist application, revealed a nicotine-induced membrane current that was blocked by α-BnTx; however, unlike previously described responses, the effect of α-BnTx was readily reversible. Moreover, the kinetics of the nicotinic response differed, with activation and desensitization rates much slower than previously observed (Figure 4.4). The involvement of α7 subunits in mediating this response was confirmed by blocking the effect of nicotine with α7 mAb. Subsequently, it has been demonstrated that rat SCG neurons also express functional α-BnTx receptors; in fact, two types of α-BnTx receptor were described (Cuevas et al., 2000). One type corresponds to α7-containing receptors in intracardiac ganglion neurons, as the responses are slowly desensitizing and reversibly blocked by α-BnTx. The other receptor, seen less frequently, mediates rapidly desensitizing responses and is irreversibly blocked by α-BnTx. Thus, rat peripheral neurons do, in fact, have functional α-BnTx-sensitive

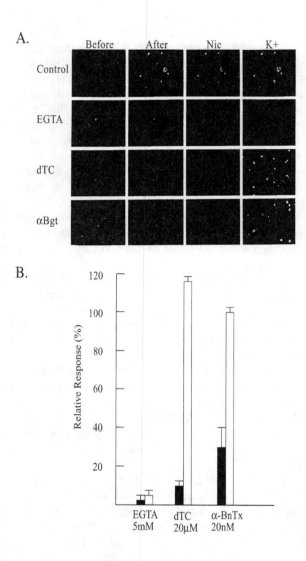

FIGURE 4.7 Calcium imaging of chick ciliary cervical ganglion neurons to nicotine and selective neuronal nAChR antagonists. A. The horizontal rows of 4 columns (panels) show the images obtained BEFORE the application of 1 μM nicotine, 2–4 seconds AFTER its application, the difference between the BEFORE and AFTER panels (NIC) and the net increase produced by 50 mM KCl (K+). The 4 rows indicate the CONTROL in the sense that nothing was added, in the presence of 5 mM EGTA (to remove extracellular calcium), in the presence of 20 μM d-Tubocurarine (dTC), and in the presence of 50 nM α-BnTx (α-BnTx). B. The selected results of experiments shown in A presented as the relative fluorescence produced by nicotine (stippled bars) and 50 mM KCl (open bars adapted from Vijaraghavan et al., 1992).

α7-containing nAChRs and do not, after all, contain a uniform population of neuronal nAChRs.

There have been relatively few studies of the functional types of neuronal nAChRs in DRG or TG ganglion neurons (Sucher et al., 1990; Liu et al., 1993; 1996). In these cells it was found that currents activated by nicotine or other nAChR agonists (e.g., DMPP) could be blocked by α-BnTx, hexamethonium, mecamylamine, and even atropine. These suggest that many functional types of nAChRs exist in primary sensory neurons. There have, however, been more extensive studies done regarding the functional types of nAChRs in epithelial cells. In patch clamp studies using mature human keratinocytes it was found that ACh activated a 32 pS channel that was inhibited by κ-BnTx or mecamylamine, suggesting the presence of α3 or α4 subunits (Grando et al., 1995). In BES cells, ACh or nicotine activated 12 pS channels (Maus et al., 1998). In these same cells whole cell currents activated by nicotine were inhibited by κ-BnTx. Based on a variety of considerations and experiments, it was concluded that BES cells contain α3α5 β2 and α3α5 β4 complexes.

4.2.6 PROBING FUNCTION WITH MOLECULAR DELETIONS: "ANTISENSE" AND GENE "KNOCK-OUTS"

A novel and exciting approach to determining the contribution of individual subunits to native nAChRs is to remove specific subunits selectively and measure the effects of the deletion on nAChR physiology and pharmacology. This can be achieved acutely, by using antisense oligonucleotides — short nucleic acid sequences that bind to mRNA — to prevent synthesis of new subunit protein. Chronic removal of proteins can be achieved by removal of a gene, or a critical part of the gene, encoding the protein, to produce a "knock-out" or "knock-down" model, typically a mouse. Besides the inherent technical difficulties of both types of experiment, one major drawback is the biological tendency to compensate for what is missing, thus producing potentially misleading results.

4.2.6.1 Autonomic Nervous System

Antisense oligonucleotides to the α3 subunit were used by Listerud et al. (1991) to illustrate that functional nAChRs in chick sympathetic neurons are largely composed of α3-containing receptors. Removal of surface receptors using bromoacetylcholine, followed by incubation with α3 antisense for 48 hours, led to a >90% reduction in channel activity compared with control cells. Antisense to α4 had no affect on nAChR channel conductance or kinetics. Removal of the α3 subunit with antisense also led to a compensatory expression of the α7 subunit.

Interestingly, α5 antisense caused the selective loss of a 52 pS conductance channel in single-channel recordings from chick lumbar sympathetic neurons, indicating that α5 subunits make a functional contribution to nAChRs in these cells (Ramirez-Latorre et al., 1996). Deletion of α5 from chick sympathetic neurons using antisense modified the agonist potency profile, but also led to compensatory changes in α7 subunit expression (Yu and Role, 1998a). Antisense deletion of α7 subunits from chick sympathetic neurons removed the ACh-activated, whole-cell current components sensitive to α7 receptor antagonists, α-BnTx, and methyllycaconitine. In single channel recordings, three different conductance states were found to be sensitive to α7 antagonists and/or α7 antisense, suggesting a heterogeneous population of α7-containing nAChRs in chick sympathetic ganglion neurons (Yu and Role, 1998b).

An α3 null mouse (homozygous α3 -/-) has been developed by an exon replacement that deletes three of the transmembrane domains in the α3 subunit (Xu et al., 1999a). The mice survive to birth, but have impaired growth, increased postnatal mortality, and bladder defects, although no other significant peripheral or CNS defects are apparent. In dissociated SCG neurons from the α3 null mice, half the patches tested lacked ACh-activated channels, compared with all patches being responsive in wildtype mice. Patches from the SCG neurons in the α3 null mice responsive to ACh showed low opening probabilities and fewer conductance classes than wildtype mice. These data support the results from antisense deletion experiments, suggesting α3 subunits make a substantial contribution to the functional nAChRs expressed in sympathetic ganglion neurons.

The development of mice lacking specific β subunits (β2 -/-; β4 -/-; β2-/-β4-/-) enabled the contribution of the β subunits to functional nAChRs to be investigated (Xu et al., 1999b). Mice with a single mutation for either β2 or β4 grow to adulthood with no obvious phenotypic abnormalities, while the β2-/-β4-/- double mutant shows similar problems to the α3 null mouse: increased postnatal mortality and bladder defects. Dissociated SCG neurons from the double mutant mice show no responsiveness to nicotine in whole-cell recordings; responses to nicotine are substantially reduced in SCG neurons from the β4 -/- mutant, while in the β2 -/- mutant, responses are similar to those seen in wildtype mice. These findings suggest that the main functional nAChR in sympathetic neurons is composed of α3β4 subunits, although β2 subunits might contribute to function, as the double mutant shows more marked attenuation of responses than the single β4 null mouse.

4.2.6.2 Sensory Systems

Few studies have tested the effects of knocking out (KO) neuronal nAChRs in sensory neurons. Deletion of α4 or β2 subunits reduces nicotine–induced antinociception (Cordero-Erausquin et al., 2000), although these effects may arise from CNS rather than peripheral changes.

4.3 SUMMARY AND CONCLUSIONS

Peripheral neuronal nAChRs are found in the autonomic nervous system, in sensory neurons, and in non-neuronal tissue. Classical pharmacological approaches gave early indications of the diversity of functional nAChRs in autonomic ganglia, which was later confirmed by the detection of mRNA and proteins for multiple nAChR subunits. The contribution of these subunits to the receptors that function in synaptic transmission in the ganglia is now being revealed by comparisons with cloned receptors, by antisense deletions, and by gene knock-outs. In the autonomic nervous system neuronal nAChRs play a role in fast synaptic transmission as well as in modulating synaptic strength and efficacy through their relatively high calcium permeabilities. In sensory neurons, nAChRs have been found in nociceptors and have a role in evoking nociception, pain, and inflammation. In non-neuronal cells nAChRs have many effects, including cell viability, and adhesion.

ACKNOWLEDGMENTS

We thank Mr. Doug Buchacek for help with the figures and Drs. Berg, Grando, Liu, Loring, Vijayaraghavan, and Zigmond who were so generous in sending them to us. Dr. Sid A. Simon also acknowledges support from the Philip Morris External Research Program.

REFERENCES

Alimohammadi, H. and Silver, W.L. (1999), Evidence for nicotinic acetylcholine receptors on nasal trigeminal nerve endings of the rat, *Chem. Senses,* 25, 61–66.

Arias, H.R. (2000), Localization of agonist and competitive antagonist binding sites on nicotinic acetylcholine receptors, *Neurochem. Int.,* 36, 595–645.

Baccaglini,P.I., and Cooper, E. (1982), Influences on the expression of acetylcholine receptors on rat nodose neurons in cell culture, *J. Physiol.,* 324, 441–452.

Ballivet, M., Nef, P., Couturier, S., Rungger, D., Bader, C.R., Bertrand, D., and Cooper, E. (1988), Electrophysiology of a chick neuronal acetylcholine receptor expressed in Xenopus oocytes after cDNA injection, *Neuron,* 1, 847–852.

Boulter, J., Evans, K., Goldman, D., Martin, G., Treco, D., Heinemann, S., and Patrick, J. (1986), Isolation of a cDNA clone coding for a possible neural nicotinic acetylcholine receptor α-subunit, *Nature,* 319, 368–374.

Boulter, J., Connolly, J., Deneris, E., Goldman, D., Heinemann, S., and Patrick, J. (1987), Functional expression of two neuronal nicotinic acetylcholine receptors from cDNA clones identifies a gene family, *PNAS,* 84, 7763–7767.

Boulter, J., O'Shea-Greenfield, A., Duvoisin, R.M., Connolly, J.G., Wada, E., Jensen, A., Gardner, P.D., Ballivet, M., Deneris, E.S., McKinnon, D., Heinemann, S., and Patrick, J. (1990), α3, α5, and β4: three members of the rat neuronal nicotinic acetylcholine receptor-related gene family form a gene cluster, *J. Biol. Chem.,* 265, 4472–4482.

Boyd, R.T., Jacob, M.H., Couturier, S., Ballivet, M., and Berg, D.K. (1988), Expression and regulation of neuronal acetylcholine receptor mRNA in chick ciliary ganglia, *Neuron,* 1, 495–502.

Cachelin, A.B. and Jaggi, R. (1991), β subunits determine the time course of desensitization in rat α3 neuronal nicotinic acetylcholine receptors, *Pfluegers Arch.,* 419, 579–582.

Chiapinelli, V.A. (1983), Kappa-bungarotoxin: a probe for the neuronal nicotinic receptor in the avian ciliary ganglion, *Brain Res.,* 277, 9–21.

Chiapinelli, V.A. and Dryer, S.E. (1984), Nicotinic transmission in sympathetic ganglia: blockade by the snake venom neurotoxin kappa-bungarotoxin, *Neurosci. Lett.,* 50, 239–244.

Chiapinelli, V.A. (1985), Actions of snake venom toxins on neuronal nicotinic receptors and other neuronal receptors, *Pharmac. Ther.,* 31, 1–32.

Clarke, P.B.S. (1992), The fall and rise of neuronal α-bungarotoxin binding proteins, *Trends Pharmacol. Sci.,* 13, 407–413.

Clarke, P.B.S., Schwartz, R.D., Paul, S.M., Pert, C.B., and Pert, A. (1985), Nicotinic binding in rat brain: Autoradiographic comparison of [^{3}H] Acetylcholine, [^{3}H] nicotine, and [^{125}I]-α-bungarotoxin, *J. Neurosci.,* 5, 1307–1315.

Conroy, W.G., Vernallis, A.B., and Berg, D.K. (1992), The α5 gene product assembles with multiple acetylcholine receptor subunits to form distinctive receptor subtypes in brain, *Neuron,* 9, 679–691.

Conroy, W.G. and Berg, D.K. (1995), Neurons can maintain multiple classes of nicotinic acetylcholine receptors distinguished by different subunit compositions, *J. Biol. Chem.,* 270, 4424–4431.

Cordero-Erausquin, M., Marbuio, L.M., Klink, R., and Changeux, J. (2000), Nicotinic receptor function: new perspectives from knockout mice, *TIPS,* 21, 211–217.

Corriveau, R.A. and Berg, D.K. (1993), Coexpression of multiple acetylcholine receptor genes in neurons: quantification of transcripts during development, *J. Neurosci.,* 13, 2662–2671.

Couturier, S., Erkman, L., Valera, S., Rungger, D., Bertrand, S., Boulter, J., Ballivet, M., and Bertrand, D. (1990a), α5, α3 and non-α3. Three clustered avian genes encoding neuronal nicotinic acetylcholine receptor-related subunits, *J. Biol. Chem.,* 265, 17560–17567.

Couturier, S., Bertrand, D., Matter, J.-M., Hernandez, M.-C., Bertrand, S., Millar, N., Valera, S., Barkas, T., and Ballivet, M. (1990b), A neuronal nicotinic acetylcholine receptor subunit (α7) is developmentally regulated and forms a homo-oligomeric channel blocked by α-BTX, *Neuron,* 5, 847–856.

Covernton, P.J.O., Kojima, H., Sivilotti, L.G., Gibb, A.J., and Colquhoun, D. (1994), Comparison of neuronal nicotinic receptors in rat sympathetic neurons with subunit pairs expressed in *Xenopus* oocytes, *J. Physiol.,* 481, 27–34.

Crabtree, G., Rameriz-Latorre, J., and Role, L.W. (1997), Assembly and Ca^{2+} regulation of neuronal nicotinic receptors including the α7 and α5 subunits, *Soc. Neurosci.,* 23, 331.

Cuello, A.C. (1993), *Immunohistochemistry, II. International Brain Research Organisation,* Chichester, U.K.

Cuevas, J. and Berg, D.K. (1998), Mammalian nicotinic receptors with α7 subunits that slowly desensitize and rapidly recover from α-bungarotoxin blockade, *J. Neurosci.,* 18, 10335–10344.

Cuevas, J., Roth, A.L., and Berg, D.K. (2000), Two distinct classes of functional α7-containing nicotinic receptor on rat superior cervical ganglion neurons, *J. Physiol.,* 525, 735–746.

Deneris, E.S., Boulter, J. Swanson, L.W., Patrick, J., and Heinemann, S. (1989), β3: a new member of nicotinic acetylcholine receptor gene family is expressed in brain, *J. Biol. Chem.,* 264, 6268–6272.

Deneris, E.S., Connolly, J., Rogers, S.W., and Duvoisin, R. (1991), Pharmacological and functional diversity of neuronal nicotinic acetylcholine receptors, *Trends Pharmacol. Sci.,* 12, 34–40.

Derkach, V.A., North, R.A., Selyanko, A.A., and Skok, V.I. (1987), Single channels activated by acetylcholine in rat superior cervical ganglion, *J. Physiol.,* 388, 141–151.

Dressier, J.M., O'Mahoney, M., Sieffermann, J.M., and Carstens, E. (1998), Mecamylamine inhibits nicotine but not capsaicin irritation on the tongue: psychophysical evidence that nicotine and capsaicin activate separate receptors, *Neurosci. Lett.,* 240, 65–68.

Dryer, S.E. (1994), Functional development of the parasympathetic neurons of the avian ciliary ganglion: a classic model system for the study of neuronal differentiation and development, *Prog. Neurobiol.,* 43, 281–322.

Duner-Engstrom, M., Fredholm, B.B., Larsson, O., Lundberg, J.M., and Saria, A. (1986), Autonomic mechanisms underlying capsaicin induced oral sensations and salivation in man, *J. Physiol. Lond.,* 373, 87–95.

Duvoisin, R.M., Deneris, E.S., Patrick, J., and Heinemann, S. (1989), The functional diversity of the neuronal nicotinic acetylcholine receptors is increased by a novel subunit: β4, *Neuron,* 3, 487–496.

Elgoyhen, A.B., Johnson, D.S., Boutler, J., Vetter, D.E., and Heinemann, S. (1994), α9: acetylcholine receptor with novel pharmacological properties expressed in rat cochlear hair cells, *Cell,* 79, 705–715.

Elsele, J., Bertand, S., Galzi, J.-L., Devillers-Thiery, A., Changeux, J., and Bertand, D. (1993), Chimeric nicotinic-serotonic receptor combines distinct ligand binding and channel specificities, *Nature,* 366, 479–483.

Flores, C.M., Decamp, R.M., Kilo, S., Rodgers, S.W., and Hargraves, J.M. (1996), Neuronal nicotinic receptor expression in sensory neurons of the rat trigeminal ganglion: demonstration of α3β4, a novel subtype in the mammalian nervous system, *J. Neurosci.,* 16, 7892–7901.

Fumagalli, L. and De Renzis, G. (1984), Extrasynaptic localization of α-bungarotoxin receptors in the rat superior cervical ganglion, *Neurochem. Int.,* 6, 355–364.

Gerzanich, V., Anand, R., and Lindstrom, J. (1994), Homomers of α8 and α7 subunits of nicotinic receptors exhibit similar channel but contrasting binding site properties, *Mol. Pharmacol.,* 45, 212–220.

Goldman, D., Deneris, E., Luyten, W., Kochhar, A., Patrick, J., and Heinemann, S. (1987), Members of a nicotinic acetylcholine receptor gene family are expressed in different regions of the mammalian central nervous system, *Cell,* 48, 965–973.

Goodman, A.G. and Gilman, L. S. (1985), *The Pharmacological Basis of Therapeutics.* ed. VII, A.G. Gilma, L.S. Goodman, T.W. Rall, and F. Murad, Eds., Macmillan Publishing Co., New York, Chap. 5.

Gotti, C., Hanke, W., Maury, M., Ballivet, M., Clementi, F., and Bertrand, D. (1994), Pharmacology and biophysical properties of α7 and α7-α8, α-bungarotoxin receptor subtypes immunopurifies from the chick optic lobe, *Eur. J. Neurosci,* 6, 1281–1291.

Grando, S. (1997), Biological functions of keratinocyte cholinergic receptors, *J. Invest. Dermatol.,* 2, 41–48.

Grando, S. and Horton, R.M. (1997), The keratinocyte cholinergic system with acetylcholine as an epidermal cytotransmitter, *Curr. Opin. Dermatol.,* 4, 262–268.

Grando, S.A., Horton, R.M., Pereria, E.F.R., Diethelm-Okita, B.M., George, P.M., Aldskogius, H., and Conti-Fine, B.M. (1995), A nicotinic acetylcholine receptor regulating cell adhesion and motility is expressed in human keratinocytes, *J. Invest. Dermatol.,* 105, 774–781.

Gross, A., Ballivet, M., Rungger, D., and Bertrand, D. (1991), Neuronal nicotinic acetylcholine receptors expressed in *Xenopus* oocytes: role of the α subunit in agonist sensitivity and desensitization, *Pfluegers Arch.,* 419, 545–551.

Halvorsen, S.W. and Berg, D.K. (1987), Affinity labeling of neuronal acetylcholine receptor subunits with an α-neurotoxin that blocks receptor function, *J. Neurosci.,* 7, 2547–2555.

Halvorsen, S.W. and Berg, D.K. (1990), Subunit composition of nicotinic acetylcholine receptors from chick ciliary ganglia, *J. Neurosci.,* 10, 1711–1718.

Hamill, O.P., Marty, A., Neher, E. Sakmann, B., and Sigworth, F.J. (1981), Improved patch clamp techniques for high resolution current recording from cells and cell-free membrane patches, *Pfluegers Arch.,* 391, 85–100.

Izumi, H. and Karita, K. (1993), Reflex vasodilation in the cat lip elicited by stimulation of nasal mucosa by chemical irritants, *Am. J. Physiol.,* 265, R733–788.

Jacob, M.H. and Berg, D.K. (1983), The ultrastructural localization of α-bungarotoxin binding sites in relation to synapses on chick ciliary ganglion neurons, *J. Neurosci.,* 3, 260–271.

Jacob, M.H., Berg, D.K., and Lindstrom, J.M. (1984), Shared antigenic determinant between the Electrophorus acetylcholine receptor and a synaptic component on chicken ciliary ganglion neurons, *Proc. Natl. Acad. Sci. USA,* 81, 3223–3227.

Keele, C.A. (1962), The common chemical sense, *Arch. Int. Pharmacodyn.,* CXXXIX, 547–557.

Keiger, C.H.J. and Walker, J.C. (2000), Individual variation in the expression profiles of nicotinic receptors in the olfactory bulb and trigeminal ganglion and identification of α2,α6 and α9 transcripts, *Biochem. Pharmacol.,* 59, 233–240.

Lewis, T.M., Harkness, P.C., Sivilotti, L.G., Colquhoun, D., and Millar, N.S. (1997), The ion channel properties of a rat recombinant nicotinic receptor are dependent on the host cell type, *J. Physiol.,* 505, 299–306.

Lindstrom, J. (1996), Neuronal nicotinic acetylcholine receptors, in *Ion Channels,* T. Narahashi, Ed., Plenum Press, NY, 377–450.

Listerud, M., Brussaard, A.B., Devay, P. Colman, D.R., and Role, L.W. (1991), Functional contribution of neuronal AChR subunits revealed by antisense oligonucleotides, *Science,* 254, 1518–1521.

Liu, L., Ma, H., Pugh, W., and Simon, S.A. (1993), Identification of acetylcholine receptors in adult rat trigeminal ganglion neurons, *Brain Res.,* 617, 37–42.

Liu, L. and Simon, S.A. (1996), Capsaicin and nicotine both activate a subset of rat trigeminal ganglion neurons, *Am. J. Physiol.* 270, C1807–1814.

Liu, L., Chang, G., Jiao, Y., and Simon, S.A. (1998), Neuronal nicotinic acetylcholine receptors in rat trigeminal ganglia, *Brain Res.,* 809, 238–245.

Loring, R.H., Chiapinelli, V.A., Zigmond, R.E., and Cohen, J.B. (1984), Characterization of a snake venom neurotoxin which blocks nicotinic transmission in the avian ciliary ganglion, *Neurosci.,* 11, 989–999.

Loring, R.H., Dahm, L.M., and Zigmond, R.E. (1985), Localization of α-bungarotoxin binding sites in the ciliary ganglion of the embryonic chick: an autoradiographic study at the light and electron microscopic level, *Neuroscience,* 14, 645–660.

Loring, R.H. and Zigmond, R.E. (1987), Ultrastructural distribution of -^{Tox}in F binding sites on chick ciliary neurons: synaptic localization of a toxin that blocks ganglionic nicotinic receptors, *J. Neurosci.,* 7, 2153–2162.

Loring, R.H. and Zigmond, R.E. (1988), Characterization of neuronal nicotinic receptors by snake venom neurotoxins, *Trends Neurosci.,* 11, 73–78.

Luetje, C.W. and Patrick, J. (1991), Both α and β subunits contribute to the agonist sensitivity of neuronal nicotinic acetylcholine receptors, *J. Neurosci.,* 11, 837–845.

MacIver, M.B. and Tanelian, D.L. (1993), Structural and functional specialization of Ad and C fiber free nerve endings innervating rabbit corneal epithelium, *J. Neurosci.,* 13, 4511–4524.

Mandelzys, A., Pié, B., Deneris, E.S., and Cooper, E. (1994), The developmental increase in ACh current densities on rat sympathetic neurons correlates with changes in nicotinic ACh receptor α-subunit gene expression and occurs independent of innervation, *J. Neurosci.,* 14, 2357–2364.

Mandelzys, A., De Koninck, P., and Cooper, E. (1995), Agonist and toxin sensitivities of ACh-evoked currents on neurons expressing multiple nicotinic ACh receptor subunits, *J. Neurophysiol.,* 74, 1212–1221.

Mathie, A., Cull-Candy, S.G., and Colquhoun, D. (1987), Single–channel and whole-cell currents evoked by acetylcholine in dissociated sympathetic neurons of the rat, *Proc. Roy. Soc. Lond. B.,* 232, 239–248.

Mathie, A., Cull-Candy, S.G., and Colquhoun, D. (1991), Conductance and kinetic properties of single nicotinic acetylcholine receptor channels in rat sympathetic neurones, *J. Physiol.,* 439, 717–750.

Maus, A.D.J., Pereria, E.F.R., Karachunski, P.I., Horton, R.M., Navaneetham, D., Macklin, K., Cortes, W.S., Albuquerque, E.X., and Conti-Fine, B.M. (1998), Human and rodent bronchial epithelial cells express functional nicotinic acetylcholine receptors, *Mol. Pharm.,* 54, 779–788.

McGehee, D.S. and Role, L.W. (1995), Physiological diversity of nicotinic acetylcholine receptors expressed by vertebrate neurons, *Ann. Rev. Physiol.,* 57, 521–546.

Morita, K. and Katayama, Y. (1989), Bullfrog dorsal root ganglion cells having tetrodotoxin-resistant spikes are endowed with nicotine receptors, *J. Neurophysiol.,* 62, 657–664.

Moss, B.L., Schuetze, S.M., and Role, L.W. (1989), Functional properties and developmental regulation of nicotinic acetylcholine receptors on embryonic chicken sympathetic neurons, *Neuron,* 3, 597–607.

Nef, P., Oneyser, C., Alliod, C. Couturier, S., and Ballivet, M. (1988), Genes expressed in the brain define three distinct neuronal nicotinic acetylcholine receptors, *EMBO J.,* 7, 595–601.

Nguyen, V.T., Ndoye, A., and Grando, S.A. (2000a), Novel human $\alpha 9$ acetylcholine receptor regulating keratinocyte adhesion is targeted by pemphigus vulgaris autoimmunity, *Am. J. Pathology,* 157, 1–15.

Nguyen, V.T., Hall, L.L., Gallacher, G., Ndoye, A., Jalkovsky, D.L., Webber, R.J., Buchli, R., and Grando, S.A. (2000b), Choline acetyltransferase, acetylcholinesterase, and nicotine acetylcholine receptors of human gingival and espohageal epithelia, *J. Dental Res.,* 79, 929–949.

Ogden, D.C., Gray, P.T.A., Colquhoun, D., and Rang, H.P. (1984), Kinetics of acetylcholine activated ion channels in chick ciliary ganglion neurones grown in tissue culture, *Pfluegers Arch.,* 400, 44–50.

Papke, R.L., Boulter, J., Patrick, J., and Heinemann, S. (1989), Single-channel currents of rat neuronal nicotinic acetylcholine receptors expressed in *Xenopus* oocytes, *Neuron,* 3, 589–596.

Papke, R.L. and Heinemann, S.F. (1991), The role of the $\beta 4$ subunit in determining the kinetic properties of rat neuronal nicotinic acetylcholine $\alpha 3$-receptors, *J. Physiol.,* 440, 95–112.

Patrick, J. and Stallcup, W.B. (1977), Immunological distinction between acetylcholine receptor and the α-bungarotoxin-binding component on sympathetic neurons, *Proc. Natl. Acad. Sci. USA,* 74, 4689–4692.

Poth, K., Nutter, T.J., Cuevas, J., Parker, M.J., Adams, D.J., and Luetje, C.W. (1997), Heterogeneity of nicotinic receptor class and subunit mRNA expression among individual parasympathetic neurons from rat intracardiac ganglia, *J. Neurosci.,* 17, 586–596.

Puttfarcken, P.S., Manelli, A.M., Arneric, S.P., and Donnelly-Roberts, D.L. (1997), Evidence for nicotinic receptors potentially modulating nociceptive transmission at the level of the primary sensory neuron: studies with F11 cells, *J. Neurochem.,* 69, 930–938.

Ramirez-Latorre, J., Yu, C.R., Qu, X., Perin, F., Karlin, A., and Role, L. (1996), Functional contributions of $\alpha 5$ subunit to neuronal acetylcholine receptor channels, *Nature,* 380, 347–351.

Ravdin, P.M. and Berg, D.K. (1979), Inhibition of neuronal acetylcholine sensitivity by α-toxins from Bungarus multicinctus venom, *Proc. Natl. Acad. Sci. USA,* 76, 2072–2076.

Rust, G., Burgunder, J.-M., Lauterburg, T.E., and Cachelin, A.B. (1994), Expression of neuronal nicotinic acetylcholine receptor subunit genes in the rat autonomic nervous system, *Eur. J. Neurosci.,* 6, 478–485.

Sargent, P.B. and Pang, D.Z. (1989), Acetylcholine receptor-like molecules are found in both synaptic and extrasynaptic clusters on the surface of neurons in the frog cardiac ganglion, *J. Neurosci.,* 9, 1062–1072.

Sargent, P.B. (1993), The diversity of neuronal nicotinic acetylcholine receptors, *Ann. Rev. Neurosci.,* 16, 403–443.

Sargent, P.B. and Garrett, E.N. (1995), The characterization of α-bungarotoxin receptors on the surface of parasympathetic neurons in the frog heart, *Brain Res.,* 680, 99–107.

Schechter M.D. and Rosecrans J.A. (1971), Behavioral evidence for two types of cholinergic receptors in the C.N.S, *Eur. J. Pharmacol.,* 15(3):375–8.

Schoepfer, R., Conroy, W.G., Whiting, P., Gore, M., and Lindstrom, J. (1990), Brain α-bungarotoxin binding protein cDNAs and mAbs reveal subtypes of this branch of the ligand-gated ion channel gene superfamily, *Neuron,* 5, 35-48.

Séguéla, P., Wadiche, J., Dineley-Miller, K., Dani, J.A., and Patrick, J.W. (1993), Molecular cloning, functional properties, and distribution of rat brain α7: a nicotinic cation channel highly permeable to calcium, *J. Neurosci.,* 13, 596–604.

Sivilotti, L.G., McNeil, D.K., Lewis, T.M., Nassar, M.A., Schoepfer, R., and Colquhoun, D. (1997), Recombinant nicotinic receptors, expressed in *Xenopus* oocytes, do not resemble native rat sympathetic ganglion receptors in single-channel behaviour, *J. Physiol.,* 500, 123–138.

Smith, M.A., Stollberg, J., Lindstrom, J.M., and Berg, D.K. (1985), Characterization of a component in chick ciliary ganglia that cross-reacts with monoclonal antibodies to muscle and electric organ acetylcholine receptor, *J. Neurosci.,* 5, 2726–2731.

Stott, D. I. (1994), Immunoblotting, dot-blotting, and ELISPOT assays: methods and applications, in *Immunochemistry,* C. J. van Oss and M. H. V. van Regenmortel, Eds., Marcel Dekker, Inc., New York, 925–948.

Sucher, N.J., Cheng, T.P.O., and Lipton, S.A. (1990), Neural nicotine acetylcholine responses in sensory neurons from postnatal rat, *Brain Res.,* 533, 248–254.

Tecott, L.H., Eberwine, J.H., Barchas, J.D., and Valentino, K.L. (1987), Methodological considerations in the utilization of in situ hybridization, in *In Situ Hybridization, Applications to Neurobiology,* K. L. Valentino, J. H. Eberwine, and J. D. Barchas, Eds., Oxford University Press, New York, 3–24.

Trouslard, J., Marsh, S.J., and Brown, D.A. (1993), Calcium entry through nicotinic receptor channels and calcium channels in cultured rat superior cervical ganglion cells, *J. Physiol.,* 468, 53–71.

Ullian, E.M., McIntosh, J.M., and Sargent, P.B. (1997), Rapid synaptic transmission in the avian ciliary ganglion is mediated by two distinct classes of nicotinic receptors, *J. Neurosci.,* 17, 7210–7219.

Verino, S., Amador, M., Luetje, C.W., Patrick, J., and Dani, J.A. (1992), Calcium modulation and high calcium permeability of neuronal nicotinic acetylcholine receptors, *Neuron,* 8, 127–134.

Vernallis, A.B., Conroy, W.G., and Berg, D.K. (1993), Neurons assemble acetylcholine receptors with as many as three kinds of subunits while maintaining subunit segregation among receptor subtypes, *Neuron,* 10, 451–464.

Vijayaraghavan, S., Pugh, P.C., Zhang, Z.-W., Rathouz, M.M., and Berg, D.K. (1992), Nicotinic receptors that bind α-bungarotoxin on neurons raise intracellular free Ca^{2+}, *Neuron,* 8, 353–362.

Wada, K., Ballivet, M., Boulter, J., Connolly, J., Wada, E., Deneris, E.S., Swanson, L.W., Heinemann, S., and Patrick, J. (1988), Functional expression of a new pharmacological subtype of brain nicotinic acetylcholine receptor, *Science,* 240, 330–334.

Walker, J.C., Kendal-Reed, M., Keiger, C.J., Bencherif, M., and Silver, W.L. (1996), Olfactory and trigeminal responses to nicotine, *Drug Dev. Res.,* 38, 160–168.

Wang, F., Gerzanich, V., Wells, G.B., Anand, R., Peng, X., Keyser, K., and Lindstrom, J. (1996), Assembly of human neuronal nicotinic receptor α5 subunits with α3, β2 and β4 subunits, *J. Biol. Chem.,* 271, 17656–17665.

Wessler, I., Kirkpatrick, C.J., and Racke, K. (1998), Non-neuronal acetylcholine locally acting molecule, widely distributed in biological systems: expression and function in humans, *Pharmacol. Ther.,* 77, 59–79.

Wessler, I., Kirkpatrick, C.J., and Racke, K. (1999), The cholinergic "pitfall" acetylcholine, a universal cell molecule in biological systems, including humans, *Clin. Exp. Pharm. Physiol.,* 26, 198–205.

Wilson-Horch, H.L. and Sargent, P.B. (1995), Perisynaptic surface distribution of multiple classes of nicotinic acetylcholine receptors on neurons in the chicken ciliary ganglion, *J. Neurosci.,* 15, 7778–7795.

Wisden, W., Morris, B. J., and Hunt, S. P. (1991), *In situ* hybridization with synthetic DNA probes, in *Molecular Neurobiology, A Practical Approach,* J. Chad and H. Wheal, Eds., Oxford University Press, 205–225.

Xu, W., Gelber, S., Orr-Urtreger, A., Armstrong, D., Lewis, R.A., Ou, C.-N., Patrick, J., Role, L., De Biasi, M. and Beaudet, A.L. (1999a), Megacystis, mydiasis, and ion channel defect in mice lacking the α3 neuronal nicotinic acetylcholine receptor, *Proc. Natl. Acad. Sci. USA,* 96, 5746–5751.

Xu, W., Orr-Urtreger, A., Nigro, F., Gelber, S., Sutcliffe, S.B., Armstrong, D., Patrick, J., Role, L., Beaudet, A.L., and De Biasi, M. (1999b), Multiorgan autonomic dysfunction in mice lacking the β2 and the β4 subunits of neuronal nicotinic acetylcholine receptors, *J. Neurosci.,* 19, 9298–9305.

Yu, C.R. and Role, L.W. (1998a), Functional contribution of the α5 subunit to neuronal nicotinic channels expressed by chick sympathetic ganglion neurones, *J. Physiol.,* 509, 667–681.

Yu, C.R. and Role, L.W. (1998b), Functional contribution of the α7 subunit to multiple subtypes of nicotinic receptors in embryonic chick sympathetic neurones, *J. Physiol.,* 509, 651–665.

Zhang, Z.W., Coggan, J.S., and Berg, D.K. (1996), Synaptic currents generated by neuronal acetylcholine receptors sensitive to α-bungarotoxin, *Neuron,* 17, 1231–1240.

Zhang, Z.-W., Vijayaraghavan, S., and Berg, D.K. (1994), Neuronal acetylcholine receptors that bind α-bungarotoxin with high affinity function as ligand-gated ion channels, *Neuron,* 12, 167–177.

Section 2

5 Nicotinic Cholinergic Receptors as Targets in Pain Control

M. Imad Damaj and Christopher M. Flores

CONTENTS

5.1 INTRODUCTION

Pain can be defined as a sensory experience evoked by stimuli that injure or threaten to destroy tissue. The management and treatment of pain is probably one of the commonest and yet most difficult aspects of medical practice. Analgesic therapy is currently dominated by the two major classes of analgesic drugs, namely opiates and nonsteroidal antiinflammatory drugs (NSAIDs). Although there have been several improvements to these analgesic mainstays in terms of formulation, selectivity, etc., there has been little conceptual innovation. Most acute types of pain usually, though not always, arise from an identifiable injury or disease and, most importantly, are self-limiting and of a relatively short duration, because they either respond readily to treatment or subside on their own. It is the more persistent pain syndromes (e.g., migraine, cancer pain, and neuropathic pains) that pose the greatest challenge in terms of analgesic drug therapy. Thus, NSAIDs exhibit little if any efficacy for many types of persistent pain, and the long term use of opioids in patients with nonmalignant pains, for example, remains controversial. Moreover, both classes of analgesic drugs upon long term use produce serious side effects, such as gastrointestinal disturbances, gastric ulcerations, and renal damage (with NSAIDs), and nausea, respiratory depression, and possible dependence (with opioids). Accordingly, safer, more efficacious analgesic agents are needed. One novel strategy for the management of pain involves the use of nicotinic cholinergic agonists (see Figure 5.1).

As originally demonstrated by Davis et al.,[1] nicotine and its congeners have been shown to reduce responses to noxious stimuli in experimental animals using a variety of nociceptive tests. Typically these effects are blocked by nicotinic (e.g., mecamylamine) but not opioid (e.g., naloxone) receptor antagonists. However, such antinociceptive effects of nicotinic compounds are typically of rather short duration (5 to 30 min) and occur at doses that reduce locomotor activity and body temperature. Studies on humans suggest that smokers have a significantly elevated pain threshold and that cigarette smoking can have analgesic effects which may depend on factors associated with smoking but not necessarily nicotine, and may be difficult to resolve from withdrawal relieving effects. On the other hand, when nicotine proper is administered to human nonsmokers, it produces analgesia independent of withdrawal relief; however, these analgesic effects are observed in men

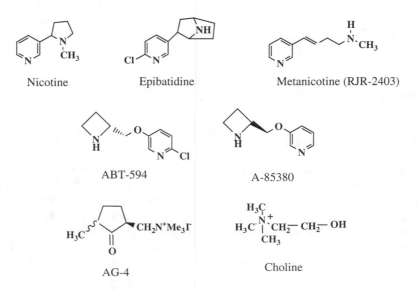

FIGURE 5.1 Structures of some nicotinic agonists tested as antinociceptive agents.

but not women and often require relatively high doses. Thus, there are significant, unresolved issues to be addressed before nicotinic receptor agonists may be considered viable analgesic drug candidates. Nonetheless, interest in a nicotinic cholinergic approach to pain control has been rekindled by recent discoveries of epibatidine, a novel nicotinic agonist isolated from the skin of *Edpipedobates tricolor,* an Ecuadoran frog[2] and its synthetic analog ABT-594,[3] both of which exhibit antinociceptive activity with a potency at least two orders of magnitude greater than and efficacy equal to that of morphine. Furthermore, rapid advances in the understanding of the molecular biology, physiology, and biochemistry of nAChRs, which have occurred over the last decade, will likely lead to creation of more selective nicotinic agonists with safer therapeutic profiles.

This chapter (1) highlights some of the recent contributions providing key insights on mechanisms of nicotinic receptor-mediated antinociception in animal models; (2) reviews the distribution of nicotinic cholinergic receptors along the neuraxis, emphasizing the identities and locations of those that appear most important in terms of pain and analgesia; (3) discusses the genetic factors, including gender, that may determine variability in analgesic responses to nicotinic drugs, integrating the relevant clinical literature; and (4) describes the latest generation of nicotinic agents under investigation, with an eye toward prospects for the ultimate development of one or more clinically viable nicotinic analgesic drugs.

5.2 ANALGESIC EFFECTS OF NICOTINE AND nAChRS AGONISTS IN DIFFERENT PAIN MODELS

Perhaps the most compelling revelation to emerge from the collective preclinical studies on nicotinic receptor-mediated antinociception is the remarkable breadth of animal pain models in which nicotinic agonists exhibit efficacy (Table 5.1). Thus, at least ten, biochemically and pharmacologically distinct agents have shown antinociceptive as well as antihyperalgesic/antiallodynic activity in tests of mechanical, thermal, and chemical nociception. Interestingly, these agents are effective in models of both inflammatory (e.g., carrageenan) and neuropathic (e.g., Chung) pain. This is particularly significant because persistent pain of neuropathic origin in humans is often extremely difficult to treat by any means. Taken together, animal studies to date indicate that nicotinic agonists display a profile of antinociceptive activity at least as broad as or broader than that of any other known analgesic drug class. Moreover, this spectrum of action is characterized by very high potency and by efficacy equal to that of opioids in each case in which the two drug classes have been directly compared. What remains to be seen is to what extent such effects will translate to analgesia in humans, for which there is very little and conflicting data.

In addition to the synoptic generalizability of the nicotinic agonist effects mentioned previously, other attributes are worth noting. For instance, the vast majority of behavioral studies on nicotinic receptor-mediated antinociception to date (see Table 5.1) have assessed responses in models of spinal, somatic pain. Given the widespread incidence and prevalence of orofacial pain,[4] it would be important to know whether nicotinic cholinergic drugs are likely to have any efficacy

TABLE 5.1
Summary of Nicotinic Agonist Effects on Nociception in Animals

Modality/Model	Nociceptive Test	Nicotinic Drug	Species
		Effect	

Antinociception[1]

Modality/Model	Nociceptive Test	Nicotinic Drug	Species
Chemical	Acetic Acid	NIC	Mouse
	Formalin (Acute Phase)	ABT-594,[111] EPI[5]	Rat
		META,[156] NIC[156]	Mouse
	Para-phenylquinone	ABT-594,[157] EPI,[16] META,[156] NIC[156]	Mouse
Electrical	Tooth Pulp Stimulation	DMPP, NIC	Rabbit
Mechanical	Gall Bladder Distension	NIC	Cat
	Paw Withdrawal (M)	META,[156] NIC	Rat
Thermal	Cold Plate	ABT-594[157]	Mouse
	Hot Plate	ABT-418, ABT-594, EPI, EPX, META[156]	Mouse
		NIC	Mouse, Rat
		NMCC	Rat
	Paw Withdrawal (T)	A-85380,[47] ABT-594, CYT, META,[10] NIC	Rat
	Skin Twitch	NIC	Dog
	Tail-Flick	ABT-418, META[156]	Mouse
		NMCC, CYT, DMPP	Rat
		EPI, NIC	Mouse, Rat
	Tail Withdrawal	EPI[8]	Mouse
		NIC	Rat

Antihyperalgesia/Antiallodynia[2]

Modality/Model	Nociceptive Test	Nicotinic Drug	Species
Capsaicin	Paw Withdrawal (T)	META,11 NIC	Rat
Carrageenan	Paw Withdrawal (M)	EPI, NIC	Rat
	Paw Withdrawal (T)	EPI	Rat
CFA	Paw Withdrawal (M)	META,[156] NIC[156]	Rat
Chung	Paw Withdrawal (M-vF)	ABT-594, META[11]	Rat
Formalin (Tonic Phase)	Nocifensive Behavior	A-85380,[129] ABT-594, EPI[5]	Rat
		META[156], NIC[156]	Mouse

Adapted from Flores and Hargreaves.[51]

Abbreviations: CFA, complete Freund's adjuvant; CYT, cytisine; DMPP, dimethylphenylpiperazinium; EPI, epibatidine; EPX, epiboxidine; M, mechanical (Randall-Selitto); M-vF, mehcanical (von Frey); META, metanicotine (RJR-2403); NIC, nicotine; NMCC, N-methylcarbamylcholine; T, thermal (Hargreaves).

[1]antinociceptive: relating to an increase in nociceptive threshold in naïve animals.

[2]antihyperalgesic/antiallodynic: relating to a reversal of the exaggerated responsiveness to noxious stimuli or of the reduction in nociceptive threshold, respectively, following injury.

in treating painful conditions of trigeminal origin. Recently, Gilbert et al.,[5] showed that, in the orofacial adaptation of the formalin model,[6] epibatidine exhibited dose- and time-dependent antinociception in both phases of the test. This is important because the effects in the acute and tonic phases were abolished by mecamylamine, indicating that epibatidine was exerting its antinociceptive and antihyperalgesic effects, respectively, via nicotinic receptors, thereby further extending the breadth of action for this drug class. In another example, Lawand et al.,[7] showed that, in the kaolin/carrageenan model of knee joint inflammation, not only did spinally administered epibatidine reduce spontaneous pain-related behaviors as well as edema and hyperthermia of the joint, but also significantly blocked the secondary thermal hyperalgesia that developed in the ipsilateral paw. Moreover, epibatidine also was capable of reversing these effects when administered 4.5 hours after the induction of experimental arthritis. This latter finding is particularly relevant from the standpoint of human behavior that most commonly involves treatment with analgesic and/or anti-inflammatory drugs following rather than preceding tissue damage. It is important to note that this is one of only a relatively few rat studies[8–11] showing any direct antinociceptive effects of nicotinic agents at the level of the spinal cord. Indeed, whereas some have shown that intrathecally administered nicotinic agonists have little or no antinociceptive effect,[12–14] others have suggested a pronociceptive effect.[15] Conversely, antinociception following intrathecal injection of nicotinic agonists is more commonly and uniformly observed in the mouse.[16–18] Yet to be determined is the extent to which such discrepancies arise as a function of the particular species or pain model/test employed, or indicate an acute dependence on the precise route and/or site of drug application (e.g., direct intraspinal drug compared with intrathecal injections) or, more likely, on the drug itself.

Another promising avenue of research involves the evaluation of combinatorial antinociceptive effects produced by the coadministration of nicotine and opioids. For example, intrathecal or intracerebroventricular nicotine, at doses which had little effect by themselves, potentiated the antinociception produced by morphine.[19,20] Similarly, Zarrindast et al.,[21] found that doses of nicotine as low as 0.0001 mg/kg, which had no antinociceptive effects alone, potentiated morphine-induced antinociception in the tail-flick assay. It is interesting that whereas naloxone predictably reduced the response to morphine in the presence or absence of nicotine, atropine — but not mecamylamine — blocked the potentiating effects of nicotine. Whether such a result is idiosyncratic, indicates a complex interaction between nicotinic and muscarinic receptors or implicates the involvement of an atropine-sensitive nicotinic receptor subtype (e.g., α9[22]) remains to be determined. Nonetheless, this line of investigation provides a rationale for the relatively attractive approach of using a nicotinic analgesic as an opioid-sparing adjunct or vice versa. Indeed, the major limitation of currently utilized analgesics agents, aside from a lack of efficacy against certain types of pain, is dose-limiting toxicity. For the opioids, in particular, such untoward effects include a potential abuse liability. This concern over chemical dependence increases as tolerance develops because of the consequent necessity to escalate the dose and frequency of administration, raising apprehensions that can result in underutilization and undertreatment.

5.3 NICOTINIC RECEPTOR SUBTYPES INVOLVED IN THE ANTINOCICEPTIVE EFFECTS OF NICOTINE AND NICOTINIC AGONISTS

The ultimate development of one or more viable nicotinic analgesic compounds, at least as a result of any rational drug design strategy, will depend initially on the precise localization and identification of those neuronal nicotinic receptor subtypes whose activation produces analgesia. Additionally, these discoveries will lead to a more comprehensive view of the mechanisms from which this analgesia derives, further improvements in the formulation, scheduling, and possible adjunction of nicotinic analgesics and, ultimately, greater insights on nociceptive processing itself. Considerable progress has been made in the past decade regarding the characterization of nicotinic receptors along the neuraxis. Molecular cloning, mRNA mapping, and radioligand binding studies have revealed the existence and location of multiple neuronal nicotinic receptor subtypes in brain and spinal cord as well as in the peripheral nervous system. Even so, it has been difficult to determine which subtypes of nAChRs mediate the various effects of nicotinic agonists and the anatomic location at which these actions occur. Most of the information available to date on nAChR subtypes involved in nicotinic analgesia has been generated using a combination of pharmacological and anatomical approaches. In addition, gene deletion and antisense strategies have been used more recently to examine the role of nAChRs subtypes in nicotine-induced antinociception. Nonetheless, there continue to be substantial gaps in knowledge; therefore, an examination of neuronal nicotinic receptor populations involved in nociception is warranted.

5.3.1 SUPRASPINAL SITES

The distribution of cholinergic somata, projections and terminals in mammalian brain has been well described (for reviews, see Lewis and Shute[23] and Woolf[24]). Briefly, the cerebral mantle receives innervation from the cholinergic basal complex, comprising the medial septum, diagonal band of Broca, and nucleus basalis of Meynert, while the subcortical mass receives cholinergic innervation from the pontomesencephalon. The few cholinergic interneurons appear to be restricted to the striatum (as well as the cortex in rat). In terms of the various nicotinic receptors that are the targets of this cholinergic input, distribution has been extensively mapped through radioligand binding analyses. Thus, following the initial demonstration of high affinity, saturable and sterospecific binding of [3H]-nicotine in rat brain homogenates,[25] the regional distribution of neuronal nicotinic receptors has been elucidated using [3H]-nicotine,[26] [3H]-Ach,[27] [3H]-N-methylcarbamylcholine,[28] and [3H]-cytisine.[29] All of these agents label a single class of sites, albeit potentially consisting of different affinity states depending on the conditions used. This receptor class appears to be predominantly, if not entirely, composed of α4 and β2 subunits, and exhibits varying densities of expression that are highest in thalamus, intermediate in cortex, striatum, and colliculi, and lowest in hippocampus, brainstem, cerebellum, and hypothalamus.[30] Autoradiographic analyses have localized these sites with a higher degree of resolution to discrete regions that include, *inter alia,* the medial habenula, interpeducular nucleus, several motor and sensory nuclei of the thalamus, the basal ganglia, layers I, II, and IV of cerebral cortex, and the molecular layer of the dentate gyrus.[31–35] More recently, [3H]-epibatidine has been shown to label at least two sites in rodent and human brain homogenates.[36] Higher resolution has been achieved autoradiographically, indicating that the distribution of [3H]-epibatidine-labeled sites, which includes the α4β2 subtype as well as other subtype(s) that most likely contain α3, is very similar to that revealed by the aforementioned ligands, albeit of generally higher density.[37]

In addition, certain nicotinic receptors that include α7-containing subtypes may be labelled with [125I]-α-bungarotoxin.[38] Compared with the aforementioned reagents, radioligand binding with [125I]-α-bungarotoxin exhibits a distinct distribution pattern primarily associated with brain areas receiving direct sensory inputs, such as olfactory bulbs, superior colliculus, ventral lateral geniculate nucleus, cochlear nuclei and olivary nuclei, as well as limbic areas in hippocampus, amygdala, olfactory tubercle, medial mammillary nucleus, and the dorsal tegmental nucleus of Gudden.[31, 35, 39] Nicotinic receptor labeling has also been achieved with [125I]-κ-bungarotoxin, a second component isolated from the venom of the Formosan krait, *Bungarus multicinctus*.[40] κ-bungarotoxin, also called toxin F , toxin 3.1, or neuronal bungarotoxin (owing to its selectivity as an antagonist at nicotinic receptors located on neurons rather than on myocytes) appears to label subsets of receptors that are measured with many of the other nicotinic radioligands. This manifests as two saturable, high affinity components, the first of which is displaced by unlabelled nicotine and most densely distributed in fasciculus retroflexus, the lateral geniculate nucleus, the medial terminal nucleus of the accessory optic tract, and the olivary pretectal nucleus. The second component is displaced by α-bungarotoxin and localized to the lateral geniculate nucleus, the subthalamic nucleus, the dorsal tegmental nucleus, and the medial mammillary nucleus. In particular, κ-bungarotoxin has been shown to antagonize nicotine-induced dopamine release from rat striatum,[41] a function widely ascribed to an α3β2-containing subtype.

It is as yet unclear, however, precisely which nicotinic receptor population(s) is(are) responsible for the plethora of antinociceptive effects depicted in Table 5.1. In large part, this lack of knowledge results from a relative paucity of subtype-specific agonists and antagonists. Nonetheless, multiple experimental approaches have proved insightful. The earliest and most exploited strategy involved microinjecting various nicotinic agonists into discrete CNS loci and assessing the effect on nociceptive status using a variety of tests. Thus, several "centers" of nicotinic antinociception were identified. These included the periaqueductal gray,[42, 43] the rostral ventromedial medulla, particularly the nucleus raphe magnus[44–48] and the dorsolateral mesopontine tegmentum, including the locus coeruleus.[46, 49, 50] Principally, these studies have provided the preponderance of fundamental knowledge regarding those sites in brain at which nicotine and certain of its congeners produce antinociception, at least in the rat. Taken together, these investigations tend to suggest that there are multiple CNS loci involved in nicotinic antinociception. Interestingly, but not coincidentally, these loci represent the origin of opioidergic, serotonergic, and noradrenergic fibers that descend from the brain stem to the spinal/medullary dorsal horn. It is known that these descending, pain inhibitory pathways modulate nociceptive input at the level of the dorsal spinal cord; this would seem to be the frontrunning mechanism whereby brain nicotinic receptor activation leads to antinociception (for review, see Flores and Hargreaves[51]).

At this point, it is difficult to predict which nicotinic receptor subtypes and which descending pathways are the most important and/or what factors would make one or another a better target in terms of nicotinic analgesia. However, two independent investigations using knockout and antisense approaches have shed some light on these questions. First, Marubio et al.,[52] demonstrated that mice lacking either the α4 or β2 nicotinic receptor subunit gene exhibited a significantly reduced antinociceptive response to nicotine in the hot plate assay, and this was correlated to a loss in high affinity [3H]-nicotine binding sites (presumably α4β2 receptors) throughout brain. Moreover, patch clamp recordings in areas implicated in supraspinal nicotinic antinociception (e.g., nucleus raphe magnus and thalamus) showed a loss of nicotine-elicited currents in α4 or β2 knockout mice.

Consistent with this finding, Bitner et al.,[53] have shown that i.c.v. infusion of α4 subunit antisense oligonucleotide led to a significant attenuation of the antinociception produced by A-85380 (a nicotinic receptor agonist) in the rat paw withdrawal test compared with mismatch oligonucleotide treated animals. This change in nicotinic receptor-mediated antinociception in the α4 antisense animals was associated with approximately 50% diminution in immunohistochemical staining for this subunit in the dorsal raphe nucleus. In contrast, there was no change in the staining of the serotonin neuronal marker tryptophan hydroxylase, the expression of which almost nearly completely overlaps with the α4 subunit in control animals. In a previous study, Bitner et al.,[54] showed that, in the nucleus raphe magnus (NRM), there is also a nearly complete overlap of α4 subunit and tryptophan hydroxylase immuostaining. Moreover, the microinjection of ABT-594,[55] epibatidine, or A-85380[47] into the NRM produces antinociception in the paw withdrawal assay, and both α4 staining as well as antinociception in response to nicotinic agonist treatment were attenuated by pre-treatment with the serotonin neurotoxin 5.7-dihydroxytryptamine.[54] Collectively, these studies provide strong evidence that at least one mechanism of nicotinic receptor-mediated antinociception involves α4-containing receptors located on serotonergic neurons in one or more raphe nuclei.

The role of non-α4β2 nAChRs in nicotinic analgesia is still largely unknown. However, recent studies have reported that the α7 antagonists α-bungarotoxin (α-BGTX) and methyllycaconitine (MLA) failed to block the antinociceptive effects of various nicotinic agonists after i.c.v. or i.t. administration in rats and mice[18, 56] using the tail-flick test. Although these results indicate little involvement of α7 subtypes in the antinociceptive effects of nicotinic agonists, none of the agonists used in these studies have high selectivity and/or affinity for the α7 nAChR subtypes. On the other hand, Damaj et al.,[57] assessed the involvement of α7 nicotinic receptors in nicotinic analgesia after spinal (i.t.) or intraventricular (i.c.v.) administration in mice using the tail-flick and hot-plate tests. Dose-dependent antinociceptive effects were seen with the α7 agonist choline after spinal or supraspinal injection. Furthermore, MLA or α-BGTX significantly blocked the effects of choline, while dihydro-β-erythroidine or mecamylamine did not. These results strongly support the involvement of α7 subunits in choline's antinociceptive effects. Interestingly, the α7 partial agonists DMXB or 4-OH-DMXB failed to elicit a significant antinociceptive effect but did block choline-induced antinociception in a dose-dependent manner following i.t. injection. Taken together, these results suggest that activation of α7 receptors in the CNS elicits antinociception in an acute thermal pain model.

It is worth noting that the vast majority of mechanistic studies conducted to date have largely focused on tests of acute, thermal nociception. Thus, it is entirely possible that the mechanisms, and, thus, the nAChRs subtypes involved in the antinociceptive effects of nicotinic agonists may vary as a function of the type of pain model used. Nonetheless, the continuing acquisition of the kind of information cited previously would rank among the most important research objectives impacting the future potential success of nicotinic analgesic drug development.

5.3.2 SPINAL SITES

As alluded to earlier, the dorsal spinal cord is a primary locus for the neuromodulation of sensory input, including nociception and the initial point of connectivity between the central and peripheral nervous systems. With respect to nociception, it comprises interneurons and the somata of ascending, second-order projection neurons of the anterolateral system as well as terminals of both the central projections of primary sensory neurons and the descending modulatory neurons of the dorsolateral funiculus. As such, the spinal cord would seem an opportune site at which to direct pain modulating medicaments, including nicotinic analgesics. Indeed, the repertoire of nicotinic receptors localized to the mammalian spinal cord is equally heterogeneous as in the brain. However, the functional role that these receptors play is controversial. Essentially, the historical description of nicotinic receptor binding sites in the spinal cord paralleled that in brain, evolving primarily as a function of the radioligands developed to measure these sites.

Thus, Schecter et al.,[86] first described [125I]-α-bungarotoxin binding in both thoracic and lumbar spinal cord homogenates of the rat. More detailed autoradiographic analyses revealed that these sites were numerous within the substantia gelatinosa (equivalent to Rexed's lamina II) in rat[58] and human[59] spinal cord, and, in the rat, these sites were reduced following dorsal rhizotomy.[60] Nicotinic acetylcholine receptors that could be labeled by [3H]-acetylcholine and that are not sensitive to α-bungarotoxin were localized to laminae II, III, IX, and X of rat, cat, monkey and human cervical spinal cord.[61] It should be noted that lamina X neurons in slices of rat thoracolumbar spinal cord can be depolarized by the nicotinic receptor agonists DMPP, nicotine or cytisine, and these responses are blocked by mecamylamine but not methyllycaconitine.[62] The molecular identity of at least some of these receptors was elucidated by Flores et al.,[30] who used immunoprecipitation of [3H]-cytisine binding sites to demonstrate the existence of spinal α4β2 receptors. Interestingly, these spinal nicotinic receptors, as detected by [3H]-nicotine binding, are among the first to appear ontogeneticallly (rat gestational day 12), according to a caudo-rostral developmental pattern of expression.[63] More recently, several investigators have used [3H]-epibatidine to demonstrate that nicotinic cholinergic receptors in rodent spinal cord comprise at least two classes of binding sites.[64–66] Apparently, there are approximately equal numbers of each site in dorsal lumbar spinal cord, as there were approximately twice as many receptors bound by [^{3}H]-epibatidine as by [^{3}H]-cytisine.[64] In addition, $\alpha2$, $\alpha3$, $\alpha4$, $\alpha7$ and $\beta2$ mRNAs were found in the spinal cord.[67–69] However, no signal was detected for $\beta4$ mRNA.[70]

From a functional standpoint, nicotinic receptors in spinal cord are perhaps the most enigmatic. Ironically, one of the first functional nicotinic receptors of central neuronal origin was described on the Renshaw cell.[71] However, related to nociceptive transmission, widely conflicting reports originally refuted the idea of spinal nicotinic antinociception, then adopted a pronociceptive action of nicotinic agonists administered intrathecally and, most recently, have affirmed the concept of spinally-mediated nicotinic analgesia. As mentioned previously, the antinociceptive effects of intrathecally administered nicotinic agonists are consistently and reproducibly observed in the mouse tail-flick test.[72] However, this is not the case in the rat. Thus, Yaksh et al.,[12] showed that intrathecally administered muscarinic (oxotremorine) but not nicotinic (nicotine) agonists were antinociceptive in the rat hot-plate and tail-flick assays and that the antinociceptive effects of nonselective cholinomimetics (carbachol or acetylcholine) were blocked by pretreatment with atropine but not curare. Similarly, while nicotine had virtually no effect in the rat tail-flick test (however, see later comments), either carbachol or neostigmine produced dose-dependent antinociception with the latter being reversed by atropine.[13] It is interesting that at the 5 µg dose, however, nicotine produced a statistically significant but rapid and transient antinociceptive effect. Although this observation was attributed to a paralytic action of the drug, no such action was noted at the 10 µg dose. In addition, no effect of i.t. N-methyl-carbamylcholine, a selective nicotinic receptor agonist, was detected in the rat hot-plate or tail-flick tests.[46] Finally, Gillberg et al.,[14] showed that, in the rat-tail withdrawal test, intrathecal physostigmine produced antinociception that was completely antagonized by atropine but not mecamylamine. Moreover, nicotine was without effect, leading to the conclusion that nicotinic receptors are not involved in spinal antinociception.

On the other hand, it has been suggested that nicotinic receptors may mediate a pronociceptive action at the level of the spinal cord. Thus, Khan et al.,[15] have shown that a variety of nicotinic agonists, including nicotine, cytisine, DMPP, and N-methyl-carbamylcholine, produce a constellation of motoric and vocalization responses, called an irritation index, that have been interpreted as nociceptive. These effects of i.t. nicotinic agonists are antagonized to varying degrees by mecamylamine, hexamethonium, dihydro-β-erythroidine, α-lobeline, and methyllycaconitine, indicating the involvement of one or (probably) more neuronal nicotinic receptor subtypes. However, the relationship of this devised behavioral index to pain is questionable. Primarily because, although the irritation response of either nicotine or cytisine was blocked by i.t. administration of the antinociceptive agents, morphine, or MK-801, these drugs also significantly reduced the chronotropic (nicotine) and pressor (cytisine) responses as well.[15] In a follow-up study Khan et al.,[73] showed that epibatidine, as well as nicotine and cytisine, was capable of producing the irritation response and that, with the same rank order of potency, these three compounds dose dependently and mecamylamine reversibly induced the spinal release of the excitatory amino acids glutamate and aspartate. The ability of MK-801 to inhibit the irritation response to nicotine, therefore, is consistent with the hypothesis that nicotinic agonists produce this behavioral response, at least in part, via the release of excitatory amino acids and the subsequent acitvation of NMDA receptors. Similarly, i.t. administration in rat of A-85380, a novel nicotinic agonist, induced a pronociceptive effect as measured by a decrease in the latency to paw withdrawal to an acute thermal pain.[74] This hyperalgesia was blocked by mecamylamine but not by α-bungarotoxin or dihydro-β-erythroidine.

More recently, a slowly growing number of studies suggest that certain nicotinic agonists, in fact, may be capable of producing antinociception via nicotinic receptors in the spinal cord. For example, i.t. epibatidine, in addition to producing an irritation response, caused a thermal antinociception in the paw withdrawal test, although the latter effect required higher doses and was of shorter duration than the former and could not be replicated with nicotine or cytisine.[74–76] Interestingly, this antinociceptive effect of epibatidine was blocked by mecamylamine or α-lobeline, but not by methyllycaconitine or dihydro-β-erythroidine.[75,76]

It is also worth noting several diverse lines of investigation that implicate a role for spinal nicotinic receptors in neurogenic inflammatory processes. Thus, it was shown that ABT-594 could inhibit capsaicin-evoked calcitonin, gene-related peptide release from slices of rat spinal dorsal horn *in vitro*.[77] Moreover, i.t. nicotine dose dependently inhibited bradykinin-induced plasma extravasation into the synovium of the rat knee joint.[78] Finally, in a model of arthritis, i.t. epibatidine blocked or reversed the secondary hyperalgesia, edema and hyperthermia produced by the injection of kaolin/carrageenan into the knee joint of the rat.[7] Abundantly clear is that spinal cholinergic systems, either nicotinic or muscarinic, are not playing a tonic role in the modulation of nociception (e.g., contributing to basal nociceptive thresholds), as the i.t. injection of nicotinic (e.g., mecamylamine) or muscarinic (e.g., atropine) antagonists are without effect.[46, 79, 80] Nonetheless, it would seem that spinal nicotinic receptors also might be emerging as potential analgesic and anti-inflammatory drug targets.

5.3.3 PERIPHERAL SITES

Peripheral nicotinic cholinergic receptors located on autonomic ganglia[81] as well as at the neuromuscular junction[82] were the first and perhaps best studied of all acetylcholine receptors (for reviews, see references[83–85]). However, much less is known about the structure and function of nicotinic receptors located on sensory neurons, including nociceptors. Nonetheless, the prospect of targeting those neurons responsible for the initial detection of noxious environmental stimuli and the subsequent transduction and transmission is immediately attractive, and this attraction becomes stronger still given that these neurons exist outside the blood-brain barrier. Nicotinic receptor binding sites were originally demonstrated in sensory neurons of the rat trigeminal and dorsal root ganglia, first in membrane homogenates[86] and later using autoradiography.[39] These initial findings were later replicated in rat, monkey, cat, and human dorsal root ganglia and extended to show that these receptors were trafficked to both the central and peripheral processes of the sensory nerve axon.[87,88] Pursuant to the molecular biology windfall that began in the 1980s and has since produced nearly a dozen nicotinic receptor subunit clones, extensive mapping of their distribution in the central and peripheral nervous systems has been undertaken. Consequently, it was shown that, similar but not identical to the subunit expression in autonomic ganglia, sensory ganglia expressed $\alpha3$, $\alpha4$, $\alpha5$, $\beta2$ and $\beta4$ subunit mRNA in rat[67, 89] and chick.[90] Direct evidence for $\alpha3\beta4$ and $\alpha4\beta2$ subtypes in rat trigeminal ganglia was subsequently provided, indicating a relative density of approximately 3 to 1, respectively.[91] Importantly, at least some of these receptors appear to be localized on nociceptors, as the destruction of a subpopulation of polymodal nociceptors via neonatal treatment with capsaicin led to a decrease in [3H]-nicotine binding sites in mixed homogenates of spinal cord, dorsal roots, and DRG.[92] Taken together with the earlier work on α-bungarotoxin sensitive binding sites, currently at least three major nicotinic receptor subtypes are expressed by sensory neurons at the protein level, namely $\alpha7$, $\alpha3\beta4$ and $\alpha4\beta2$, although others are likely to be identified. In addition, more recent investigations using RT-PCR have suggested that the diversity of subunit expression may be greater still and include $\alpha2$, $\alpha6$, $\alpha9$, and $\beta3$ subunits at the mRNA level.[93,94]

This heterogeneity of spinal nicotinic receptor expression appears to be paralleled at the functional level. Thus, nearly a half century ago, Armstrong et al.,[95] implicated an algogenic role for acetylcholine in human skin, an effect which was attributed, at least in part, to a nicotinic receptor mechanism.[96] Similarly, the application of nicotine to the eye of the rat produced blepharospasm, which the authors represented as "nicotinic" pain.[97] Physiologically, nicotinic agonists can 1) activate nociceptors *in vivo*[97–99] or *in vitro*;[100] 2) depolarize primary sensory neurons in culture;[101–105]; and 3) induce the efflux of ^{86}Rb+ as well as the release of immunoreactive substance P from cultured F11 cells[.106] a DRG-neuroblastoma hybrid cell line used to model nociceptors.[107]

Based on the impressive list above, it would not appear immediately tenable to propose a peripherally mediated mechanism of nicotinic antinociception. And the prospect is hardly enhanced by the demonstration that quaternary methiodide derivatives of nicotine, presumably not brain penetrant, were without effect in the rat and mouse tail-flick assays.[108] However, there has been an intriguing if not promising accretion of studies tending to support the concept of peripherally-mediated nicotinic analgesia. For example, Aceto et al.,[108] also showed that nicotine pyrrolidine methiodide as well as nicotine were antinociceptive in the mouse phenylquinone test and that these effects were partially blocked by the peripherally active, ganglionic, nicotinic receptor antagonist hexamethonium. In 1995, Caggiula et al.,[109] demonstrated that chlorisondamine, which also does not readily enter the CNS, antagonized the antinociceptive effects of nicotine in the tail withdrawal assay, but not the paw withdrawal assay. Consistent with these findings, partial blockade of nicotine-induced antinociception by hexamethonium was described in the rat[56,110] and mouse[110] tail-flick assays.

In addition, more recent studies have yielded similar results using $\alpha4$ or $\beta2$ null mutant mice. Thus, while there was an almost complete abolition of nicotine-induced antinociception in the hot-plate test among mice in which either the $\alpha4$ or $\beta2$ nicotinic receptor subunit gene had been knocked out, such genetic maniupulations led to a decrease in the antinociceptive potency but not efficacy of nicotine in the tail-flick test.[52] Moreover, in the tail-flick test, hexamethonium blocked nearly 50%

of the antinociception produced in the wildtype animals by a maximally effective dose of nicotine (2 mg/kg s.c.) and nearly all of the antinociception (approximately 60% of the maximal effect) in the α4 null mutants at this same dose. Collectively, these studies strongly support the existence of a peripheral component to nicotinic receptor-mediated antinociception, at least in the tail-flick test, and further suggest that such effects involve a nicotinic receptor subtype that does not contain the α4 subunit. Whether such findings will be restricted to a single assay of reflexive, spinally mediated nociception of uncertain clinical relevance or may be generalized to other, more germane types of human pain remains to be seen.

Regarding a mechanism that could account for the peripheral antinociceptive effects of nicotinic agonists, the direct or indirect inhibition of nociceptor activation would seem the most logical. However, insofar as agonist-mediated activation of virtually all nicotinic receptors is associated with a depolarizing event, taken together with the aforementioned literature documenting excitation of nociceptors by nicotinic agonists *in vitro*, one cannot immediately intuit how these behavioral and cellular phenomena may be most easily reconciled. On the other hand, a comprehensive understanding of the biophysical and pharmacological properties of all native nicotinic receptors does not yet exist and, almost certainly, nor has the complete repertoire of these subtypes been revealed.

Related to this incomplete data set is compelling evidence for nicotinic agonist-mediated inhibition of nociceptor function. For example, spinal neuronal responses to noxious thermal and mechanical stimulation were dose dependently inhibited by the intradermal injection of ABT-594.[111] Interestingly, not only were these effects antagonized by mecamylamine but also they could be elicited by doses of ABT-594 as low as 2.7 nmoles, strongly suggesting a peripheral site of action. Similarly, ABT-594 has been shown to inhibit the capsaicin-evoked release of calcitonin gene-related peptide from either the peripheral[112] or central[113] terminals of a subpopulation of capsaicin-sensitive primary sensory neurons. Although this latter effect was shown to be mecamylamine reversible, it is still unknown which nicotinic receptor subtype(s) is(are) responsible for any of the inhibitory actions cited earlier. Until this information is known, it will be difficult to assess on a cellular/molecular level the mechanisms that underlie this inhibition.

However, in this regard it is worth noting the rather intriguing findings that nicotine as well as certain other nicotinic agonists is capable of direct inhibitory actions on cerebellar Purkinje cells[114] or dorsolateral septal neurone[115] as a result of potassium-mediated hyperpolarization. Subsequent studies show that these inhibitory effects of nicotine, recorded intracellularly *in vitro*; 1) were sensitive to mecamylamine, κ-bungarotoxin, hexamethonium, and chlorisondamine but not to α-bungarotoxin, dihydro-β-erythroidine, or curare; 2) occured via calcium-dependent, potassium-mediated hyperpolarization; 3) were not blocked by treatment with dithiothreitol or iontophoresis of excess Mg^{2+} ion; and 4) were blocked by intraneuronal injections of $GTP\gamma S$.[114, 116–118] Taken together, these data indicate that nicotinic agonists may engage a novel, possibly G-protein-coupled, nicotinic receptor to directly inhibit neuronal firing via calcium-dependent potassium channel activation on the postsynaptic membrane. Whether or not these or similar nicotinic receptors exist on sensory neurons will require further examination, but could potentially account for some of the nicotinic agonist actions mentioned previously and, therefore, be considered a viable analgesic drug target.

5.4 FACTORS MODIFYING NICOTINIC ANALGESIA

5.4.1 GENDER STUDIES

There is now considerable and incontrovertible evidence documenting differences in the experience of experimental and clinical pain between males and females (for reviews, see Fillingim and Maixner[119] and Berkley[120]). Indeed, research efforts in this area have begun to focus on the elucidation of the biological and genetic mechanisms as well as the environmental and psychosocial factors that lead to these differences. In particular, attention has been directed at one of the more publicized examples of this phenomenon in humans, wherein it was reported that kappa opioids manifested greater postoperative analgesia in women compared with men following third molar extraction.[121] This particular result, however, should be interpreted cautiously, because subjects received only a single dose of either nalbuphine or butorphanol; subsequent experiments using different doses of nalbuphine showed biphasic effects in women and antianalgesic effects in men.[122] Nonetheless, experimentation in animals is fast leading to a wealth of data to suggest that sex-based differences in antinociceptive responsiveness to a variety of analgesic drug classes may be less the exception than the rule. Based on such precedents, it is reasonable to hypothesize that similar differences might exist in response to nicotinic agents.

In fact, Craft and Milholland[123] demonstrated that, in the rat hot-plate test, the intracerebroventricular injection of nicotine (0.1 to 1.0 µg) produced significantly greater antinociception in females compared with males. Similarly, the nicotinic receptor agonist metanicotine or the acetylcholinesterase inhibitor neostigmine injected intrathecally was more efficacious and/or potent in producing antinociception in the paw withdrawal test[10] or antiallodynia in the Chung nerve injury model[11] in female rats compared with their male counterparts. Interestingly, whereas mecamylamine partially antagonized the effects of neostigmine among females, it was without effect among males. This apparent differential sensitivity supports the concept that differences in responses between males and females, in fact, may be a function of sex-based differences in antinociceptive mechanisms and/or pathways, as suggested in the case of kappa opioid analgesia.[123a]

In contrast, Damaj[124] demonstrated using the tail-flick assay that female ICR mice were two- to three-fold less sensitive than were males to the antinociceptive effects of nicotine or epibatidine administered subcutaneously and of nicotine or neostigmine administered intrathecally. Similar differences in nicotine sensitivity were observed in the hot-plate assay. Interestingly, while males were also 2.5 times more sensitive than females to the anxiolytic effects of nicotine, there were no sex differences in the hypothermic, antimotoric or proconvulsive effects of nicotine. These data are remarkable, insofar as they suggest that 1) the sex-based differences in antinociception and anxiolysis observed likely are a function of pharmacodynamic rather than pharmacokinetic factors; and 2) the receptors (and perhaps the downstream mechanisms they engage) which underlie antinociception and anxiolysis are distinct from those which underlie hypothermia, hypomotility, and epileptogenesis. This constitutes compelling evidence in support of a rationale for the development of nicotinic analgesic drugs that are relatively devoid of certain untoward side effects.

Regarding potential mechanisms to account for the gender differences in nicotine sensitivity, there is good evidence that ovarian sex steroids may play a role. Thus, Valera et al.,[125] demonstrated that progesterone or 17-β-estradiol was capable of noncompetitively inhibiting α4β2 or α3β4 receptors expressed in *Xenopus* oocytes; this work has been confirmed.[126] Moreover, it was shown that progesterone and its A-ring metabolites could noncompetitively inhibit ^{86}Rb+ efflux from mouse thalamic synaptosomes[127] as well as from TE671 or SH-SY5Ycells.[128] Importantly, inhibitory effects are observed at concentrations of hormone levels (low μM) at or near those that can be achieved in plasma during the late follicular (17-β-estradiol) and midluteal (progesterone) phases of the menstrual cycle or in response to stress. More directly relevant to understanding sex-based differences in nicotinic analgesia were the *in vivo* studies of Damaj[124] showing that pretreatment with progesterone or 17-β-estradiol led to a rightward shift in the antinociceptive response to nicotine in female mice. While the effects of these sex hormones in male mice were not assessed, testosterone was without effect. It is not known whether such differences in sensitivity to nicotine and/or effects of ovarian sex steroid would be observed in humans, the relevant literature for which is discussed later. Finally, among eight different inbred mouse strains, Flores et al.,[129] observed no significant sex differences in the antinociceptive effects of epibatidine in the tail withdrawal assay, although there were significant genetic-based but gender-unrelated differences between strains (see the next section).

5.4.2 Genetic Studies

Recent advances in the understanding of pain behavior in humans indicate that genetic factors play an important role in individual variation to pain response and to analgesic drugs. For example, allelic differences in the hepatic P450 enzyme CYP2D6 underlie profound variability in the analgesic response to the pro-drug codeine, which requires conversion to morphine by this enzyme.[130] Similarly, differences in nicotine metabolism (for review, see Benowitz and Jacob[131]) have been related to genetic polymorphisms[158] and homozygous deletions[130,132] in CYP2A6, the enzyme primarily responsible for nicotine oxidation to cotinine in humans.[133]

Data demonstrating strain differences in a variety of mouse behaviors following nicotine administration (for review see Collins and Marks[134]) have been reported, although the factors underlying these differences are unknown, and relatively few studies have assessed whether genetic-based variability in antinociceptive response to nicotinic agonists exists. Using the tail-flick assay, Seale et al.,[135,136] demonstrated differences in antinociceptive sensitivity to i.p. nicotine or cytisine between two mouse stocks, CD-1 and CF-1, so that the CF-1 displayed less sensitivity than CD-1. Interestingly, other behavioral effects of nicotine or cytisine, such as hypomotility and seizures, did not differ between these strains, most likely excluding a pharmacokinetic explanation for the data. Unfortunately, however, these studies do not directly implicate the potential allelic basis of this variability, because genetically heterogeneous, outbred populations were used. More recently, Flores et al.,[129] have investigated the antinociceptive effect of epibatidine in eight inbred mouse strains using the tail-immersion withdrawal assay. They found that these strains differed significantly in their sensitivity to epibatidine, with DBA/2, BALB/c, and A strains showing much greater sensitivity than all others. In particular, the A strain exhibited 20-fold higher antinociceptive potency compared with the C3H/He strain. Importantly, the effect of epibatidine was blocked by mecamylamine in both strains, indicating the involvement of neuronal nAChRs in mediating the observed antinociception. Based on these data, Flores et al.,[129] suggested that such pharmacogenetic differences most probably derive from pharmacodynamic rather than pharmacokinetic mechanisms, and this view is supported by findings that similar strain differences in antinociceptive sensitivity using the mouse tail-flick and hot-plate tests were observed for nicotine (Damaj and Collins, unpublished observations). Additional studies, including quantitative trait locus mapping analyses, are needed to elucidate the specific genes or sets of genes that subserve this pharmacogenetic variability in nicotinic receptor-mediated antinociception. It will be of particular interest to discover whether such differential sensitivity will be observed in humans as well.

5.5 HUMAN STUDIES

Unfortunately, data from well controlled clinical trials on the analgesic efficacy of any nicotinic receptor agonist in humans are extremely scarce. It is likely that this paucity of information stems, at least in part, from actual and perceived untoward effects of the prototype drug, nicotine, and from an absence of any other pharmacologically similar compound approved for use in humans. (However, according to the June 7, 1999 Drug Report Issue of the Investigational Drugs Database, ABT-594 is in phase II clinical trials for pain.) What few studies have been performed suffer predominantly from use of tobacco products as the independent variable rather than nicotine proper, leading to significant confounds in interpretations of the data. In other cases, evaluations of analgesic effects of nicotine were performed only in smokers who were in varying stages of abstinence and/or they lacked appropriate controls.

To compound matters, the literature on the subject to date is rather conflicting and, for reasons stated earlier, seldom comparable. Thus, whereas some studies found that smoking cigarettes was associated with increases in pain threshold and tolerance, still others found the opposite or no effect. To a great extent, such limitations and contradictions focused early attention on and perhaps overemphasized the anxiolytic effects of nicotine in preventing withdrawal and the tendency to "misinterpret" such effects as being analgesic in nature. This is consistent with the view espoused by Schacter,[137] who believed that, at least among smokers, the tolerance to a painful stimulus was not increased by smoking but rather was reduced by not smoking. Indeed, Silverstein[138] was in concurrence that the major effect of nicotine, delivered in the form of a cigarette, was to end withdrawal, although he did measure a significantly higher threshold to electrical shock among individuals smoking a high-nicotine cigarette compared with the low-nicotine cigarette, or smoking deprived or nonsmoking groups. On the other hand, Pomerleau[139] has argued against this model of withdrawal relief, demonstrating that withdrawal-free exsmokers as well as smokers exhibited significantly increased endurance to cold-pressor pain following high-nicotine cigarettes or snuff compared with sham-tobacco-use groups. Partially supporting these findings are those of Nesbitt,[140] who found that smoking a high-nicotine cigarette produced a significant increase in pain tolerance to electrical shock compared with smoking a low-nicotine cigarette or simulating smoking with an unlit cigarette. However, this effect was observed only among smokers and not among nonsmokers.

In stark contrast, Milgrom-Friedman et al.,[141] reported that following ischemia with a blood pressure cuff, smokers had the shortest onset to pain and the lowest tolerance to pain, significantly different from nonsmokers or deprived smokers (1 or 12 h). In fact, smokers deprived of cigarettes for 12 h exhibited the latest onset and greatest tolerance to pain of any group tested, leading the authors to conclude that smoking deprivation is analgesic in chronic smokers. On the contrary, Lane[142] reported an increase in pain tolerance to thermal stimulation among smokers who were allowed to smoke compared with a group of deprived smokers. Moreover, Pauli et al.,[143] observed no differences in pain threshold to thermal stimulation between deprived (12 h) and minimally deprived (30 min) smokers prior to smoking, and a significant increase in pain threshold after smoking, but only in the deprived individuals. Consistent with this report, no significant difference in pain thresholds of deprived smokers (overnight) and nonsmokers was found.[144] Interestingly, no significant differences in either pain threshold or tolerance were observed between smokers who smoked 2 cigarettes 75 and 15 min prior to testing and smokers who chewed 2 pieces of nicotine containing gum (2 mg) 1 h and just prior to commencement of testing.[141] This latter finding clearly indicates an effect of smoking, in contrast to sham smoking (as reported by Fertig et al.)[145] that is independent of nicotine, highlighting the confounds that continually plague this type of research. Similarly, a retrospective epidemiological analysis of over 60,000 subjects from a multiphasic healthcare program indicated that, in the Achilles tendon pressure test, a significantly lower pain tolerance was associated with white male and female smokers compared with nonsmokers. Finally, several studies have found no effect of smoking cigarettes on pain responses in the cold-pressor[146] or electrical stimulation[146–149] tests.

To date, there have been only two studies to evaluate the analgesic effects of nicotine *per se* in smokers and, most importantly with respect to proof of concept, non-smokers. Thus, Perkins et al.,[150] found that, compared with placebo, nicotine (5 to 20 μg/kg) administered in a nasal spray produced a small but significant increase in pain detection latencies among smokers and nonsmokers. Similarly, Jamner et al.[151] described a significant (approximately 20%) increase in pain threshold and pain tolerance to electrical stimulation following treatment with a nicotine patch compared with placebo patch. Interestingly, the study observed this effect only in male smokers and nonsmokers, providing the initial evidence for gender-based differences in the pain modulating effects of nicotine in humans. This finding lends strong support to the rather astute conclusion of the authors that previous studies which reported little or no analgesic effects of nicotine (albeit in the form of tobacco) consisted of mixed gender, predominantly female, or entirely female subject populations. It is important to note that the females evaluated by Jamner et al.[151] were staged and tested in the midluteal phases of their menstrual cycles when progesterone levels are maximal and 17-β-estradiol levels are elevated (approximately half-maximal). Although the authors invoked an explanation involving opposing cardiovascular actions of estrogen and nicotine, these results are perhaps more easily explained by the investigations mentioned earlier demonstrating a direct, noncompetitive inhibitory action of ovarian sex steroids on nicotinic receptor function *in vitro*[125–128] and *in vivo*.[124] Alternatively, such differences, paralleled by elevated basal pain threshold and tolerance levels among only male smokers compared with nonsmokers, could also be explained by differences between men and women in nociceptive mechanisms and/or pathways engaged by nicotinic agonists. In any case, gender-based differences in analgesic sensitivity of the magnitude described by Jamner et al.[151] would have profound implications for future development and potential use of nicotinic analgesics in humans.

5.6 NEW NICOTINIC AGONISTS AS ANALGESICS

Until very recently, relatively few selective ligands were available to study neuronal nicotinic receptors. These included a wide range of agonists (e.g., nicotine and DMPP), antagonists (e.g., mecamylamine and dihydro-β-erythroidine), and mixed agonist/antagonists or partial agonists (e.g., cytisine and lobeline). Unfortunately, none of these exhibited any subtype specificity, with the exception of the α7 selective antagonists MLA and α-BGTX. Medicinal and natural product chemistry efforts over the past decade have expanded considerably the number of nicotinic ligands available. However, the lack of clearly defined molecular structures for the various nicotinic acetylcholine receptor subtypes has precluded comprehensive molecular modeling studies and complicated the search for additional subtype-selective agonists and antagonists.

In this regard, two complementary approaches have been typically used for the identification of useful agents. The first is to evaluate candidate compounds on defined human or animal (usually rat or chick) subunit combinations exogenously expressed in *in vitro* cell systems (e.g., *Xenopus* oocytes). Drugs identified as being subtype-selective are then targeted for further behavioral studies. Although this strategy may be exploited for high throughput screening of many compounds, it suffers from the potential lack of generalizability between *in vitro* and *in vivo* systems. Another approach is to assess selectivity of various compounds, *in vivo*, in terms of the desired behavioral effect (e.g., antinociception). Although less efficient, this method ensures that the ligands under study will be interacting with native nAChRs and that various pharmacokinetic properties, such as CNS permeation of the compounds, are immediately considered. Compounds showing good behavioral selectivity then would be evaluated for selectivity on different subunit combinations *in vitro*.

The analgesic potential of nicotinic receptor agonists has been demonstrated with structurally diverse molecules such as nicotine, epibatidine, and ABT-594 (see Table 5.1). Mechanistic studies strongly implicate α4-containing nAChRs subtypes, and likely the α4β2 subtype, as being critically important to the analgesic activity of nicotinic agonists. However, the selectivity for this subtype over other nicotinic subtypes, which may function to counteract the analgesic activity of a given agonist, must be considered (see Section 5.3.2). As newer compounds are evaluated in more sophisticated molecular systems, it becomes increasingly clear that some ligands have mixed pharmacological properties at the various nAChRs. Thus, full agonist activity of a ligand at one receptor subtype does not necessarily predict its activity at other subtypes; in fact, partial agonists at one subtype may have full antagonistic activity at other subtypes. Thus, extensive effort is being directed toward designing novel structural classes with high affinity and selectivity for the $\alpha4\beta2$ subtype.

5.6.1 ABT-594

Recently, a derivative of epibatidine, ABT-594 ((R)-5-2-azetidinylmethoxy)-2-chloropyridine), has been developed by Abbott Laboratories as a nicotinic analgesic and, as mentioned earlier, is currently in phase II clinical trials. ABT-594 shows similar affinity to epibatidine at the human $\alpha4\beta2$ subtype (K_i of 37 and 42 pM for ABT-594 and epibatidine, respectively), but has 4000 times less activity at the neuromuscular receptor.[77] The systemic administration of ABT-594 produced broad-spectrum antinociceptive activity in preclinical assays of acute and persistent pain with equal efficacy to that of morphine, but did not cause opioid-like withdrawal, physical dependence or decreased gastric motility.[3] In addition, its antinociceptive effects were blocked by mecamylamine but not by naloxone. Moreover, Bitner et al.[54] suggested that the analgesic activity of ABT-594 is mediated by selective modulation of the $\alpha4\beta2$ nAChR subtype, possibly located on serotonergic neurons in one or more raphe nuclei. More recently, two other groups reported that the antinociceptive effect of ABT-594 in acute thermal pain models in rats occurs at doses that induce adverse effects such as hypothermia, hypotension, motor impairment, and convulsions.[152,153] In addition, the authors reported that ABT-594 was less efficacious than epibatidine in the hot-plate and tail-flick tests. Such discrepancies in the response of rodents to the effects of ABT-594 in acute pain models may be due to differences in species of test animal, route of drug administration, time of evaluation and/or pain test employed. However, Kesingland et al.[152] found that ABT-594 displayed a clear separation between its antimotoric and antihyperalgesic effects in models of persistent pain. Studies with ABT-594 support the hypothesis that nicotinic agonists with improved selectivity for certain nAChRs subtypes (e.g., $\alpha4\beta2$) could provide significant antinociception with a much improved therapeutic window compared with nicotine.

5.6.2 Metanicotine

Metanicotine (known also as RJR-2403), N-methyl-4-(3-pyridinyl)-3-butene-1-amine, is another novel nicotinic agonist with increased receptor selectivity.[154,155] This compound displaces high affinity [³H]-nicotine binding in rat brain with a K_i value of 26 nM, while possessing relatively weak affinity ($K_i = 36$ µM) for the [¹²⁵I]- α-BGTX-sensitive nAChR subtype. *In vitro* functional assays indicate that metanicotine does not stimulate ion flux in rat autonomic ganglia (PC 12 cells) or human muscle (TE671 cells). On the other hand, metanicotine is equal to or better than nicotine as a cognitive enhancer in rats. Metanicotine was 10 to 30-fold less potent than nicotine in eliciting changes in blood pressure, heart rate, and temperature in rats.[155] Recently, Damaj et al.[156] found that metanicotine produced significant antinociceptive effects in mice and rats subjected to either thermal (tail-flick), mechanical (paw-pressure), chemical (PPQ or formalin), or persistent (complete Freund's adjuvant) noxious stimuli. Metanicotine was about five-fold less potent than nicotine in the tail-flick test after s.c administration, but slightly more potent after central administration, and its duration of action was longer than that of nicotine. Nicotinic antagonists, mecamylamine, and dihydro-β-erythroidine (but not naloxone) blocked metanicotine-induced antinociception in the different pain models. In contrast to nicotine, antinociceptive effects of metanicotine were observed at doses that had virtually no effect on spontaneous activity or body temperature in mice. These data indicate that metanicotine is a centrally acting neuronal nicotinic agonist with preferential antinociceptive effects in animals. The analgesic activity of metanicotine in preclininal acute and persistent pain models was also reported in rats after systemic and intrathecal injections.[10,11] Thus, metanicotine and related nicotinic agonists may have great potential for development as a new class of analgesics.

5.7 SUMMARY

This review reflects a substantial and growing body of preclinical evidence that the concept of developing therapeutically useful nicotinic analgesic drugs is a viable one. As reviewed in this chapter, nicotinic agonists show a remarkable spectrum of efficacy in various animal models of acute and persistent pain. Clearly, it will be of primary importance to ascertain whether nicotinic agonists prove to be effective analgesics in humans as well. Based on the equivocal results to date from human experimental pain studies, it is fairly obvious that nicotine may not be the ideal prototype for a nicotinic analgesic drug. This observation highlights the necessity to clarify certain unresolved issues relating to mechanisms and sites of action, variability in analgesic response, and adverse effects. It is likely that such efforts will be greatly facilitated by the discovery of compounds with greater pharmacological and physiological selectivity. In turn, this should help to elucidate the functional role of native neuronal nAChRs about which relatively little is known. It is hoped that, through this process, the potential of this drug class as safe and effective analgesic therapeutics will be fully realized.

ACKNOWLEDGMENTS

This work was supported by National Institute on Drug Abuse grant # DA 05274 (M.I.D.) and DA 10510 (C.M.F.).

REFERENCES

1. Davis, L., Pollock, L.J., and Stone, T.T., Visceral pain, *Surg. Gynecol. Obstet.,* 55, 418–426, 1932.
2. Spande, T., Garraffo, H., Edwards, M., Yeh, H., Pannell, L., and Daly, J., Epibatidine: A novel (chloropyridyl) azabicycloheptane with potent analgesic activity from an Ecuadoran poison frog, *J. Am. Chem. Soc.,* 114, 3475–3478, 1992.
3. Bannon, A.W., Decker, M.W., Holladay, M.W., Curzon, P., Donnelly-Roberts, D., Puttfarcken, P.S., Bitner, R.S., Diaz, A., Dickenson, A.H., Porsolt, R.D., Williams, M., and Arneric, S.P., Broad-spectrum, non-opioid analgesic activity by selective modulation of neuronal nicotinic acetylcholine receptors, *Science,* 279, 77–81, 1998.
4. Lipton, J.A., Ship, J.A., and Larach-Robinson, D., Estimated prevalence and distribution of reported orofacial pain in the U.S., *J. Am. Dent. Assoc.,* 124, 115–121, 1993.
5. Gilbert, S.D., Clark, T.M., and Flores, C.M., Antihyperalgesic activity of Epibatidine in the formalin model of facial pain, *Pain, in press,* 2000.
6. Clavelou, P., Pajot, J., Dallel, R., and Raboisson, P., Application of the formalin test to the study of orofacial pain in the rat, *Neurosci. Lett.,* 103, 349–353, 1989.
7. Lawand, N.B., Lu, Y., and Westlund, K.N., Nicotinic cholinergic receptors: potential targets for inflammatory pain relief, *Pain,* 80, 291–299, 1999.
8. Aceto, M.D., Bagley, R.S., Dewey, W.L., Fu, T.-C., and Martin, B.R., The spinal cord as a major site for the antinociceptive action of nicotine in the rat, *Neuropharmacology,* 25, 1031–1036, 1986.
9. Christensen, M.-K. and Smith, D.F., Antinociceptive effects of the stereoisomers of nicotine given intrathecally in spinal rats, *J. Neural. Transm.,* 80, 189–194, 1990.
10. Chiari, A., Tobin, J.R., Pan, H.-L., Hood, D.D., and Eisenach, J.C., Sex differences in cholinergic analgesia I, *Anesthesiology,* 91, 447–454, 1999.
11. Lavand'homme, P.M. and Eisenach, J.C., Sex differences in cholinergic analgesia II, *Anesthesiology,* 91, 455–461, 1999.
12. Yaksh, T.L., Dirksen, R., and Harty, G.J., Antinociceptive effects of intrathecally injected cholinomimetric drugs in the rat and cat, *Eur. J. Pharmacol.,* 117, 81–88, 1985.
13. Smith, M.D., Yang, X., Nha, J.-Y., and Buccafusco, J.J., Antinociceptive effect of spinal cholinergic stimulation: interaction with substance P, *Life Sci.,* 45, 1255–1261, 1989.
14. Gillberg, P.G., Hartvig, P., Gordh, T., Sottile, A., Jansson, I., Archer, T., and Post, C., Behavioral effects after intrathecal administration of cholinergic receptor agonists in the rat, *Psychopharmacology,* 100, 464–469, 1990.
15. Khan, I.M., Taylor, P., and Yaksh, T.L., Cardiovascular and behavioral responses to nicotinic agents administered intrathecally, *J. Pharmacol. Exp. Ther.,* 270, 150–158, 1994.
16. Damaj, M.I., Creasy, K.R., Grove, A.D., Rosecrans, J.A., and Martin, B.R., Pharmacological effects of epibatidine optical enantiomers, *Brain Res.,* 664, 34–40, 1994.
17. Damaj, M.I., Creasy, K., Welch, S.P., Rosecrans, J., Aceto, M., and Martin, B.R., Comparative pharmacology of nicotine and ABT-418, a new nicotinic agonist, *Psychopharmacology,* 120, 483–490, 1995.
18. Damaj, M.I., Fei-Yin, M., Dukat, M., Glassco, W., Glennon, R.A., and Martin, B.R., Antinociceptive responses to nicotinic acetylcholine receptor ligands after systemic and intrathecal administration in mice, *J. Pharmacol. Exp. Ther.,* 284, 1058–1065, 1998.

19. Suh, H.-W., Song, D.K., Lee, K.J., Choi, S.R., and Kim, Y.H., Intrathecally injected nicotine enhances the antinociception induced by morphine but not b-endorphin, D-Pen2,5-enkephalin and U50,488H administered intrathecally in the mouse, *Neuropeptides,* 30, 373–378, 1996.

20. Suh, H.W., Song, D.K., Choi, S.R., Chung, K.M., and Kim, Y.H., Nicotine enhances morphine- and b-endorphin-induced antinociception at the supraspinal level in the mouse, *Neuropeptides,* 30, 479–484, 1996.

21. Zarrindast, M., Nami, A.B., and Farzin, D., Nicotine potentiates morphine antinociception: a possible cholinergic mechanism, *Eur. Neuropsychopharmacol,* 6, 127–133, 1996.

22. Elgoyhen, A., Johnson, D.S., Boulter, J., Vetter, D., and Heinemann, S., α9: An acetylcholine receptor with novel pharmacological properties expressed in rat cochlear hair cells, *Cell,* 79, 705–715, 1994.

23. Lewis, P.R. and Shute, C.C.D., Cholinergic pathways in CNS, in *Handbook of Psychopharmacology,* Vol. 9, L.L. Iversen, S.D. Iversen and S.D. Synder, Eds., Plenum Press, New York, 1978, 315–355.

24. Woolf, N.J., Cholinergic systems in mammalian brain and spinal cord, *Prog. Neurobiol.,* 37, 475–524, 1991.

25. Romano, C. and Goldstein, A., Stereospecific nicotine receptors on rat brain membranes, *Science,* 210, 647–650, 1980.

26. Marks, M.J. and Collins, A.C., Characterization of nicotine binding in mouse brain and comparison with the binding of alpha-bungarotoxin and quinuclidinyl benzilate, *Mol. Pharmacol.,* 22, 554–564, 1982.

27. Schwartz, R.D., McGee, R., JR., and Kellar, K., Nicotinic cholinergic receptors labeled by [^{3}H] acetylcholine in rat brain, *Mol. Pharmacol.,* 22, 56–62, 1981.

28. Abood, L.G. and Grassi, S., [^{3}H]methylcarbamylcholine, a new radioligand for studying brain nicotinic receptors, *Biochem. Pharmacol.,* 35, 4199–4202, 1986.

29. Pabreza, L.A., Dhawan, S., and Kellar, K.J., [^{3}H]cytisine binding to nicotinic cholinergic receptors in brain, *Mol. Pharmacol.,* 39, 9–12, 1990.

30. Flores, C.M., Rogers, S.W., Pabreza, L.A., Wolfe, B.B., and Kellar, K.J., A subtype of nicotinic cholinergic receptor in the rat brain is composed of alpha-4 and beta-2 subunits and is up-regulated by chronic nicotine treatment, *Mol. Pharmacol.,* 41, 31–37, 1992.

31. Clarke, P.B.S. and Pert, A., Autoradiographic evidence for nicotine receptors on nigrostriatal and mesolimbic dopaminergic neurons, *Brain Res.,* 348, 355–358, 1985.

32. Clarke, P.B.S., Schwartz, R.D., Paul, S.M., Pert, C.B., and Pert, A., Nicotinic binding in rat brain: autoradiographic comparison of [^{3}H]acetylcholine, [3H]nicotine, and [125I]-α-bungarotoxin, *J. Neurosci.,* 5, 1307–1315, 1985.

33. London, E.D., Connolly, R.J., Szikszay, M., and Wamsley, J.K., Distribution of cerebral metabolic effects of nicotine in the rat, *Eur. J. Pharmacol.,* 110, 391–392, 1985.

34. Boksa, P. and Quirion, R., [3H]N-methyl-carbamylcholine, a new radioligand specific for nicotinic acetylcholine receptors in brain, *Eur. J. Pharmacol.,* 139, 323–333, 1987.

35. Harfstrand, A., Adem, A., Fuxe, K., Agnati, L., Anderson, K., and Nordberg, A., Distribution of nicotinic cholinergic receptors in the rat tel- and diencephalon: a quantitative receptor autoradiographical study using [^{3}H]acetylcholine, [^{125}I]-α-bungarotoxin and [^{3}H]nicotine, *Acta. Physiol. Scand.,* 132, 1–14, 1988.

36. Houghtling, R., Davila-Garcia, M., and Kellar, K., Characterization of ($\pm$)-[^{3}H]Epibatidine binding to nicotinic cholinergic receptors in rat and human brain, *Mol. Pharmacol.,* 48, 280–287, 1995.

37. Perry, D. and Kellar, K., [^{3}-H]Epibatidine labels nicotinic receptors in rat brain: An autoradiographic study, *J. Pharmacol. Exp. Ther.*, 275, 1030–1034, 1995.
38. Salvaterra, P.M. and Moore, W.J., Binding of (125I)-alpha bungarotoxin to particulate fractions of rat and guinea pig brain, *Biochem. Biophys. Res. Commun.*, 55, 1311–1318, 1973.
39. Polz-Tejera, G. and Schmidt, J., Autoradiographic localisation of alpha-bungarotoxin-binding sites in the central nervous system, *Nature*, 258, 349–351, 1975.
40. Schulz, D.W., Loring, R.H., Aizenman, E., and Zigmond, R.E., Autoradiographic localization of putative nicotinic receptors in the rat brain using 125I-neuronal bungarotoxin, *J. Neurosci.*, 11, 287–297, 1991.
41. Schulz, D.W. and Zigmon, R.E., Neuronal bungarotoxin blocks the nicotinic stimulation of endogenous dopamine release from rat striatum, *Neurosci. Lett.*, 98, 310–316, 1989.
42. Llewelyn, M.B., Azami, J., Grant, C.M., and Roberts, M.H.T., Analgesia following microinjection of nicotine into the periaqueductal gray matter, *Neurosci. Lett., Suppl.* 7, S277, 1981.
43. Guimaraes, A. and Prado, W., Antinociceptive effects of carbachol microinjected into different portions of the mesencephalic periaqueductal gray matter of the rat, *Brain Res.*, 647, 220–230, 1994.
44. Hamann, S.R. and Martin, W.R., Thermally evoked tail avoidance reflex - input-output relationships and their modulation, *Brain Res. Bull.*, 29, 507–509, 1992.
45. Rogers, D. and Iwamoto, E., Multiple spinal mediators in parenteral nicotine-induced antinociception, *J. Pharmacol. Exp. Ther.*, 267, 341–349, 1993.
46. Iwamoto, E.T. and Marion, L., Adrenergic, serotonergic and cholinergic components of nicotinic antinociception in rats, *J. Pharmacol. Exp. Ther.*, 265, 777–789, 1993.
47. Curzon, P., Nikkel, A.L., Bannon, A.W., Arneric, S.P., and Decker, M.W., Differences between the antinociceptive effects of the cholinergic channel activators A-85380 and (+-)-epibatidine in rats, *J. Pharmacol. Exp. Ther.*, 287, 847–853, 1998.
48. Bitner, R.S., Nikkel, A.L., Curzon, P., Arneric, S.P., Bannon, A.W., and Decker, M.W., Role of the nucleus raphe magnus in antinociception produced by ABT-594: Immediate early gene responses possibly linked to neuronal nicotinic acetylcholine receptors on serotonergic neurons, *J. Neurosci.*, 18, 5426–5432, 1998.
49. Iwamoto, E.T., Antinociception after nicotine administration into the mesopontine tegmentum of rats: evidence for muscarinic actions, *J. Pharmacol. Exp. Ther.*, 251, 412–421, 1989.
50. Parvini, S., Hamann, S.R., and Martin, W.R., Pharmacologic characteristics of a medullary hyperalgesic center, *J. Pharmacol. Exp. Ther.*, 265, 286–293, 1993.
51. Flores, C.M. and Hargreaves, K.M., Neuronal nicotinic receptors: new targets in the treatment of pain, in *Neuronal Nicotinic Receptors: Pharmacology and Therapeutic Opportunities,* S.P. Arneric and J.D. Brioni, Eds., Wiley-Liss, New York, 1998, 359–378.
52. Marubio L. M, Arroyo-Jimenez M. M, Cordero-Erausquin M., Léna C., Le Novère N., de Kerchove d'Exaerde A., Huchet M., Damaj M. I., and Changeux J-P., Reduced antinociception in mice lacking neuronal nicotinic receptor subunits, *Nature*, 398, 805–810, 1999.
53. Bitner, R.S., Nikkel, A.L., Curzon, P., Donnelly-Roberts, D.L., Puttfarcken, P.S., Namovic, M., Jacobs, I.C., and Meyer, M.D., Reduced nicotinic receptor-mediated antinociception following *in vivo* antisense knock-down in rat, *Brain Res.*, 871, 66–74, 2000.

54. Bitner, R.S., Nikkel, A.L., Curzon, P., Arneric, S.P., Bannon, A.W., and Decker, M.W., Role of the nucleus raphe magnus in antinociception produced by ABT-594: immediate early gene responses possibly linked to neuronal nicotinic acetylcholine receptors on serotonergic neurons, *J. Neurosci.*, 18, 5426–5432, 1998.

55. Bannon, A.W., Decker, M.W., Curzon, P., Buckley, M.J., Kim, D.J.B., Radek, R.J., Lynch, J.K., Wasicak, J.T., Lin, N.-H., Arnold, W.H., Holladay, M.W., Williams, M., and Arneric, S.P., ABT-594 [(R)-5-(2-azetidinylnethozy)-2-chloropyridine]: a novel, orally effective antinociceptive agent acting via neuronal nicotinic acetylcholine receptors: II. *In vivo characterization, J. Pharmacol. Exp. Ther.*, 285, 787–794, 1998.

56. Rao, T.S., Correa, L.D., Reid, R.T., and Lloyd, G.K., Evaluation of anti-nociceptive effects of neuronal nicotinic acetylcholine receptor (NAChR) ligands in the rat tail-flick assay, *Neuropharmacology*, 35, 393–405, 1996.

57. Damaj, M.I., Meyer, E.M., and Martin, B.R., The antinociceptive effects of α7 nicotinic agonists in an acute pain model, *Neuropharmacology*, 39, in press, 2000.

58. Hunt, S. and Schmidt, J., Some observations on the binding patterns of alpha-bungarotoxin in the central nervous system of the rat, *Brain Res.*, 157, 213–232, 1978.

59. Gillberg, P. and Aquilonius, S., Cholinergic, poioid and glycine receptor binding sites localized in human spinal cord by *in vitro autoradiography, Acta Neurol. Scand.*, 72, 299–306, 1985.

60. Gillberg, P.G. and Wiksten, B., Effects of spinal cord lesions and rhizotomies on cholinergic and opiate receptor binding sites in rat spinal cord, *Acta Physiol. Scand.*, 126q, 575–582, 1986.

61. Gillberg, P.G., d'Argy, R., and Aquilonius, S.M., Autoradiographic distribution of [3H]acetylcholine binding sites in the cervical spinal cord of man and some other species, *Neurosci. Lett.*, 90, 197–202, 1988.

62. Bordey, A., Feltz, P., and Trouslard, J., Nicotinic actions on neurons of the central autonomic area in rat spinal cord slices, *J. Physiol.*, 497, 175–187, 1996.

63. Naeff, B., Schlumpf, M., and Lichtensteiger, W., Prenatal and postnatal development of high-affinity [3H]nicotine binding sites in rat brain regions: an autoradiographic study, *Brain Res. Dev. Brain Res.*, 68, 163–174, 1992.

64. Flores, C.M., Davila-Garcia, M.I., Ulrich, Y.M., and Kellar, K.J., Differential regulation of neuronal nicotinic receptor binding sites following chronic nicotine administration, *J. Neurochem.*, 69, 2216–2219, 1997.

65. Khan, I.M., Yaksh, T.L., and Taylor, P., Epibatidine binding sites and activity in the spinal cord, *Brain Res.*, 753, 269–282, 1997.

66. Ulrich, Y.M., Hargreaves, K.M., and Flores, C.M., A comparison of multiple injections vs. continuous infusion of nicotine for producing up-regulation of neuronal [3H]-epobatidine binding sites, *Neuropharmacology*, 36, 1119–1125, 1997.

67. Wada, E., Wada, K., Boulter, J., Deneris, E., Heinemann, S., Patrick, J., and Swanson, L., Distribution of alpha2, alpha3, alpha4, and beta2 neuronal nicotinic receptor subunit mRNAs in the central nervous system: a hybridization histochemical study in the rat, *J. Comp. Neurol.*, 284, 314–335, 1989.

68. Séguéla, P., Wadiche, J., Dineleymiller, K., Dani, J.A., and Patrick, J.W., Molecular cloning, functional properties, and distribution of rat brain-alpha7 — a nicotinic cation channel highly permeable to calcium., *J. Neurosci.*, 13, 596–604, 1993.

69. Marks, M.J., Pauly, J.R., Gross, S.D., Deneris, E.S., Hermans-Borgmeyer, I., Heinemann, S.F., and Collins, A.C., Nicotine binding and nicotinic receptor subunit RNA after chronic nicotine treatment, *J. Neurosci.*, 12, 2765–2784, 1992.

70. Dineley-Miller, K. and Patrick, J., Gene transcripts for the nicotinic acetylcholine receptor subunit, beta4, are distributed in multiple areas of the rat central nervous system, *Mol. Brain Res.,* 16, 339–344, 1992.

71. Eccles, J.C., Fatt, P., and Koketsu, K., Cholinergic and inhibitory synapses in a pathway from motor-axon collaterals to motoneurons, *J. Physiol.* (London), 126, 524–562, 1954.

72. Damaj, M.I., Welch, S.P., and Martin, B.R., Nicotine-induced antinociception in mice: role of G-proteins and adenylate cyclase, *Pharmacol. Biochem. Behav.,* 48, 37–42, 1994.

73. Khan, M.M., Marsla, M., Printz, M.P., Taylor, P., and Yaksh, T.L., Intrathecal nicotinc agonist-elicited release of excitatory amino acids as measured by *in vivo spinal microdialysis in rats, J. Pharmacol. Exp. Ther.,* 278, 97–106, 1996.

74. Reuter, L.E., Meyer, M.D., and Decker, M.W., Spinal mechanisms underlying A-85380-induced effects on acute thermal pain, *Brain Res.,* 872, 93–101, 2000.

75. Khan, I.M., Yaksh, T.L., and Taylor, P., Epibatidine binding sites and activity in the spinal cord, *Brain Res.,* 753, 269–282, 1997.

76. Khan, I.M., Buerkle, H., Yaksh, T.L., and Taylor, P., Nociceptive and antinociceptive responses to intrathecally administered nicotinic agonists, *Neuropharmacology,* 37, 1515–1525, 1998.

77. Donnelly-Roberts, D.L., Puttfarcken, P.S., Kuntzweiler, T.A., Briggs, C.A., Anderson, D.J., Campbell, J.E., Piattoni-Kaplan, M., McKenna, D.G., Wasicak, J.T., Holladay, M.W., Williams, M., and Arneric, S.P., ABT-594 [(R)-5-(2-azetidinylmethoxy)-2-chloropyridine]: a novel, orally effective analgesic acting via neuronal nicotinic acetylcholine receptors: I. *In vitro* characterization, *J. Pharmacol. Exp. Ther.,* 285, 777–786, 1998.

78. Miao, F.J.-P., Dallman, M.F., Benowitz, N.L., Basbaum, A.I., and Levine, J.D., Adrenal medullary modulation of the inhibition of Bradykinin- induced plasma extravasation by intrathecal nicotine, *J. Pharmacol. Exp. Ther.,* 264, 839–844, 1993.

79. Naguib, M. and Yaksh, T.L., Antinociceptive effects of spinal cholinesterase inhibition and isobolographic analysis of the interaction with μ and α2 receptor systems, *Anesthesiology,* 80, 1338–1348, 1994.

80. Zhuo, M. and Gebhart, G.F., Spinal serotonin receptors mediate descending facilitation of a nociceptive reflex from the nuclei reticularis gigantocellularis and gigantocellularis pars alpha in the rat, *Brain Res.,* 550, 35–48, 1991.

81. Langley, J.N. and Dickenson, W.L., On the local paralysis of peripheral ganglia and on the connection of different classes of nerve fibers with them, *Proc. Roy Soc. Lond,* 46, 423–431, 1889.

82. Langley, J.N., On the reaction of cells and of nerve-endings to certain poisons, chiefly as regards the reaction of striated muscle to nicotine and to curare., *J. Physiol.,* 33, 374–413, 1905.

83. Conti-Tronconi, B.M. and Raftery, M.A., The nicotinic cholinergic receptor: correlation of molecular structure with functional properties, *Ann. Rev. Biochem.,* 51, 491–530, 1982.

84. Popot, J.L. and Changeux, J.P., Nicotinic receptor of acetylcholine: structure of an oligomeric integral membrane protein, *Physiol. Rev.,* 64, 1162–1239, 1982.

85. McCarthy, M.P., Earnest, J.P., Young, E.F., Choe, S., and Stroud, R.M., The molecular neurobiology of the acetylcholine receptor, *Ann. Rev. Neurosci.,* 9, 383–413, 1986.

86. Schechter, N., Handy, I.C., Pezzementi, L., and Schmidt, J., Distribution of α-bungarotoxin binding sites in the central nervous system and peripheral organs of the rat, *Toxicon*, 16, 245–251, 1978.

87. Polz-Tejera, G., Hunt, S.P., and Schmidt, J., Nicotinic receptors in sensory ganglia, *Brain Res.*, 195, 223–230, 1980.

88. Ninkovic, M. and Hunt, S.P., Alpha-bungarotoxin binding sites on sensory neurons and their axonal transport in sensory afferents, *Brain Res.*, 272, 57–69, 1983.

89. Wada, K., Ballivet, M., Boulter, J., Connolly, J., Wada, E., Deneris, E.S., Swanson, L.W., Heinemann, S., and Patrick, J., Functional expression of a new pharmacological subtype of brain nicotinic acetylcholine receptor, *Science*, 240, 330–334, 1988.

90. Boyd, R.T., Jacob, M.H., McEachern, A.E., Caron, S., and Berg, D.K., Nicotinic acetylcholine receptor mRNA in dorsal root ganglion neurons, *J. Neurobiol.*, 22, 1–14, 1991.

91. Flores, C.M., DeCamp, R.M., Kilo, S., Rogers, S.W., and Hargreaves, K.M., Neuronal nicotinic receptor expression in sensory neurons of the rat trigeminal ganglion: demonstration of α3β4, a novel subtyoe in the mammaliam nervous system, *J. Neurosci.*, 16, 7892–7901, 1996.

92. Roberts, R.G.D., Stevenson, J.E., Westerman, R.A., and Pennefather, J., Nicotine acetylcholine receptors on capsaicin-sensitive nerves, *NeuroReport*, 6, 1578–1582, 1995.

93. Liu, L. and Simon, S.A., Responses of cultured rat trigeminal ganglion neurons to bitter tastants, *Chem. Senses*, 23, 125–130, 1998.

94. Keiger, C.J. and Walker, J.C., Individual variation in the expression profiles of nicotinic receptors in the olfactory bulb and trigeminal ganglion and identification of alpha2, alpha6, alpha9, and beta3 transcripts, *Biochem. Pharmacol.*, 59, 233–240, 2000.

95. Armstrong, D., Dry, R.M.L., Keele, C.A., and Markham, J.W., Observations on chemical excitants of cutaneous pain in man, *J. Physiol.*, 120, 326–351, 1953.

96. Keele, C.A. and Armstrong, D., *Substances Producing Pain and Itch*, Edward Arnold, London, 1964, 107–123.

97. Jancsò, N., Jancsò-Gabor, S., and Takats, I., Pain and inflammation induced by nicotine, acetylcholine and structurally related compounds and their prevention by desensitizing agents, *Acta Physiol.*, 19, 113–132, 1961.

98. Juan, H., Nicotinic nociceptors on perivascular sensory nerve endings, *Pain*, 12, 259–264, 1982.

99. Tanelian, D.L., Cholinergic activation of a population of corneal afferent nerves, *Exp. Brain Res.*, 86, 414–420, 1991.

100. Steen, K.H. and Reeh, P.W., Actions of cholinergic agonists and antagonists on sensory nerve endings in rats, *in vitro*, *J. Neurophysiol.*, 70, 397–405, 1993.

101. Baccaglini, P.I. and Cooper, E., Influences on the expression of acetylcholine receptors on rat nodose neurones in cell culture, *J. Physiol.*, 324, 441–451, 1982.

102. Mandelzys, A., Cooper, E., Verge, V.M., and Richardson, P.M., Nerve growth factor induces functional nicotinic acetylcholine receptors on rat sensory neurons in culture, *Neuroscience*, 37, 523–50, 1990.

103. Sucher, N.J., Cheng, T.P., and Lipton, S.A., Neural nicotinic acetylcholine responses in sensory neurons from postnatal rat, *Brain Res.*, 533, 248–254, 1990.

104. Liu, L., Identification of acetylcholine receptors in adult rats trigeminal ganglion neurons, *Brain Res*, 617, 37–42, 1993.

105. Liu, L. and Simon, S.A., Similarities and differences in the currents activated by capsaicin, piperine, and zingerone in rat trigeminal ganglion cells, *J. Neurophysiol.,* 76, 1858–1869, 1996.

106. Puttfarcken, P.S., Manelli, A.M., Arneric, S.P., and Donnelly-Roberts, D.L., Evidence for nicotinic receptors potentially modulating nociceptive transmission at the level of the primary sensory neuron: studies with F11 cells, *J. Neurochem.,* 69, 930–938, 1997.

107. Francel, P.C., Harris, K., Smith, M., Fishman, M.C., Dawson, G., and Miller, R.J., Neurochemical characteristics of a novel dorsal root ganglion X neuroblastoma hybrid cell line, F-11, *J. Neurochem.,* 48, 1624–1641, 1987.

108. Aceto, M.D., Awaya, H., Martin, B.R., and May, E.L., Antinociceptive action of nicotine and its methiodide derivatives in mice and rats, *Br. J. Pharmacol.,* 79, 869–876, 1983.

109. Caggiula, A.R., Epstien, L.H., Perkins, K.A., and Saylor, S., Different methods of assessing nicotine-induced antinociception may engage different neural mechanisms, *Psychopharmaacology,* 122, 301–306, 1995.

110. Tripathi, H.L., Martin, B.R., and Aceto, M.D., Nicotine-induced antinociception in rats and mice: correlation with nicotine brain levels, *J. Pharmacol. Exp. Ther.,* 221, 91–96, 1982.

111. Bannon, A.W., Decker, M.W., Curzon, P., Buckley, M.J., Kim, D.J.B., Radek, R.J., Lynch, J.K., Wasicak, J.T., Lin, N.-H., Arnold, W.H., Holladay, M.W., Williams, M., and Arneric, S.P., ABT-594 [(R)-5-(2-azetidinylmethoxy)-2-chloropyridine]: a novel, orally effective antinoceiceptive agent acting via neuronal nicotinic acetylcholine receptors: II. *In vivo characterization, J. Pharmacol. Exp. Ther.,* 285, 787–794, 1998.

112. Dussor, G.O., Leong, A.S., Gracia, N.B., Hargreaves, K.M., Arneric, S.P., and Flores, C.M., Differential effects of neuronal nicotinic receptor agonists on capsaicin-evoked CGRP release from peripheral terminals of primary sensory neurons., *Soc. Neurosci. Abst.,* 24, 1625, 1998.

113. Donnelly-Roberts, D.L., Puttfarcken, P.S., Kuntzweiler, T.A., Briggs, C.A., Anderson, D.J., Campbell, J.E., Piattoni-Kaplan, M., McKenna, D.G., Wasicak, J.T., Holladay, M.W., Williams, M., and Arneric, S.P., ABT-594 [(R)-5-(2-azetidinylmethoxy)-2-chloropyridine]: a novel, orally effective analgesic acting via neuronal nicotinic acetylcholine receptors: I. *In vitro characterization, J. Pharmacol. Exp. Ther.,* 285, 777–786, 1998.

114. de la Garza, R., Bickford-Wimer, P.C., Hoffer, B.J., and Freedman, R., Heterogeneity of nicotine actions in the rat cerebellum: an *in vivo electrophysiologic study, J. Pharmacol. Exp. Ther.,* 240, 689–695, 1987.

115. Wong, L.A. and Gallagher, J.P., A direct nicotinic receptor-mediated inhibition recorded intracellularly *in vitro, Nature,* 341, 439–442, 1989.

116. Wong, L.A. and Gallagher, J.P., Pharmacology of nicotinic receptor-mediated inhibition in rat dorsolateral septal neurones, *J. Physiol.,* 436, 325–346, 1991.

117. Sorenson, E.M. and Gallagher, J.P., The reducing agent dithiothreitol (DTT) does not abolish the inhibitory nicotinic response recorded from rat dorsolateral septal neurons, *Neurosci. Lett.,* 152, 137–140, 1993.

118. Sorenson, E.M. and Gallagher, J.P., The membrane hyperpolarization of rat dorsolateral septal nucleus neurons is mediated by a novel nicotinic receptor, *J. Pharmacol. Exp. Ther.,* 277, 1733–1743, 1996.

119. Fillingim, R.B. and Maixner, W., The influence of resting blood pressure and gender on pain responses, *Psychosom. Med., 58, 326–332, 1996.*

120. Berkley, K.J., On the dorsal columns: translating basic research hypotheses to the clinic, *Pain,* 70, 103–107, 1997.

121. Gear, R.W., Gordon, N.C., Heller, P.H., Paul, S., Miaskowski, C., and Levine, J.D., Gender difference in analgesic response to the kappa-opioid pentazocine, *Neurosci. Lett.,* 205, 207–209, 1996.

122. Gear, R.W., Miaskowski, C., Gordon, N.C., Heller, P.H., Paul, S., and Levine, J.D., The kappa opioid nalbuphine produces gender- and dose-dependent analgesia and antianalgesia in patients with postoperative pain, *Pain,* 83, 339–345, 1999.

123. Craft, R. and Milholland, R., Sex Differences in cocaine- and nicotine-induced antinociception in the rat, *Brain Res.,* 809, 137–140, 1998.

124. Damaj, M.I., Influence of gender and sex hormones on nicotine acute pharmacological effects in mice, *J. Pharmacol. Exp. Ther.,* in press, , 2000.

125. Valera, S., Ballivet, M., and Bertrand, D., Progesterone modulates a neuronal nicotinic acetylcholine receptor, *Proc. Natl. Acad. Sci. USA,* 89, 9949–9953, 1992.

126. Buisson, B. and Bertrand, D., Allosteric modulation of neuronal nicotinic acetylcholine receptors, *J. Physiol. (Paris),* 92, 89–100, 1998.

127. Bullock, A., Clark, A., Grady, S., Robinson, S., Slobe, B., Marks, M., and Collins, A., Neurosteriods modulate nicotinic receptor function in mouse striatal and thalamic synaptosomes, *J. Neurochem.,* 68, 2412–2423, 1997.

128. Ke, L. and Lukas, R.J., Effects of steroid exposure on ligand binding and functional activities of diverse nicotinic acetylcholine receptor subtypes, *J. Neurochem.,* 67, 1100–1111, 1996.

129. Flores, C.M., Wilson, S.G., and Mogil, J.S., Pharmacogenetic variability in neuronal nicotinic receptor-mediated antinociception, *Pharmacogenetics,* 9, 619–625, 1999.

130. Poulsen, L., Brosen, K., Arendt-Nielsen, L., Gram, L.F., Elbaek, K., and Sindrup, S.H., Codeine and morphine in extensive and poor metabolizers of sparteine: pharmacokinetics, analgesic effect and side effects., *Eur. J. Clin. Pharmacol.,* 51, 289–295, 1996.

131. Benowitz, N.L. and Jacob, P.I., Individual differences in nicotine kinetics and metabolism in humans, *NIDA Res. Monogr.,* 173, 48–64, 1997.

132. Nakajima, M., Yamagishi, S., Yamamoto, H., Yamamoto, T., Kuroiwa, Y., and Yokoi, T., Deficient cotinine formation from nicotine is attributed to the whole deletion of the CYP2A6 gene in humans, *Clin. Pharmacol. Ther.,* 67, 57–69, 2000.

133. Yamazaki, H., Inoue, K., Hashimoto, M., and Shimada, T., Roles of CYP2A6 and CYP2B6 in nicotine C-oxidation by human liver microsomes, *Arch. Toxicol.,* 73, 65–70, 1999.

134. Collins, A.C. and Marks, M.J., Genetic studies of nicotinic and muscarinic agents. in *The Genetic Basis of Alcohol and Drug Actions,* J.C. Crabbe Jr., and R.A. Harris, Eds., Plenum Press, New York, 1991, 323–352.

135. Seale, T.W., Nael, R., and Basmadjian, G., Inherited, selective hyporesponsiveness to the analgesic action of nicotine in mice, *NeuroReport,* 8, 191–195, 1996.

136. Seale, T.W., Nael, R., Singh, S., and Basmadjian, G., Inherited, selective hypoanalgesic response to cytisine in the tail-flick test in CF-1 mice, *NeuroReport,* 9, 201–205, 1998.

137. Schachter, S., Pharmacological and psychological determinants of smoking, *Ann. Int. Med.,* 88, 104–114, 1978.

138. Silverstein, B., Cigarette smoking, nicotine addiction, and relaxation, *J. Pers. Soc. Psychol.,* 42, 946–950, 1982.

139. Pomerleau, C.S., Nicotine as a psychoactive drug: anxiety and pain reduction, *Psychopharmacol. Bull.,* 22, 865–869, 1986.

140. Nesbitt, P.D., Smoking, physiological arousal, and emotional response, *J. Pers. Soc. Psychol.,* 25, 137–144, 1973.
141. Milgrom-Friedman, J., Penman, R., and Meares, R., A preliminary study on pain perception and tobacco smoking, *Clin. Exp. Pharmacol. Physiol., 10,* 161–169, 1983.
142. Lane, J., Rose, J., Lefebvre, J., and Keefe, F., Effects of cigarette smoking on perception of thermal pain, *Exp. Clin. Psychopharmacol.,* 3, 140–147, 1995.
143. Pauli, P., Rau, H., Zhuang, P., Brody, S., and Birbaumer, N., Effects of smoking on thermal pain thresholds in deprived and minimally-deprived habitual smokers, *Psychopharmacology.,* 111, 472–476, 1993.
144. Perkins, K., Epstein, L., Grobe, J., and Fonte, C., Tobacco abstinence, smoking cues, and the reinforcing value of smoking, *Pharmacol. Biochem. Behav.,* 47, 107–112, 1994.
145. Fertig, J.B. and Pomerleau, O.F., Nicotine-produced antinociception in minimally deprived smokers and exsmokers, *Addic. Behav.,* 11, 239–248, 1986.
146. Sult, S.C. and Moss, R.A., The effects of cigarette smoking on the perception of electrical stimulation and cold pressor pain, *Addic. Behav.,* 11, 447–451, 1986.
147. Waller, D., Schalling, D., Levander, S., and Edman, G., Smoking, pain tolerance, and physiological activation, *Psychopharmacology,* 79, 193–198, 1983.
148. Shiffman, S. and Jarvik, M.E., Cigarette smoking, physiological arousal, and emotional response: Nesbitt's paradox re-examined, *Addic. Behav.,* 9, 95–98, 1984.
149. Knott, V.J. and De Lugt, D., Subjective and brain-evoked responses to electrical pain stimulation: effects of cigarette smoking and warning condition, *Pharmacol. Biochem. Behav.,* 39, 889–893, 1991.
150. Perkins, K.A., Grobe, J.E., Stiller, R.L., Scierka, A., Goettler, J., Reynolds, W.A., and Jennings, J.R., Effects of nicotine on thermal pain detection in humans, *Exp. Clin. Psychopharmacol.,* 2, 95–106, 1994.
151. Jamner, L., Girdler, S., Shapiro, D., and Jarvik, M., Pain Inhibition, Nicotine, and Gender, *Exp. Clin. Psychopharmacol.,* 6, 96–106, 1998.
152. Kesingland, A.C., Gentry, C.T., Panesar, M.S., Bowes, M.A., Vernier, J.-M., Cube, R., Walker, K., and Urban, L., Analgesic profile of the nicotinic acetylcholine receptor agonists, (+)-epibatidine and ABT-594 in models of persistent inflammatory and neuropathic pain, *Pain,* 86, 113–118, 2000.
153. Boyce, S., Webb, J.K., Shepheard, M.G.N., Russell, R.G., Hill, R.G., and Rupniak, N.M.J., Analgesic and toxic effects of ABT-594 resemble epibatidine and nicotine in rats, *Pain, 85,* 443–450, 2000.
154. Bencherif, M., Lovette, M.E., Fowler, K.W., Arrington, S., Reeves, L., Caldwell, W.S., and Lippiello, P.M., RJR-2403: a nicotinic agonist with CNS selectivity I. *In vitro characterization, J. Pharmacol Exp. Ther.,* 279, 1413–1421, 1996.
155. Lippiello, P.M., Bencherif, M., Gray, J.A., Peters, S., Grigoryan, G., Hodges, H., and Collins, A.C., RJR-2403: a nicotinic agonist with CNS selectivity II. *In vivo characterization, J. Pharmacol. Exp. Ther.,* 279, 1422–1429, 1996.
156. Damaj, M.I., Glassco, W., Aceto, M.D., and Martin, B.R., Antinociceptive and pharmacological effects of metanicotine, a selective nicotinic agonist, *J. Pharmacol. Exp. Ther.,* 291, 390–398, 1999.
157. Decker, M.W., Bannon, A.W., Buckley, M.J., Kim, D.J.B., Holladay, M.W., Ryther, K.B., Lin, N.-H., Wasicak, J.T., Williams, M., and Arneric, S.P., Antinociceptive effects of the novel neuronal nicotinic acetylcholine receptor agonist, ABT-594, in mice, *Eur. J. Pharmacol.,* 346, 23–33, 1998.
158. Inoue, K., Yamazaki, H., Shimada, T., CYP2A6 genetic polymorphisms and liver microsomal coumarin and nicotine oxidation activities in Japanese and Caucasians, *Arch. Toxicol.,* 73, 532–539, 2000.

6 Mouse Models to Evaluate Genetic Influences on Responses to Nicotine

Michael J. Marks and Allan C. Collins

CONTENTS

6.1   INTRODUCTION

Genetic factors influence tobacco use in humans as first illustrated by Fisher,[1,2] who reported that monozygotic twins displayed significantly more concordance for smoking behavior than did dizygotic twins. This observation has been repeatedly confirmed in twin and family studies of smoking (see Heath and Madden[3] for a recent review). More recent studies have been extended to evaluate the genetic influence of coexpression of tobacco use and several personality traits as well as the use of other drugs, including alcohol.[4–6] The data suggest genetic factors play an important role in regulating vulnerability to becoming a smoker, but only limited progress has been made in humans in terms of identifying genes critical in regulating nicotine dependence. However, some progress has been made in understanding the regulation of responses to nicotine in the mouse. This chapter will outline strategies that have been, or could be, used to identify genes that regulate behaviors. Specific examples of how these strategies have been used to gain an understanding of genetic influences on nicotine-related behaviors will be provided.

Virtually all of the studies of genetic regulation of responses to nicotine have used the mouse. Historically, the mouse has been the species of choice for genetic studies largely because of the availability of large numbers of genetically-homogeneous inbred strains. An inbred strain is derived by brother-sister mating and, at least theoretically, all members of a given inbred strain are identical genetically. Thus, an inbred mouse strain can provide multiple copies of the mouse equivalent of monozygotic twins. Inbred mouse strains have been derived from multiple sources and many are commercially available. Some of the currently available inbred strains are closely related to one another (for example, the C57BL/6, C57BL/10, C57BR strains); others are only distantly related. An excellent source of information concerning inbred mouse strains is provided by the Jackson Laboratories on their home page (www.informatics.jax.org). The large number of inbred strains, some of which are closely related and others not, offers the scientist some very powerful tools to use to characterize genetic regulation of a phenotype, as well as ultimately identifying genes that regulate a phenotype.

Recently, the mouse has become the species of choice for gene mapping experiments because of the rapid progress that has been and is being made in mapping the mouse genome. The mouse is also the species of choice for gene targeting experiments such as the production of transgenic and null mutant (gene knockout) mice.

6.2   GENERAL APPROACHES: FORWARD AND REVERSE GENETICS

Understanding the role of heritable factors in any biological system, including response to nicotine, can be approached from two complementary directions. The first approach, which has been termed forward genetics, generally starts with identifying genetic regulation of a phenotype, such as a specific response to nicotine. This often involves screening inbred strains for variability in a phenotype of interest. Inbred mouse strain differences can be either qualitative or quantitative. The differences across inbred mouse strains (i.e., strain mean differences) are determined by genetic factors, whereas the variability within an inbred strain arises because of environmental influences on the phenotype. The identification of inbred strain differences is the first step. Other strategies, such as studying animals derived by breeding disparate strains together, are required to determine the source of the genetic variance and, ultimately, to localize the chromosomal sources of the variability and identify the genes themselves.

The second and more recently developed genetic approach is sometimes called reverse genetics. This approach involves using modern molecular methods to alter or eliminate a gene. Once an animal is produced that carries the modified gene, the effects of genetic manipulation on phenotypes of interest is studied. For studies of nicotine actions, the most important example of this approach has been the development of nicotinic receptor null mutant (gene knockout) mice. Null mutant mice exhibit a loss of gene function. A gain in function can be achieved by developing mice that overexpress a gene (transgenic mice) and, for something like nicotinic receptors, animals which express a nicotinic receptor subunit that desensitizes more slowly can be developed. The latter approach has been used for the α4 and α7 nicotinic receptor subunit genes.[7,8]

Both forward and reverse genetics approaches have advantages and disadvantages and each can provide valuable information about the genetic control of a complex phenotype. The ultimate goal of the genetic strategies should be to obtain converging evidence using both forward and reverse genetics to provide a comprehensive description of the complex polygenic influences on nicotine response and nicotine dependence. Neither approach will, in all likelihood, yield answers that do not require circumspect interpretation, and one approach cannot be used without taking into consideration the other. For example, two α4 nicotinic receptor subunit null mutant strains have been produced:[9,10] the French-made α4 null mutants[9] were derived from a 129-C57BL/6 cross, whereas the Australian-made α4 mice[10] were derived from a BALB/c-CF1 cross. An example of the potential effect of this difference is found when the effects of nicotine on locomotor activity were measured in the two α4 null mutant mice. Marubio et al.[9] reported that the locomotor activities of their α4 null mutants did not differ from the wildtype controls following injection with 1 and 2 mg/kg nicotine. In contrast, Ross et al.,[10] reported that their α4 null mutants showed a reduced sensitivity to the locomotor depressant effects of nicotine produced by injection with 0.9 and 2.7 mg/kg nicotine. In both cases, the α4 gene had been knocked out with the same biochemical consequences (loss of high affinity nicotine binding in most brain regions), but the behavioral effects of gene deletion were markedly different. One possible explanation for the discrepant results is that differential genetic background influences the animal's ability to cope with loss of the α4 gene product. Several recent reports provide in depth discussions of the factors that should be considered when evaluating data obtained with transgenic mice and the approaches that should be used in the behavioral phenotyping of these mice.[11–14]

6.3 CHOICE OF PHENOTYPE

The effects of nicotine are varied and complex. Part of this complexity arises because nicotine interacts with specific and distinct nicotinic acetylcholine receptors found on skeletal muscle, in the autonomic nervous system, on secretory tissue, and throughout the central nervous system. This complexity is compounded by the fact that nicotinic receptor activation subsequently stimulates additional biological responses such as membrane depolarization, muscle contraction, hormone secretion and neurotransmitter release. Partially as a consequence of the multiple sites of expression of nicotinic receptors, many different behavioral and physiological responses to nicotine have been observed. Each of the responses is amenable to study and many may have consequences either to reinforce or diminish the use of nicotine. Each of the responses may be under specific genetic control.

Nicotine affects many measures in mice:

1. Locomotion measured in open field arena[15] and in the Y-maze[16]
2. Body temperature[15]
3. Acoustic startle response[15]
4. Heart rate[15]
5. Respiratory rate[15]
6. Convulsive activity[17,18]
7. Antinociception[19]
8. Anxiety[20-23]
9. Memory retrieval[24]
10. Depression[25]
11. Discrimination learning and consolidation[26]

In addition, several behaviors that may be directly related to nicotine abuse have been measured in mice. These include:

1. Development of tolerance following chronic administration[27]
2. Oral nicotine consumption[28]
3. Intravenous self-administration and drug discrimination[29]
4. Conditioned place preference[30]
5. Withdrawal from chronic nicotine[31]

Once tests have been chosen and the appropriate nicotine dose or doses have been determined for a representative strain, the investigation of the effects of genotype can proceed. In general, mice require higher doses of nicotine to achieve a particular behavioral response than do rats. This difference arises, in part, because the mouse metabolizes nicotine more rapidly than the rat; the half-life for nicotine is approximately 6 minutes in the mouse[32] and nearly 60 minutes in the rat.[33]

Quite frequently it is of both scientific and practical value to develop methods that will allow the evaluation of several responses in the same mouse. Caution must be taken to avoid distorting responses through the introduction of intertest interactions that may occur when more than a single response is measured. Multitest batteries are particularly valuable when expensive mice (such as knockout mice) or genetically unique mice (such as mice from a segregating population) are being evaluated.

6.4 GENETIC MODELS FOR THE ANALYSIS OF THE EFFECTS OF NICOTINE

Various mouse models that have been used, or can be used in the future, to study the genetic influences on nicotine responses will be discussed in the sections that follow. Specific examples of experimental results will be included.

6.4.1 INBRED STRAINS

The initial screen for genetic effects almost universally involves the measurement of the effects of nicotine in several inbred mouse strains. Inbred mice can be obtained from several suppliers including Jackson Laboratories, which maintains a large number of inbred strains. Inbred mice are a stable breeding population and, barring mutations leading to genetic drift, maintain a constant genotype over time. Consequently, it is not necessary to measure all phenotypes of interest at once. Comparison of results obtained with inbred mice collected at different times and different places is feasible. However, care must be taken to consider that environmental factors such as housing conditions, season, and experimental differences among laboratories can affect the phenotype measured.[34]

Many experiments have been conducted using a screen of two or more inbred strains, and in virtually every study, the effects of nicotine varied with genotype. Several examples are cited here.

The C57BL/6 and DBA/2 strains of mice are among the most commonly studied of all inbreds; these two strains have been examined for the effects of nicotine on learning,[26] locomotor activity,[16,35] nicotine metabolism,[32] seizure sensitivity,[17] oral nicotine intake[28,36] and nicotine discrimination.[29] Although the C57BL/6 and DBA/2 strains are excellent choices for an initial strain comparison, the information obtained may be applicable to only these two strains. In order to obtain more generalizable data, screening of several inbred strains is advisable. Several such screens have been conducted and will be briefly discussed here to illustrate the progression of such strain screens. In one study, four inbred mouse strains (BALB/cByJ, C3H/2Ibg, C57BL/6J, and DBA/2) were screened for their responses to nicotine for several behavioral or physiological tests, including body temperature, rotarod performance, open field activity, respiratory rate, heart rate, and acoustic startle response.[15] Each test was administered independently and dose-response curves were constructed for each of the responses to nicotine. In this study, DBA and C57BL/6 mice displayed similar responses to nicotine, while responses of C3H mice differed qualitatively; C3H mice exhibited an increase in open field activity and startle response following nicotine whereas most other strains showed a decrease in these measures following nicotine administration. Inbred strains can also differ quantitatively from one another. C3H mice, for example, differ from C57BL/6, DBA/2, and BALB/cByJ mice in the nicotine dose required to decrease body temperature by 2° and in the dose required to decrease rotarod performance by 50%. Thus, even with this relatively modest comparison of four inbred strains, the advantage of examining strains other than C57BL/6 and DBA was illustrated.

A subsequent analysis of nicotine response used 19 strains.[37] The strains studied included sets of closely related mice, such as the C57 family, as well as strains with no obvious genetic relationship with other strains, such as RIIIS/J, SWR/J, and BUB/BnJ. Before beginning these analyses, a multicomponent test battery was developed and validated so that several measures of nicotine response could be made with one mouse (respiratory rate, acoustic startle response, Y-maze crossing and rearing activities, heart rate, and body temperature),[38] thereby reducing the number of animals required. Nicotine affected all 6 responses in each of the 19 inbred strains.[37] Some responses differed quantitatively (same direction of response, but differences in the dose of nicotine required to elicit the response, e.g., locomotor activity, body temperature) while others differed qualitatively (different direction of response, e.g., acoustic startle). In a parallel study with these same strains, clonic seizures elicited by either ip or iv administration of nicotine were also evaluated.[39]

When several measurements are made within a large population — in this case strains of inbred mice — responses measured for each test can be compared to those measured for other tests. An initial comparison of responses can be made using regression analysis to evaluate the relationship between two variables. Regression analysis was used to compare the responses among the strains and revealed a wide range of relationships.[40] For example, the relative sensitivities of the strains to nicotine-induced changes in Y-maze crosses and rears were highly correlated ($r = 0.93$), while relative sensitivities to nicotine's effects on Y-maze crossing activities and seizure sensitivity were not related ($r = 0.07$). A potential explanation for this observation is that the sets of genes influencing Y-maze activities and seizure responses are not the same. This conclusion is based on the assumption that these responses are mediated by multiple genes. It is possible, however, that one or two common genes modulate these responses to nicotine.

When regression analysis detects significant correlations among several variables, further analyses, such as factor analysis (also termed principal components analysis), may be applied. In general, a factor analysis assumes that relationships among several variables occur because these variables are influenced by a relatively few underlying common variables (factors). The goal of a factor analysis is to describe a set of experimental variables with as few underlying factors as possible. The resulting simplification of a complex set of observations may be useful in categorizing observations and improving understanding of the possible relationships among them. Of course, any relationship among variables and calculation of potential underlying factors does not prove a common biological basis for a set of responses, but only suggests such a relationship. When applied to responses to nicotine measured in the inbred mouse strains,[37] factor analysis suggested a relatively simple relationship among the responses. Two major groups of responses (two factors), which accounted for about 70% of the variance for the interstrain differences for eight tests, were indicated: one for Y-maze crosses, Y-maze rears, body temperature and, to a lesser extent, heart rate and a second for ip and iv seizures. Respiratory rate and startle response shared properties in common with both of the other groups. Thus, the measurement of many nicotine effects in a relatively large number of strains suggested that several of the responses measured may have similar underlying influences reflecting common genetic influences.

When a large number of strains are tested, it may be useful to group mouse strains by their relative sensitivity, that is, to organize the strains into groups displaying similar characteristics. One such method of organization is cluster analysis, in which individuals or groups are combined with similar individuals. Cluster analysis of the data obtained in the 19-strain study[37] suggested four clusters of strains that have similar sensitivity to nicotine's effects on Y-maze activity and temperature measures. The strains ranged from very sensitive (A, C57BL/6, C57BL/10) to very sensitive (ST/b) to very resistant (DBA/1, DBA/2). The value of clustering is that mouse strains can be selected, for future experiments, from each of the clusters to provide a range of responses without testing a large number of inbred mouse strains. It should be noted that strains in the same cluster do not necessarily have similar sensitivity because they are genetically identical, or nearly identical, at relevant loci.

One underlying goal of genetic studies is to evaluate whether a relationship exists between a phenotype measured in the whole animal and a particular molecular, biochemical, or cell physiological process. Such comparisons were made among several behavioral responses to nicotine and the levels of [^{3}H]nicotine binding and [^{125}I]α-bungarotoxin binding,[40] which measure two major nicotinic receptor subtypes.[41,42] Regression analysis revealed a relationship between overall [^{3}H]nicotine binding and Y-maze crossing and rearing activities and body temperature (r = –0.62) and between overall [^{125}I]α-bungarotoxin binding and nicotine-induced seizures (r = –0.63). These results suggest that approximately 30 to 40% of the variance for the effects of nicotine on the Y-maze and body temperature responses may be explained by variability in the number of [^{3}H]nicotine binding sites (i.e., most likely $\alpha4\beta2$-type nAChR[9,10,43–45]) and 30 to 40% of the variance in sensitivity to nicotine-induced seizures may be due to variability in [^{125}I]α-bungarotoxin binding (i.e., $\alpha7$-type nAChR[46,47]). In both cases, high basal expression of these receptor subtypes is associated with greater sensitivity to the effects of nicotine (lower ED50-like values). The finding that differences in receptor numbers "explain" only 30 to 40% of the variance implies that variables in addition to the density of binding sites contribute to the differential responses to nicotine measured in the 19 inbred mouse strains. Genetically-based variability in response to nicotine could also arise as a consequence of variation in nicotine distribution and metabolism as well as variability at any step between receptor binding and the ultimate manifestation of the response.

The cause of the strain differences in [^{3}H]nicotine binding is unknown, but analyses for $\alpha7$ nicotinic receptor differences revealed the presence of a restriction fragment linked polymorphisms (RFLP).[48] RFLPs for this gene were identified initially in a screen of ten inbred strains. DBA/1 and DBA/2 mice displayed a different pattern from all other mice tested (including C3H/2 and ST/b). Subsequent analyses (unpublished results) determined that the polymorphism detected by the RFLP occurs in noncoding regions (intron 9), indicating that the polymorphisms did not affect the primary structure of the receptor protein. However, the RFLP may influence binding since an analysis of the association between the RFLP and brain [^{125}I]α-bungarotoxin binding done in F2 mice derived from a DBA × C3H cross indicated that those F2 animals that were homozygous for the C3H-derived RFLP had higher levels of hippocampal [^{125}I]α-bungarotoxin binding than did those animals that were homozygous for the DBA-derived RFLP; heterozygotes had intermediate levels of binding.

Studies with rats indicate that α-bungarotoxin sensitive nicotinic receptors may be important in the regulation of auditory gating. Given the observation that [^{125}I]α-bungarotoxin binding varies among inbred strains, auditory gating was evaluated in nine inbred mouse strains that differ in the number of hippocampal [^{125}I]α-bungarotoxin binding sites.[49] Auditory gating varied among the inbred strains and a robust correlation (r = 0.72) was seen between test/conditioning ratio and the amount of hippocampal α-[^{125}I]bungarotoxin binding. This is consistent with the proposal that mice expressing higher levels of α7-nAChR show higher auditory gating. This study is an example of using genetic comparisons to test a hypothesis originally proposed based on results obtained in pharmacological studies.

Genetic influences on additional responses potentially important for nicotinic pharmacotherapy and nicotine dependence have also been investigated using inbred mice. For example, eight inbred mouse strains differed in antinociception elicited by the potent nicotinic agonist, epibatidine.[19] Three of the strains (A, BALB/c, and DBA/2) were much more sensitive than the other five strains. The relative rank order of sensitivity differed from that observed for either the locomotor/temperature or seizure effects of nicotine,[37,39] suggesting that genetic influences on epibatidine-mediated antinociception differed from those of the other responses. Inasmuch as nicotinic agonists are being investigated for the relief of pain, information about individual differences in the antinociceptive effects of these drugs will be very critical.

A central aspect of the study of nicotine abuse is the investigation of self-administration. As noted above, Stolerman et al.[29] compared C57BL/6 and DBA/2 mice for drug discrimination. Stolerman argued that nicotine discrimination measures the cue that underlies the reinforcing effects of nicotine. Both strains readily learned to discriminate nicotine from saline, but the DBA mice were more sensitive to response-rate reducing effects of nicotine. The finding that these two strains differ only slightly in terms of nicotine discrimination is consistent with the findings of Robinson et al.,[28] who reported that the C57BL/6 and DBA/2 strains consumed the highest daily dose of six strains analyzed when either water or 0.2% saccharin was used as the vehicle. Interestingly, a significant inverse relationship between seizure sensitivity and oral intake was observed across the six inbred strains,[28] which suggests that a toxic reaction reflected by sensitivity to nicotine-induced convulsions may limit oral intake of nicotine in the mouse.

Inbred mouse strains also differ in the development of tolerance to nicotine.[27] This result was obtained in a study that compared tolerance to nicotine using five inbred mouse strains. The strains were also differentially sensitive to the acute effects of nicotine. The mice were chronically infused with nicotine doses ranging between 0 (saline infusion) and 6 mg/kg/h through jugular cannulae and tested for tolerance 2 h after infusion was stopped. Genotype clearly influenced tolerance development. The amount of tolerance depended upon the initial sensitivity of the mice to nicotine: in general, mice initially more sensitive to the acute effects of nicotine developed tolerance following infusion of lower drug doses than did mice that were initially less sensitive. When maximum tolerance had developed, the sensitive strains had approximately the same sensitivity to nicotine as did those strains with low first dose sensitivity. The strains did not differ in the drug-induced upregulation of [^{3}H]nicotine binding and [^{125}I]α-bungarotoxin binding. This finding suggests that tolerance to nicotine is not inextricably due to upregulation of either of the two major neuronal nicotinic receptor subtypes.

6.4.2 DIALLEL CROSS ANALYSIS OF F1 HYBRID MICE

The screening of inbred mouse strains will detect a genetic influence on behaviors and suggest relationships among the behaviors, but results from inbred strain comparisons do not provide information about the mode of inheritance of the traits, including information about additive or dominance components, or initial estimates of gene number. In order to obtain information concerning these issues, additional breeding experiments are required. One such breeding experiment, which is rarely used, is the diallel cross in which several inbred mouse strains are crossed to produce all possible F1 hybrids. These hybrids are isogenic (genetically homogeneous). All members of the F1 generation are homozygous at every locus where their parents carried the same allele and are heterozygous at every locus at which their parents differed. The hybrids can be tested and the subsequent results can be analyzed for the presence of additive genetic variance, dominance, and maternal effects.[50-52] Diallel crosses allow a genetic analysis to be done in one generation. However, the phenotype will reflect an overall genetic architecture which is not amenable to the identification of specific genes. Diallel crosses have been used to evaluate the effects of nicotine on body temperature[53] and open field activity.[54] In both studies, F1 hybrid mice were generated by complete reciprocal crosses of five inbred mouse strains (A/J, BALB/ByJ, C3H/2Ibg, DBA/2J, and C57BL/6J) known to differ in response to nicotine.[15] Both studies revealed that differential response to nicotine was heritable, with both additivity and dominance contributing to the heritability. Dominance following injection of a modest (for mice) dose of 0.75 mg/kg was toward an increased response to nicotine. This result may mean that intense response to nicotine provides a selective advantage, perhaps because an intense response to a toxic agent would tend to limit consumption of this agent, thereby preventing ingestion of a lethal dose. The diallel analyses also suggested that the regulation of nicotine's effects on open field activity and body temperature were, not surprisingly, polygenic with gene estimates for the nicotine effects of about seven. Such estimates should be regarded with caution because they tend to underestimate the actual gene number.

6.4.3 CLASSICAL GENETIC CROSS ANALYSIS OF F1, F2, AND BACKCROSS MICE

The classical genetic cross is amenable to both quantitative genetic analysis and the subsequent mapping of specific loci using quantitative trait locus mapping (QTL) discussed later. In this analysis the F1 hybrid is generated by crossing mice of two inbred strains; F1 mice can subsequently be crossed with F1 mice to yield F2 mice or with each of the parental strains to yield backcross mice. The F2 and each backcross are a segregating population with different overall contributions of genes from each parental strain. Quantitative genetic analysis of the responses of the populations of mice generated by the classical genetic cross can subsequently be used to assess additive, dominance, and epistatic contributions to the heritability of the response and can be used to test certain genetic models.[51] If several responses of individual mice from the segregating populations are measured, relationships among these responses can be evaluated to determine if they cosegregate. It should be emphasized that each mouse in the segregating populations is unique. As a consequence, the tests used to evaluate these animals must be of the sort where measuring one phenotype does not alter the value of any other phenotype.

Classical genetic crosses have been used to investigate the regulation of nicotine-induced seizures.[18] Mice of C3H (sensitive) and DBA (resistant) strains differ in the ED_{50} for nicotine-induced seizures and were used as the parental strains for a classical genetic cross. Following ip administration of nicotine, dose-dependent increases in seizures were noted for mice of each generation (F1, F2, backcross), thereby demonstrating that seizure sensitivity is heritable. Substantial dominance toward resistance to nicotine-induced seizures was observed. Significantly higher [^{125}I]α-bungarotoxin binding was observed in those animals obtained from F1 × DBA and F1 × C3H mice (the backcross generations) that seized following challenge with a test dose that elicited seizures in about 60% of the F2 animals. A similar trend was suggested for F2 mice, supporting the hypothesis that mice with high [^{125}I]α-bungarotoxin binding are sensitive to nicotine-induced seizures.[55] Slightly different results were obtained when nicotine was infused intravenously (iv). As was the case following ip injection, C3H mice were more sensitive than DBA mice following iv administration of nicotine; however, no significant dominance component was observed in the F1 generation.[56] Thus, although the relative sensitivity of the parental strains remained the same when the route of nicotine administration was changed, the genetic pattern of inheritance for the two routes of administration varied. These findings suggest that the rate of penetration of nicotine into the brain is important in regulating the seizure response and that different (presumably partially different) sets of genes regulate seizures depending on route of administration.

As noted previously, C3H and DBA/2 mice display an RFLP in the noncoding region of the α7 gene.[48] C3H and DBA/2 mice also differ in a polymorphism for the α5 gene.[57] This RFLP is also associated with variance in sensitivity to nicotine-induced seizures. In the segregating F2 population derived from a C3H × DBA/2 cross, F2 mice expressing the DBA variant for α5 were less sensitive to nicotine-induced seizures than were mice with the C3H variant. Heterozygotes displayed the same seizure sensitivity as DBA (i.e., there appeared to be dominance toward the DBA genotype). In this same F2 population, mice expressing the DBA variant of α7 were less sensitive to seizures and displayed lower hippocampal (but higher striatal) [^{125}I]α-bungarotoxin binding than did mice expressing the C3H variant. The α7 heterozygotes were intermediate in seizure sensitivity and binding. The α5 and α7 genotypes appeared to act independently, indicating that each contributed to the differential seizure sensitivity. It should be noted here that the α5 RFLP was detected by hybridization with a probe to the DNA for this subunit. Inasmuch as, α5, α3, and β4 form a gene cluster, the association between the α5 RFLP and seizure sensitivity may be a measure of an association with another linked gene such as the α3 or β4 nicotinic receptor genes.

Classical genetic crosses have also been used to evaluate the effects produced by lower nicotine doses. One such study used mice that were selectively bred, starting from a heterogenous stock of mice, for differences in duration of ethanol-induced loss of the righting response (sleep time).[58] These lines, designated long sleep (LS) and short sleep (SS), are also differentially affected by nicotine,[59] with LS mice generally more sensitive to both ethanol and nicotine. A classical cross analysis that used F1, F2, and backcross generations derived from the LS and SS mice[60] demonstrated that nicotine responses measured in each of five tests (respiratory rate, Y-maze crosses, Y-maze rears, heart rate, and body temperature) were heritable and polygenic. The pattern of inheritance varied among the tests suggesting that the genetic regulation of these responses is not identical. Interestingly, sensitivity to ethanol and nicotine segregated together in the F1, F2, and backcross generations. This finding suggests that one or more genes may play an important role in regulating sensitivity to both drugs. This might relate to findings that common genes may influence vulnerability to alcohol and nicotine dependence in humans.[6]

It should be noted that segregation analyses do not prove that a specific gene contributes to variability in a phenotype. The RFLP strategy, for example, is a form of linkage analysis. Thus, even though it makes sense that variability in a nicotine-induced behavior is due, at least partially, to variability associated with a nicotinic receptor gene, it is premature to draw this conclusion. The association between nicotinic receptor RFLP and phenotype may arise because another gene closely linked to the RFLP is important in regulating variability in the trait. The RFLP data elevate the nicotinic receptor gene to a candidate gene. Other strategies will be required to demonstrate unequivocally the role of the candidate gene.

6.4.4 RECOMBINANT INBRED STRAINS

Recombinant inbred strains are generated by inbreeding through brother/sister matings of mice of a segregating F2 population originally produced by the mating of two inbred strains.[61] After 20 generations of brother-sister matings, new inbred strains (recombinant inbred strains, RI) are generated that have stable, recombined genotypes. RI strains are extremely useful for quantitative genetic studies, for uncovering major gene effects, and for QTL mapping, which will be discussed later. Because RI mice are genetically homogeneous, a reliable estimate of the value of the phenotype can be obtained by testing multiple animals from each strain. Thus, RI strains have many of the advantages of inbred mice in that phenotypes can be reliably measured and all relevant data required for analysis need not be collected from a single mouse. Results using the same RI strains can be compared even though the measurements are made at different times or even in different places and, therefore, are cumulative. The analysis of RI strains has limitations: the genetic architecture expressed in the RI strains is limited by the inbreds used to establish the lines, dominance relationships are not measured, and quantitative genetic analyses have reduced power compared to F2 analyses because the number of strains is small.

Construction of new RI strains is an enormous undertaking, but fortunately several sets of RI strains have already been established. Many are available from Jackson Laboratories. Among the well-known RI strains are: BXD (originally established from a C57BL/6 [B] female DBA/2 [D] male mating), AXB, and BXA (originally established from A female × C57BL/6 male and a C57BL/6 female × A male mating, respectively). A set of recombinant inbreds derived from the LS and SS mice are maintained at our home institute, the Institute for Behavioral Genetics.[62]

RI strain analyses of nicotine responses have only recently been initiated. A recent study[63] used LS-SS RI mice to analyze the effects of acute nicotine injection on clonic seizures. An early study[59] found that LS mice are more sensitive to nicotine-induced seizures than are SS mice, but the density of both [^{3}H]-nicotine and α-[^{125}I]bungarotoxin binding sites do not differ substantially. Thus, levels of $\alpha 4\beta 2$- and $\alpha 7$-nAChR are not likely to be major factors in this difference. Stitzel et al.[63] found RFLPs for $\alpha 2$, $\alpha 3$, $\alpha 4$, $\alpha 5$, and $\alpha 6$ in LS and SS mice; of these, only the $\alpha 4$ and $\alpha 6$ polymorphisms persist in the RIs. An analysis of seizure sensitivity in the LS × SS RI strains found a wide range of sensitivity in these strains and that those RI strains with the LS-like $\alpha 4$ genotype were more sensitive to nicotine-induced seizures than were RI mice with the SS-like $\alpha 4$ genotype. Conversely, male RI mice with the SS-like $\alpha 6$ genotype were more seizure sensitive than mice with the LS-like $\alpha 6$ genotype. The effect of the $\alpha 6$ genotype was not seen in female mice.

6.4.5 SELECTED LINES

One of the best ways of establishing the heritability of a trait of interest is via selective breeding.[64] Selective breeding for a trait of interest clearly establishes a heritable component for that trait. In addition, the main phenotypic value of the lines of animals that are generated generally exceed the extremes of the foundation population. Selective breeding must be started from a heterogeneous population such as an F2 or F3 generation or, even better, a heterogeneous stock (HS) such as that used by McClearn to derive the LS and SS mice.[58] This HS stock was derived by interbreeding eight inbred strains and is maintained by outbreeding (avoiding mating animals who share a common grandparent). The goal of selective breeding is to obtain mice that display extreme responses to the drug in succeeding generations. A well-designed selection includes replicates of each selected line as well as control lines that have not been subjected to selective pressure. Selected lines are invaluable tools that can be used to identify genes that contribute to genetically-based variability in the selected phenotype.

Smolen derived replicate lines of mice that differ in locomotion in the Y-maze following nicotine injection of 0.75 mg/kg nicotine.[64] These lines were developed starting from the heterogenous stock of mice derived by McClearn.[58] Replicate lines, called nicotine-activated (NA) and nicotine-depressed (ND), were developed by within-family selection for six generations. As selection proceeded the ND lines showed a progressive increase in sensitivity to the depressant effects of nicotine. However, selection did not work in the opposite direction: the NA animals were identical in response to nicotine to the unselected controls; i.e., selection did not work in the activated direction. Smolen also measured the effects of nicotine on body temperature and found that this effect was significantly correlated with the locomotor effect in the HS mice. These responses continued to be correlated throughout the selection, suggesting that these two measures are influenced by common genes, a result consistent with the inbred strain analyses discussed.[37] Early generations of these selected lines were also used to measure conditioned place preference (CPP) following nicotine. Nicotine injection produced CPP in the NA and control lines, but the ND line did not develop CPP following nicotine treatment,[30] indicating that these lines may have been useful for studies of the reinforcing properties of nicotine. Unfortunately, these selected lines no longer exist.

6.4.6 QUANTITATIVE TRAIT LOCUS MAPPING

Most behavioral traits arise from the effects of many genes each of which contributes a small amount to variability in the phenotype. In other words, these traits are polygenic. Certainly some genes will have a larger impact on a given response than others, but complete description of the genetic architecture of any phenotype will require identification of the underlying genes. This is obviously an ambitious goal. Quantitative trait locus (QTL) mapping is one method developed which may help achieve this goal. The ever increasing identification of unique chromosomal markers throughout the mouse genome,[66] has facilitated the use of QTL mapping.

In principle, QTL analysis is relatively straightforward and has been described in detail elsewhere.[66–69] The first step is to identify strains of mice that differ quantitatively in a phenotype of interest. Subsequently, a segregating population derived from the progenitor strains is developed and tested for the phenotype. Optimally, the members of the segregating population should show widely different phenotypes so that identification of differences among mice in the population can be reliably assessed. Given that genetic differences underlie the phenotypic differences, the genes regulating the responses would be expected to differ among the mice displaying the differing phenotypes. The question, then, is where are these genes and how are they found? Initially with the discovery of RFLP differences among mouse strains and now with the availability of a large number of microsatellite markers distributed throughout the genome[66] (see also www.informatics.jax.org and www.resgen.com), it has become possible to determine whether phenotypic differences segregate with markers that differ between the progenitor strains. Cosegregation of the markers with the phenotype indicates a linkage between these two measurements and suggests the chromosomal location of the QTL. Relationships between the markers and the phenotype are analyzed by linkage analysis using a program such as MAPMAKER/QTL.[70] QTLs that appear to regulate the response can be identified by comparison of phenotypic responses and markers distributed throughout the genome. Candidate genes can be identified or cloned by mapping the area of interest more closely.

Several suggestions for similar strategies to identify QTLs have been advanced. The following progression has been proposed for conducting a QTL analysis:[68]

1. Identify selected lines or inbred strains that differ in the phenotype of interest.
2. If available, measure the phenotype in RI strains generated from the chosen founders. As noted previously, RI mice are a stable breeding population, and chromosomal marker maps have been generated for several of them and are constantly being expanded. The availability of these maps avoids the need for genotyping of the RIs. A major problem for RI analyses is that most of the RI sets available are small (20 to 40). This limits the statistical power of RI stain analyses which, in turn, limits the power to detect QTLs.
3. Once candidate QTLs have been identified in RI strains, the phenotype can be measured in segregating populations (backcross or F2) with

subsequent genotyping and mapping. Extension of the analysis to segregating populations is more labor intensive since each mouse is unique
and must be individually tested and genotyped. The F2 analysis may be
inappropriate for some studies, particularly one where the goal is to
determine whether common genes regulate two phenotypes, and measuring one phenotype affects measurement of the other. However, a
major advantage of the F2 approach is it has greater statistical power
than does the RI approach because an unlimited number of animals can
be produced and tested.

The first step in beginning a QTL analysis is choosing the strains to be analyzed.
Eventually the power of the analysis depends upon the expression of a broad
phenotype in the segregating populations. It has been suggested that the best progenitor strains are those selectively bred for differences in the trait of interest and
subsequently inbred.[68] This choice is not available for nicotine studies because no
selectively bred lines are available.

In the absence of selected lines, progenitor strains can be chosen after screening
inbred strains for the trait of interest and determining that the trait of interest is
heritable. It would seem necessary intuitively to choose progenitor strains that differ
in the phenotype of interest, but such differences are not obligatory. It may be, for
example, that two inbred strains have similar values for a phenotype of interest
because each strain has several genes (alleles) that contribute to an increase in the
phenotypic value ("plus" alleles) as well as several alleles that contribute to a
decrease in the phenotypic value ("minus" alleles). If this is the case the extreme
of the segregating population will far exceed the means of the parental strains. Inbred
strains that are phenotypically similar are especially useful if analysis of RI strains
is to be included.

The second stage of analysis must use segregating populations, such as F2
animals generated from the chosen founder strains. A major disadvantage of testing
F2 animals is that the analysis is much more labor intensive than testing RIs because
every mouse is unique and must be tested phenotypically. Every animal must also
be genotyped. The major advantage of testing an F2 population is that a large number
of mice can be tested and the power of the subsequent analysis can be extended
beyond that which is possible with an RI strain analysis. With increased power, the
number of candidate QTLs can be expanded and candidate genes that explain a
smaller fraction of the variance in the value of the phenotype can be identified. It
is important to remember that each F2 animal is unique genetically and, consequently, measurement of the phenotype must be reliable or effects, especially small
genetic effects, may be lost in measurement error. If several traits are to be evaluated,
it is imperative that measuring one phenotype does not alter the value of the others.

Although testing and genotyping of each individual in a segregating population
provides the most extensive information about the genetic basis for the behavior,
such an analysis can be massive and massively expensive. As an alternative to
genotyping each animal, strategies have been proposed that involve measuring the
phenotype in large numbers of mice and genotyping only those animals that have
the lowest and highest phenotypic values.[67,69,71] This strategy is especially applicable

if the effort involved in phenotypic testing of large numbers of subjects is less than that required to genotype all of the animals.

QTL analyses have not yet been undertaken for the effects of nicotine, but an interesting example for a drug of abuse has been obtained for morphine. A screen of several inbred mouse strains revealed marked strain differences in oral morphine intake.[72,73] Among the most extreme strains were the DBA/2 (low intake) and C57BL/6 (high intake). Consequently, oral intake of morphine was measured in the BXD RIs.[74] QTL mapping methods were not widely available when this analysis was done, but when marker maps and quantitative methods became available, the results from the BXD RI analysis of oral morphine intake study were reexamined.[75] This analysis identified several candidate QTLs (on chromosomes 2, 4, 8, and 9). A short time after this a QTL analysis of oral morphine intake of C57BL/6 × DBA/2 F2 mice was done.[76] Intriguingly, the results obtained from the F2 analysis differed markedly from those obtained with with the BXD RI strains. The F2 analysis[76] identified highly significant QTL on chromosomes 6 and 10, and a less significant QTL on chromosome 1. A potential explanation for these discrepant findings is that RI-strain QTL analyses have low statistical power. This means that any QTL identified in an RI study must be viewed as provisional. The F2 analysis has greater statistical power and the results obtained in such an analysis should be given greater credence. Another reason for giving the F2 analysis greater credence is that (at least for this example), data were obtained that seem to make sense. Specifically, the largest QTL on chromosome 10 found in the F2 study maps precisely to OPRm, the μ opioid receptor which pharmacological studies and null mutant analysis have identified as the major site of action of morphine.[77] Interestingly, the other significant QTLs found in the F2 study do not correspond to the chromosomal locations of δ-, κ-, or σ-opioid receptors. Therefore, QTL mapping provides some expected results (mapping of oral intake at the chromosomal location of a likely candidate gene, the μ-opioid receptor) but also identifies QTLs that do not correspond to a known opioid receptor.

The pattern of QTLs will likely depend on the inbred strains chosen to establish the F2 mice. It is absolutely imperative to understand that QTL mapping identifies only chromosomal sites harboring a gene that contributes to variability in response. Thus, if two progenitor strains have the same allele for a critical gene that modulates a phenotype, a QTL analysis of the phenotype using F2 mice derived from these strains will not identify a QTL near the chromosomal location of the critical gene. Another QTL analysis using a segregating population derived from inbred strains that are polymorphic at the critical gene should detect a QTL at the map site of this critical gene.

Identifying a QTL is the first step. The next, and more difficult step, is to identify the chromosomal location of the QTL. One method that shows promise for achieving this goal involves using additional segregating populations, for example, outbred mice such as the HS[78] or more extensively crossed segregating populations,[71] in which linkage between the QTL of interest and markers not tightly coupled to the region of interest have been broken. If these advanced intercross methods provide a more precise location of the QTL, the gene(s) that correspond to the QTL may be cloned from this tightly mapped region of the chromosome.

For a drug that exhibits as complex a pattern of behaviors as does nicotine, QTL analysis may be particularly valuable in determining whether and which of the most likely candidate genes, nicotinic receptor subunits, are implicated in that phenotype. In addition, QTL analyses may suggest additional candidate genes important for manifestation of various phenotypes. Possible candidates that can be envisioned include other receptors activated by neurotransmitters such as dopamine and γ-aminobutyric acid — the release of which is stimulated by nicotine — as well as hormones and hormone receptors affected by nicotine such as prolactin, ACTH, or corticosterone.

6.4.7 REVERSE GENETICS: STUDIES OF THE EFFECTS OF CANDIDATE GENES

Each of the approaches discussed earlier involved a screen of naturally occurring mouse populations for genetic influences that might explain phenotypic variability in the population. An alternative, and increasingly more common, approach is to identify a candidate gene and investigate the effect of changing the expression or structure of that gene. A very powerful and informative method with which to evaluate the effects of a candidate gene is to construct a null mutant or transgenic mouse targeted to the gene of interest. If alteration of a phenotype is achieved by gene deletion or alteration, the gene in question may be important.

The obvious candidates for primary genetic control of nicotine-mediated phenotypes are the nicotinic receptor genes. To date, nine mammalian nicotinic receptor genes expressed in the central nervous system have been cloned: $\alpha 2$,[79] $\alpha 3$,[80] $\alpha 4$,[81] $\alpha 5$,[80] $\alpha 6$,[82] $\alpha 7$,[83] $\beta 2$,[84] $\beta 3$,[85] and $\beta 4$.[86] In addition, receptor genes that seem not to be expressed in the brain have also been identified: $\alpha 9$.[87] Each of the genes encoding the subunits of the receptor at the neuromuscular junction have also been identified[88] as well as their chromosomal localization.[89] The chromosomal location of many of the neuronal genes has also been mapped in mice (gene code: chrn, e.g., chrna4 is the $\alpha 4$ gene); see www.informatics.jax.org for current map locations). Given this diverse gene family and the differential expression of its members, examination of the effects of null mutation or of expression of mutated subunits is likely to provide insights into the regulation of the diverse responses to nicotine.[90]

Several null mutants for nicotinic receptor subunits have been produced and phenotypic analyses of these mutants are in progress: $\beta 2$,[45] $\beta 4$,[91] $\alpha 3$,[92] $\alpha 4$,[9,10] and $\alpha 7$.[47] Phenotypic analysis of several of these mutants suggests potential roles for these subunits. For example, $\alpha 3$ null mutants are born, but all of the homozygous null mutants die at or near weaning.[92] The mice suffer severe deficiencies in the function of the autonomic nervous system, consistent with the crucial role of the $\alpha 3$ subunit in the receptors in the autonomic ganglia. Similarly, mice lacking both the $\beta 2$ and $\beta 4$ subunits die due to severe autonomic malfunction,[91] while animals lacking either of these subunits are viable. Studies with $\beta 2$ null mutants suggest that receptors requiring this subunit account for almost all high-affinity nicotine binding in brain and that $\beta 2$-requiring receptors are important for regulating dopamine release and maintenance of nicotine self-administration.[93] The $\beta 2$ subunit, as well as the $\alpha 4$ subunit, has also been implicated in nicotine-mediated antinociception.[9] Null mutation of the α7 subunit eliminates the fast desensitizing response to nicotinic agonists in hippocampal cells, confirming that this response is mediated by a receptor that requires this subunit.[47] Further studies of the acute and chronic effects of nicotine with these and other nicotinic receptor null mutants may help to identify the role of defined subunits in response to nicotine.

Changes in receptor function can also affect phenotype and investigations of mice expressing altered receptors may provide additional insights into the regulation of responses to nicotine. For example, variants of the α4 subunit mutated in the channel domain have been identified in human populations and linked to certain seizure disorders.[94,95] The two mutations alter the kinetics of the ion channel, leading to a change in desensitization kinetics and Ca^{2+} permeability.[96] Seizure activity is not uniquely mediated by changes in α4 subunit, since analysis of a human population revealed potential QTL near the α3, α5, and β4 gene cluster.[74] Directed mutation of the channel domains of several nicotinic receptor subunits has been demonstrated to change receptor function.[97–99] A mutation of the α7 gene has been introduced into mice by homologous recombination to generate a receptor that appears to desensitize much more slowly than the wildtype α7 gene.[7] Mice homozygous for this mutated gene do not survive, while the heterozygotes display receptors with markedly different function. Phenotypic analysis of these mice may prove interesting and also may illustrate that receptors displaying either naturally occurring or artificially introduced mutations may prove to be very useful in unraveling the genetic basis for differential response to nicotine. As is the case with all null mutants, care must be taken when interpreting the results, since failure to express a receptor subunit throughout development may have an unanticipated effect on a phenotype, owing to compensation for the missing gene.[100]

6.5 SUMMARY

Several different approaches to the investigation of the genetic basis for individual differences in response to nicotine have been outlined. Overall, the studies reinforce the idea that genetically controlled individual differences are of considerable importance in mediating responses to acute or chronic nicotine administration and will likely also be important mediators of nicotine use, abuse, and withdrawal. Uncovering genes responsible for the differences will clearly increase our understanding of the individual differences and may also provide a firmer biological basis for understanding nicotine addiction and for developing better methods to help people stop using tobacco. The genetic basis of nicotine's effects are, no doubt, very complex. Humans display considerable individual differences in response to tobacco and mice show genetically based differences that vary depending on the nicotine response under study. The application of both classical analyses (forward genetics), in which populations are evaluated to uncover the genetic basis for a phenotype, and of candidate gene approach (reverse genetics), in which a gene likely to affect a trait is altered to examine the effects of this mutation on the response, will be useful in identifying those genes important in regulating variability in response to nicotine in the mouse. Perhaps those genes that contribute to variability in response to nicotine in the mouse also play a critical role in regulating individual difference in response to nicotine in humans.

ACKNOWLEDGMENTS

Studies done in the authors' laboratory have been supported by grants from the National Institute on Drug Abuse (DA-03194, DA-10156, DA-12242) and the National Institute on Alcohol Abuse and Alcoholism (AA-11156). ACC is supported by a Research Scientist Award (DA-00197).

REFERENCES

1. Fisher, R.A., Lung cancer and cigarettes, *Nature*, 182:180, 1958a.
2. Fisher, R.A., Cancer and smoking, *Nature,* 182:596, 1958b.
3. Heath, A.C., Madden, P.A.F., Genetic influences on smoking behavior, in Turner, J.R., Cardon, L.R., and Hewitt, J.R., Eds., *Behavior Genetic Approaches in Behavioral Medicine*, New York, Plenum Press, 1995, 45–66.
4. Pomerleau, O.F., Collins, A.C., Shiffman, S., and Pomerleau, C.S., Why some people smoke and others do not: new perspectives, *J. Cons. Clin. Psych.*, 61:723–731, 1993.
5. Heath, A.C., Madden, P.A.F., Slutske, W.S., and Martin, N.G., Personality and the inheritance of smoking behavior: a genetic perspective, *Behav. Genet.*, 25:103–117, 1995.
6. True, W.R., Xian, H., Scherrer, J.F., Madden, P.A.F, Bucholz, K.K., Heath, A.C., Eisen, S.A., Lyons, M.J., Goldberg, J., and Tsuang, M., Common genetic vulnerability for nicotine and alcohol dependence in men, *Arch. Gen. Psychiatry*, 56:655–661, 1999.
7. Orr-Urtreger, A., Broide, R.S., Kasten, M.R., Dang, H., Dani, J.A., Beaudet, A.L., and Patrick, J.W., Mice homozygous for the L250T mutation in the $\alpha 7$ nicotinic acetylcholine receptor show increased neuronal apoptosis and die within 1 day of birth, *J. Neurochem.*, 74:2154–2166, 2000.
8. Labarca, C., Schwartz, J., Deshpande, P., Schwartz, S., Nowak, M.W., Fonck, C., Nashmi, R., Kofuji, P., Dang, H., Shi, W., Fidan, M., Khakh, B., Chen, Z., Bowers, B.J., Boulter, J., Wehner, J.M., and Lester, H.A., Point mutant mice with hypersensitive $\alpha 4$ nicotinic receptors show dopaminergic deficits and increased anxiety, *Proc. Nat. Acad. Sci.,* 98:2786–2791, 2001.
9. Marubio, L., del Mar Arroyo-Jimenez, M., Cordero-Erausquin, M., Lena, C., Le Novère, N., deKerchove d'Exaer, A., Huchet, M., Damaj, M.I., and Changeux, J-P., Reduced antinociception in mice lacking neuronal nicotinic receptor subunits, *Nature*, 398:805–810, 1999.
10. Ross, S.A., Wong, J.Y.F., Clifford, J.J., Kinsella, A., Massalas, J.S., Horne, M.K., Scheffer, I.E., Kola, I., Waddington, J.L., Berkovic, S.F., and Drago, J., Phenotypic characterization of an $\alpha 4$ neuronal nicotinic acetylcholine receptor subunit knock-out mouse, *J. Neurosci.*, 20:6431–6441, 2000.
11. Crawley, J.N. and Paylor, R., A proposed test battery and constellations of specific behavioral paradigms to investigate the behavioral phenotypes of transgenic and knockout mice, *Hormones Behav.*, 31:197–211, 1997.

12. Crawley, J.N., Behavioral phenotyping of transgenic and knockout mice: Experimental design and evaluation of general health, sensory functions, motor abilities, and specific behavioral tests, *Brain Res.*, 835:18–26, 1999.
13. Gold, L.H., Hierarchical strategy for phenotypic analysis in mice, *Psychopharmacology*, 147:2–4, 1999.
14. Picciotto, M.R. and Wickman, K., Using knockout and transgenic mice to study neurophysiology and behavior, *Physiol. Rev.*, 1131–1163, 1998.
15. Marks, M.J., Burch, J.B., and Collins, A.C., Genetics of nicotine response in four inbred strains of mice, *J. Pharmacol. Exp. Ther.*, 226:291–302, 1983.
16. Hatchell, P.C. and Collins, A.C., The influence of genotype and sex on behavioral sensitivity to nicotine in mice, *Psychopharmacology*, 71:45–49, 1980.
17. Tepper, J.M., Wilson, J.R., and Schlesinger, K., Relations between nicotine-induced convulsive behavior and blood and brain levels of nicotine as a function of sex and age in two inbred strains of mice, *Pharmacol. Biochem. Behav.*, 10:349–353, 1979.
18. Miner, L.L., Marks, M.J., and Collins, A.C., Classical genetic analysis of nicotine-induced seizures and nicotinic receptors, *J. Pharmacol. Exp. Ther.*, 231:545–554, 1984.
19. Flores, C.M., Wilson, S.G., and Mogil, J.S., Pharmacogenetic variability in neuronal nicotinic receptor-mediated antinociception, *Pharmacogenitics*, 9:619–625, 1999.
20. Costall, B., Kelly, M.E., Naylor, R.J., and Onaivi, E.S., The actions of nicotine and cocaine in a mouse model of anxiety, *Pharmacol. Biochem. Behav.*, 33:197–203, 1989.
21. Cao, W., Burkholder, T., Wilkins, L., and Collins, A.C., A genetic comparison of behavioral actions of nicotine in the mirrored chamber, *Pharmacol. Biochem. Behav.*, 45:803–808, 1993.
22. Brioni, J.D., O'Neill, A.B., Kim, D.J.B., and Decker, M.W., Nicotinic receptor agonists exhibit anxiolytic-like effects on the elevated plus-maze test, *Eur. J. Pharmacol.*, 238:1–8, 1993.
23. O'Neill, A.B. and Brioni, J.B., Benzodiazepine receptor mediation of the anxiolytic-like effect of (-)-nicotine in mice, *Pharmacol. Biochem. Behav.*, 49:755–757, 1994.
24. Zarrindast, M.R., Sadegh, M., and Shafaghi, B., Effects of nicotine on memory retrieval in mice, *Eur. J. Pharmacol.*, 294:1–6, 1996.
25. Dalvi, A. and Lucki, I., Murine models of depression, *Psychopharmacology*, 147:14–16, 1999.
26. Castellano, C., Effects of nicotine on discrimination learning, consolidation and learned behavior in two inbred strains of mice, *Psychopharmacology*, 48: 37–43 1976.
27. Marks, M.J., Campbell, S.M., Romm, E., and Collins, A.C., Genotype influences the development of tolerance to nicotine in the mouse, *J. Pharmacol. Exp. Ther.*, 259:392–402, 1991.
28. Robinson, S.F., Marks, M.J., and Collins, A.C., Inbred mouse strains vary in oral self-selection of nicotine, *Psychopharmacology*, 124:332–339, 1996.
29. Stolerman, I.P., Naylor, C., Elmer, G.I., and Goldberg, S.R., Discrimination and self-administration of nicotine by inbred strains of mice, *Psychopharmacology*, 141:297–306, 1999.
30. Schechter, M.D., Meehan, S.M., and Schechter, J.B., Genetic selection for nicotine activity in mice correlates with conditioned place preference, *Eur. J. Pharmacol.*, 279:59–64, 1995.
31. Isola, R., Vogelsberg, V., Wemlinger, T.A., Neff, N.H., and Hadjiconstantiou, M., Nicotine abstinence in the mouse, *Brain Res.*, 850:189–196, 1999.

32. Petersen, D.R., Norris, K.J., and Thompson, J.A., A comparative study of the disposition of nicotine and its metabolites in three inbred strains of mice, *Drug Metab. Dispos.*, 12:725–731, 1984.

33. Adir, J., Miller, R.P., and Rotenberg, K.S., Disposition of nicotine in the rat after intravenous administration, *Res. Comm. Chem. Pathol. Pharmacol.*, 13:173–183, 1976.

34. Crabbe, J.C., Wahlsten, D., and Dudek, B.C., Genetics of mouse behavior: Interactions with laboratory environment, *Science*, 284:1670–1672, 1999.

35. Hatchell, P.C. and Collins, A.C., Influences of genotype and sex on behavioral tolerance to nicotine in mice, *Pharmacol. Biochem. Behav.*, 6:25–30, 1977.

36. Meliska, C.J., Bartke, A., McGlacken, G., and Jensen, R.A., Ethanol, nicotine, amphetamine, and aspartame consumption and preferences in C57BL/6 and DBA/2 mice, *Pharmacol. Biochem. Behav.*, 50:619–626, 1995.

37. Marks, M.J., Stitzel, J.A., and Collins, A.C., Genetic influences on nicotine responses, *Pharmacol. Biochem. Behav.*, 33:667–678, 1989.

38. Marks, M.J., Romm, E., Bealer, S., and Collins, A.C., A test battery for measuring nicotine effects in mice, *Pharmacol. Biochem. Behav.*, 23:325–330, 1985.

39. Miner, L.L. and Collins, A.C., Strain comparison of nicotine-induced seizure sensitivity and nicotinic receptors, *Pharmacol. Biochem. Behav.*, 33:469–475, 1989.

40. Marks, M.J., Romm, E., Campbell, S.M., and Collins, A.C., Variation of nicotinic binding sites among inbred strains, *Pharmacol. Biochem. Behav.*, 33:679–689, 1989.

41. Clarke, P.B.S., Schwartz, R.D., Paul, S.M., Pert, C.B., and Pert, A., Nicotinic binding in rat brain: An autoradiographic comparison of [³H]acetylcholine, [³H]nicotine and [¹²⁵I]bungarotoxin, *J. Neurosci.*, 5:1307–1315, 1985.

42. Marks, M.J., Stitzel, J.A., Romm, E., Wehner, J.M., and Collins, A.C., Nicotinic binding sites in rat and mouse brain: comparison of acetylcholine, nicotine, and α-bungarotoxin, *Mol. Pharmacol.*, 30:427–436, 1986.

43. Whiting, P.J. and Lindstrom, J.M., Characterization of bovine and human neuronal nicotinic acetylcholine receptors using monoclonal antibodies, *J. Neurosci.*, 8:3395–3404, 1988.

44. Flores, C.M., Rogers, S.W., Pabreza, L.A., Wolfe, B.B., and Kellar, K.J., A subtype of nicotinic cholinergic receptor in rat brain is composed of α4 and β2 subunits and is up-regulated by chronic nicotine treatment, *Mol. Pharmacol.*, 41:31–37, 1992.

45. Picciotto, M.R., Zoli, M., Lena, C., Bessis, A., Lallemand, Y., Le Novère, N., Vincent, P., Merlo Pich, E., Brulet, P., and Changeux, J-P., Abnormal avoidance learning in mice lacking functional high-affinity nicotine receptor in brain, *Nature*, 374:65–67, 1995.

46. Schoepfer, R., Conroy, W.G., Whiting, P., Gore, M., and Lindstrom, J., Brain alpha-bungarotoxin binding protein cDNAs and MAbs reveal subtypes of this branch of ligand-gated ion channel gene superfamily, *Neuron*, 5:35–48, 1990.

47. Orr-Urtreger, A., Goldner, F.M., Saeki, M., Lorenzo, I., Goldberg, L., DeBiasi, M., Dani, J.A., Patrick, J.W., and Beaudet, A.L., Mice deficient in the α7 neuronal nicotinic acetylcholine receptor lack α-bungarotoxin binding sites and hippocampal fast nicotinic currents, *J. Neurosci.*, 17:9165–9171, 1997.

48. Stitzel, J.A., Farnham, D.A., and Collins, A.C., Linkage of strain-specific nicotinic receptor α7 subunit restriction fragment length polymorphisms with levels of α-bungarotoxin binding in brain, *Mol. Brain Res.*, 43:30–40, 1996.

49. Stevens, K.E., Freedman, R., Collins, A.C., Hall, M., Leonard, S., Marks, M.J., and Rose, G.M., Genetic correlation of inhibitory gating of hippocampal auditory evoked response and alpha-bungarotoxin-binding nicotinic cholinergic receptors in inbred mouse strains, *Neuropsychopharm.*, 15:152–162, 1996.
50. Hayman, B.I., The theory and analysis of diallel crosses, *Genetics*, 39:789–809, 1954.
51. Mather, K. and Jinks, J.L., *Biometrical Genetics*, Chapman & Hall, London, 1982.
52. Crusio, W.E., Kerbusch, J.M.L., and van Abeelen, J.H.F., The replicated diallel cross: a generalized method of analysis, *Behav. Genet.*, 14:81–104, 1984.
53. Marks, M.J., Miner, L., Burch, J.B., Fulker, D.W., and Collins, A.C., A diallel analysis of nicotine-induced hypothermia, *Pharmacol. Biochem. Behav.*, 21:953–959, 1984.
54. Marks, M.J., Miner, L.L., Cole-Harding, S., Burch, J.B., and Collins, A.C., A genetic analysis of nicotine effects on open field activity, *Pharmacol. Biochem. Behav.*, 24:743–749, 1986.
55. Miner, L.L., Marks, M.J., and Collins, A.C., Relationship between nicotine-induced seizures and hippocampal nicotinic receptors, *Life Sci.,* 37:75–83, 1985.
56. Miner, L.L., Marks, M.J., and Collins, A.C., Genetic analysis of nicotine-induced seizures and hippocampal nicotinic receptors in the mouse, *J. Pharmacol. Exp. Ther.,* 239:853–860, 1986.
57. Stitzel, J.A., Blanchette, J.M., and Collins, A.C., Sensitivity to seizure-inducing effects of nicotine is associated with strain-specific variants of the $\alpha 5$ and $\alpha 7$ nicotinic receptor subunit genes, *J. Pharmacol. Exp. Ther.,* 284:1104–1111, 1998.
58. McClearn, G.E. and Kakihana, R., Selective breeding for ethanol sensitivity: short-sleep- and long-sleep mice, in McClearn, G.E., Deitrich, R.A., and Erwin, V.G., Eds., Development of Animal Models as Pharmacogenetic Tools, DHHS Publication No. [ADM]81–113, U.S. Government Printing Office, Washington, 147–159, 1981.
59. deFiebre, C.M., Medhurst, L.J., and Collins, A.C., Nicotine response and nicotinic receptors in long-sleep and short sleep mice, *Alcohol*, 4:493–501, 1987.
60. deFiebre, C.M. and Collins, A.C., Classical genetic analyses of responses to nicotine and ethanol in crosses derived from long-sleep and short-sleep mice, *J. Pharmacol. Exp. Ther.,* 261:173–180, 1992.
61. Bailey, D.W., Recombinant-inbred strains. An aid to finding identity, linkage, and function of histocompatibility and other genes, *Transplantation*, 11:325–327, 1971.
62. DeFries, J.C., Wilson, J.R., Erwin, V.G., and Petersen, D.R., LS × SS recombinant inbred strains of mice: Initial characterization, *Alcohol Clin. Exp. Res.,* 13:196–200, 1989.
63. Stitzel, J.A., Jimenez, M., Marks, M.J., Tritto, T., and Collins, A.C., Potential role of the alpha4 and the alpha6 nicotinic receptor subunits in regulating nicotine-induced seizures, *J. Pharmacol. Exp. Ther.*, 293:67–74, 2000.
64. Falconer, D.S., *Introduction to Quantitative Genetics*, Longman House, London, 1989.
65. Smolen, A., Marks, M.J., DeFries, J.C., and Henderson, N.D., Individual differences in sensitivity to nicotine in mice: response to six generations of selective breeding, *Pharmacol. Biochem. Behav.*, 49:531–540, 1994.
66. Dietrich, W., Katz, H., Lincoln, S.E., Shin, H-S., Friedman, J., Dracopoli, N.C., and Lander, E.S., A genetic map of the mouse suitable for typing intraspecific crosses, *Genetics*, 131:423–447, 1992.
67. Lander, E.S. and Botstein, D., Mapping mendelian factors underlying quantitative traits using RFLP linkage maps, *Genetics*, 121:185–199, 1989.
68. Johnson, T.E., DeFries, J.C., and Markel, P.D., Mapping quantitative trait loci for behavioral traits in the mouse, *Behav. Genet.*, 22:635–653, 1992.

69. Tanksley, S.D., Mapping polygenes, *Ann. Rev. Genet.*, 27: 205–233, 1993.
70. Lincoln, S., Daly, M., and Lander, E., Mapping genes controlling quantitative traits with MAPMAKER/QTL 1.1, Whitehead Institute Technical Report, Cambridge, 1992.
71. Darvasi, A., Experimental strategies for the genetic dissection of complex traits in animal models, *Nature Genet.*, 18:19–24, 1998.
72. Horowitz, G.P., Whitney, G., Smith, J.C., and Stephan, F.K., Morphine ingestion: Genetic control in mice, *Psychopharmacology*, 52:119–122, 1977.
73. Belknap, J.K., Crabbe, J.C., Riggan, J., and O'Toole, L.A., Voluntary consumption of morphine in 15 inbred mouse strains, *Psychopharmacology*, 112:352–358, 1993.
74. Phillips, H.A., Scheffer, I.E., Crossland, K.M., Bhatia, K.P., Fish, D.R., Marsden, C.D., Howell, S.J.L., Stephenson, J.B.P., Tolmie, J., Plazzi, G., Eeg-Olofsson, O., Singh, R., Lopes-Cendes, I., Andermann, E., Andermann, F., Berkovic, F., and Mully, J.C., Autosomal dominant nocturnal frontal-lobe epilepsy: genetic heterogeneity and evidence for a second locus at 15q24, *Am. J. Hum. Genet.*, 63:1108–1116, 1998.
75. Gora-Maslak, G., McClearn, G.E., Crabbe, J.C., Phillips, T.J., Belknap, J.K., Plomin, R., Use of recombinant inbred strains to identify quantitative trait loci in psychopharmacology, *Psychopharmacology*, 104:413–424, 1991.
76. Berrettini, W.H., Ferraro, T.N., Alexander, R.C., Buchberg, A.M., and Vogel, W.H., Quantitative trait loci mapping of three loci controlling morphine preference using inbred mouse strains, *Nature Genet.*, 7:54–58, 1994.
77. Matthes, H.W., Maldonado, R., Simonin, F., Valverde, O., Slowe, S., Kitchen, I., Befort, K., Dietrich, A., LeMeur, M., Dolle, P., Tzavara, E., Hanoune, J., Roques, B.Pm., and Kieffer, B.L., Loss of morphine-induced analgesia, reward effect and withdrawal symptoms in mice lacking the mu-opioid-receptor gene, *Nature*, 383:819–823, 1996.
78. Talbot, C.J., Nicod, A., Cherney, S.S., Fulker, D.W., Collins, A.C., and Flint, J., High resolution mapping of quantitative trait loci in outbred mice, *Nature Genet.*, 21:305–308, 1999.
79. Wada, K., Ballivet, M., Boulter, J., Connolly, J., Wada, E., Deneris, E.S., Swanson, L.W., Heinemann, S., and Patrick, J., Functional expression of a new pharmacological subtype of brain nicotinic acetylcholine receptor, *Science*, 240:330–334, 1988.
80. Boulter, J., O'Shea-Greenfield, A., Duvoisin, R.M., Connolly, J.G., Wada, E., Jensen, A., Gardner, P.D., Ballivet, M., Deneris, E.M., McKinnon, D., Heinemann, S., and Patrick, J., $\alpha 3$, $\alpha 5$ and $\beta 4$: Three members of the rat neuronal nicotinic acetylcholine receptor-related gene family form a gene cluster, *J. Biol. Chem.*, 265:4472–4482, 1990.
81. Goldman, D., Deneris, E., Luyten, W., Kochar, A., Patrick, J., and Heinemann, S., Members of a nicotinic receptor gene family are expressed in different regions of the mammalian central nervous system, *Cell*, 48:965–973, 1987.
82. Gerzanich, V., Kuryatov, A., Anand, R., and Lindstrom, J., "Orphan" alpha6 nicotinic AChR subunit can form a functional heteromeric acetylcholine receptor, *Mol. Pharmacol.*, 51:320–327, 1997.
83. Séguéla, P., Wadiche, J., Dineley-Miller, K., Dana, J.A., and Patrick, J.W., Molecular cloning, functional properties, and distribution of rat brain $\alpha 7$: a nicotinic cation channel highly permeable to calcium, *J. Neurosci.*, 596–604, 1993.
84. Deneris, E.S., Connolly, J., Patrick, J., Swanson, L.W., Patrick, J., and Heinemann, S., Primary structure and expression of $\beta 2$: a novel subunit of neuronal nicotinic receptors, *Neuron*, 1:45–54, 1988.

85. Deneris, E.S., Boulter, J., Patrick, J., Swanson, L.W., and Heinemann, S., β3: a new member of nicotinic acetylcholine receptor gene family is expressed in the brain, *J. Biol. Chem.*, 264: 6268–6272, 1989.

86. Duvoisin, R.M., Deneris, E., Boulter, J., Patrick, J., and Heinemann, S., The functional diversity of the neuronal nicotinic acetylcholine receptors is increased by a novel subunit: β4, *Neuron*, 3:487–496, 1989.

87. Elgoyhen, A.B., Johnson, D.S., Boulter, J., Vetter, D.E., and Heinemann, S., Alpha 9: An acetylcholine receptor with novel pharmacological properties expressed in rat cochlear hair cells, *Cell*, 79:705–715, 1994.

88. Noda, M., Takashasi, H., Tanabe, T., Toyosato, M., Kikyotani, S., Furutani, Y., Hirose, T., Takashima, H., Inayama, S., Miyata, T., and Numa, S., Structural homology of Torpedo californica acetylcholine receptor subunits, *Nature*, 302:528–532, 1983.

89. Heidmann, O., Buonanno, A., Geoffrey, B., Robert, B., Guenet, J-L., Merlie, J.P., and Changeux, J-P., Chromosomal localization of muscle nicotinic acetylcholine receptor genes in the mouse, *Science*, 234:866–868, 1986.

90. Picciotto, M.R., Calderone, B.J., King, S.L., and Zachariou, V., Nicotinic receptors in brain: Links between molecular biology and behavior, *Neuropsychopharmacology*, 22:451–465, 2000.

91. Xu, W., Orr-Urtreger, A., Nigro, F., Gelber, S., Sutcliffe, C.B., Armstrong, D., Patrick, J.W., Role, L.W., Beaudet, A.L., and DeBiasi, M., Multiorgan autonomic dysfunction in mice lacking the beta2 and the beta4 subunits of neuronal nicotinic acetylcholine receptors, *J. Neurosci.*, 19:9298–9305, 1999.

92. Xu, W., Gelber, S., Orr-Urtreger, A., Armstrong, D., Lewis, R.A., Ou, C.N., Patrick, J., DeBiasi, M., and Beaudet, A.L., Megacystis, mydriasis and ion channel defect in mice lacking the alpha3 neuronal nicotinic acetylcholine receptor, *Proc. Nat. Acad. Sci.*, 96:5746–5751, 1999.

93. Picciotto, M.R., Zoli, M., Rimondini, R., Lena, C., Marubio, L.M., Merlo Pich, E., Brulet P, Fuxe K, and Changeux J-P: Acetylcholine receptors containing the beta-2 subunit are involved in the reinforcing properties of nicotine, *Nature*, 391:173–177, 1998.

94. Steinlein, O.K., Magnusson, A., Stoodt, J., Bertrand, S., Weiland, S., Berkovic, S.F., Nakken K.O., Propping, P., and Bertrand, D., An insertion mutation of the CHRNA4 gene in a family with autosomal dominant nocturnal frontal lobe epilepsy, *Hum. Mol. Genet.*, 6:943–948, 1997.

95. Steinlein, O.K., Mulley, J.C., Propping, P., Wallace, R.H., Phillips, H.A., Sutherland, G.R., Scheffer, I.E., and Brekovic, S.F., A missense mutation in the neuronal nicotinic acetylcholine receptor α4 subunit is associated with autosomal dominant nocturnal frontal lobe epilepsy, *Nat. Genet.*, 11:201–203, 1995.

96. Kuryatov, A., Gerzanich, V., Nelson, M., Olale, F., and Lindstrom, J., Mutation causing autosomal dominant nocturnal frontal lobe epilepsy alters Ca^{2+} permeability, conductance, and gating of human α4–2 nicotinic acetylcholine receptors, *J. Neurosci.*, 17:9035–9047, 1997.

97. Revah, R., Bertrand, D., Galzi, J.-L., Devillers-Thiery, A., Mulle, C., Hussy, N., Bertrand, S., Ballivet, M., and Changeux, J.-P,. Mutations in the channel domain alter desensitization of a neuronal nicotinic receptor, *Nature*, 353:846–849, 1991.

98. Bertrand, D., Devillers, Thiery, A., Revah, F., Galzi, J.-L., Hussy, N., Mulle, C., Bertrand, S., Ballivet, M., and Changeux, J.-P., Unconventional pharmacology of a neuronal nicotinic receptor mutated in the channel domain, *Proc. Nat. Acad. Sci.*, 89:1261–1265, 1992.

99. Labarca, C., Nowak, M.W., Zhang, H., Tang, L., Deshpande, P., and Lester, H.A., Channel gating governed symmetrically by conserved leucine residues in the M2 domain of nicotinic receptors, *Nature*, 376:514–516, 1995.

100. Nelson, R.J. and Young, K.A., Behavior in mice with targeted disruption of single genes, *Neurosci. Biobehav. Rev.,* 22:453–462, 1998.

7 Nicotinic Involvement in Cognitive Function of Rats

Edward D. Levin and Amir H. Rezvani

CONTENTS

7.1   INTRODUCTION

Rat models have been very useful in demonstrating the effects of nicotinic agonist and antagonist on memory performance. Experimental rat models have been critical in providing the behavioral characterization of nicotinic effects on memory, as well as important data concerning the anatomic loci, nicotinic receptor subtypes, and neurotransmitter interactions important for nicotinic effects on memory. A variety of studies with rats have shown that nicotine and other nicotinic agonists can improve memory performance, while nicotinic antagonists such as mecamylamine can impair it (see references [11,17,31, and 46] for reviews). This research provides an important bridge between studies of the biochemical studies of nicotinic receptor function and studies of potential clinical application of nicotinic therapies for memory impairment.

7.2   ACUTE NICOTINIC AGONIST EFFECTS

Acute nicotine administration has been shown to improve memory in a variety of tasks. It facilitates retention of avoidance training [10,67] and enhances Morris water maze performance in young and aged rats.[61] Also, acute nicotine treatment has been shown to reverse delay match to sample (DMTS) performance caused by aging, and facilitates performance in aged rats exhibiting deficits in spatial working memory[16,49] or poor passive avoidance performance due to a choline-deficient diet.[58] In a series of studies it has been found that the radial-arm maze is a sensitive and reliable way to assess nicotinic effects on memory function in rats.

The radial-arm maze test is a standard and sensitive measure of working memory performance. Traditionally, it consists of a central arena with eight arms extending outward. Additional arms can be added to provide a more challenging task and more response locations to assess working and reference memory separately. In the most common radial-arm maze procedure, the subjects are reinforced for one entry per arm. For optimal performance, animals must adopt a "win-shift" strategy: once rewarded, responses must be switched and not repeated to receive additional reinforcement. Errors in this task are repeated arm entries; a greater number of arm entries before a repeat indicates better performance. The task directly measures spatial working memory because animals need to remember locations recently visited in order to refrain from repeating entries.

The radial-arm maze has been used in a variety of studies to assess the effects of nicotine on memory function.[17,46,48] The following experiments were conducted to assess the effect of acute nicotine administration on both working and reference memory function on radial-arm maze in adult female rats. After a standard 18-session training in the radial 16-arm maze, rats were injected subcutaneously with either saline or a dose of 0.2 mg/kg nicotine. In 6 separate studies with a total of 71 adult female Sprague-Dawley rats, it was found that a single dose of 0.2 mg/kg nicotine injected subcutaneously 20 minutes before testing significantly improved working memory. The choice accuracy measure (entries to repeat, i.e., the number of correct entries into the arms until an error is made) indicated that working memory was significantly improved following acute nicotine administration in rats.[29]

Other nicotinic agonists have also been found to improve memory function. For example, dimethylethanolamine (DMAE),[45] epibatidine,[47] and lobeline[20] have been shown to improve working memory in rats on radial-arm maze and delayed discrimination tasks. RJR-2403 has been found to reverse scopolamine-induced amnesia of a passive-avoidance response and also to reverse impairments to both working and reference memory in the radial-arm maze in rats with forebrain ibotenic acid lesions.[51] Interestingly, RJR-2403 binds with greater selectivity than nicotine to rat cortical sites over muscle or ganglionic receptors,[9] isonicotine, norisonicotine,[39] AR-R 17779,[32] ABT-418,[18,63] or lobeline,[20,57,64] significantly and improves memory performance.

7.3   CHRONIC NICOTINE AGONIST EFFECTS

From a therapeutic point of view, it is crucial for a drug effective in an acute form to remain efficacious with repeated administration. Interestingly, and in contrast to many of the other effects of nicotine, no tolerance is seen to the memory improving effects of chronic nicotine administration. In fact, the memory improvement caused by nicotine seems to become more robust over time. A series of studies has shown that chronic infusion of either a high dose of 12 mg/kg/day of nicotine[38,40,42] or a

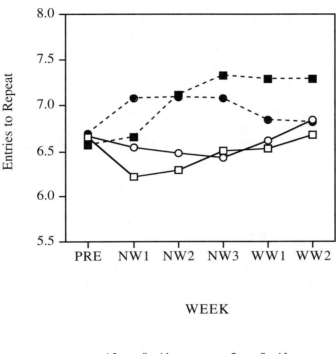

FIGURE 7.1 Chronic nicotine effects on radial-arm maze choice accuracy (entries to repeat mean±sem).

more moderate dose of 5 mg/kg/day of nicotine for 3–4 weeks significantly improves memory performance on the radial-arm maze in both male and female rats.[30,35,37,44,49] To wit: after a standard 18-session training on the radial-arm maze, rats were implanted subcutaneously with an osmotic minipump delivering 5 mg/kg/day nicotine or an equal volume of saline for 28 consecutive days, or a nicotine pellet delivering 12 mg/kg/day for 21 consecutive days. Working memory was assessed during the weeks of nicotine delivery and up to 2 weeks after termination of nicotine delivery. In 7 independent studies it was found that chronic administration of both 5 and 12 mg/kg/day nicotine significantly and consistently improved memory performance on the 8-arm radial maze in adult female rats.[37,40,49] However, only the higher dose of 12 mg/kg/day induced a persistent effect, even 2 weeks after the termination of nicotine delivery (Figure 7.1).

In a follow-up study using a 16-arm radial maze, it was found that chronic administration of 5-mg/kg/day nicotine induced facilitation that was specific for working memory, but not reference memory.[30] This selective effect of nicotine in improving working but not reference memory is consistent with the findings with acute nicotine administration. Chronic treatment with nicotinic agonists also improves memory performance in other memory tasks, such as one-way avoidance and the Lashley III maze.[6] However, chronic nicotine treatment does not appear to be effective in the T-maze alteration task.[36] This may be related to the effect of nicotine proactive interference[22] described later.

Chronic nicotine administration can also reverse working memory deficits due to lesions of fimbria and medial basalocortical projection in rats.[38] However, chronic nicotine administration does not appear to facilitate working memory performance in aged rats, possibly due to the decrease in functional nicotinic receptors in aged animals.[5,6,49]

It has been shown that memory improvements found with chronic nicotine administration can be blocked by concurrent chronic mecamylamine administration.[35] However, it has also been found that chronic nicotine administration can provide some protection against acute systemic mecamylamine challenges.[42] Thus, it seems that nicotinic stimulation of nicotinic receptors blocked by mecamylamine may be essential for induction but not for expression of the memory improvement induced by chronic nicotine administration.

While nicotine can improve cognitive performance in fully functioning animals, nicotinic drugs also have profound effects in normal organisms performing difficult tasks or in impaired organisms. For example, it has been shown that acute nicotine administration facilitates performance in aged rats exhibiting deficits in spatial working memory,[6,49,61,66] or in rats demonstrating poor passive avoidance performance due to a choline-deficient diet.[58] Acute nicotine treatment also attenuates the memory deficit caused by septal lesion in rats.[19] Chronic administration of nicotine can reverse working memory deficits due to lesions of the fimbria and medial basalocortical projection in rats.[38]

7.4 NICOTINIC ANTAGONISTS

Nicotine-induced memory improvements can be blocked by the nicotinic antagonist mecamylamine.[4,35,43] Interestingly, the muscarinic antagonist scopolamine also can reverse memory enhancement induced by nicotine in rats.[43] The effects of mecamylamine also appear to be related to task difficulty and dose, since some studies have not found mecamylamine-induced deficits,[14] and have even found paradoxical improvements in performance.[35,36,54] As discussed later, these effects may be due to the degree of proactive interference involved in the task.

7.5 BEHAVIORAL SPECIFICATION

Working memory is defined as memory with changing contents, as opposed to reference memory, which is defined as memory with fixed contents. Working memory can be differentiated from reference memory in the radial-arm maze by always baiting the same arms and not the others at the beginning of each session. Entries into baited arms are considered working memory and entries into unbaited arms are considered reference memory. Usually a 16-arm radial maze is used for assessing working and reference memory. This technique has been used to show the relative specificity of nicotine-induced improvement in working but not reference memory.[29] Baiting 12 of the 16 arms presents a difficult working memory task, while still leaving 4 arms never baited allows for the assessment of reference memory. This technique has been used to show the relative specificity of nicotine-induced improvement in working but not reference memory.[29,30]

The specificity of the cognitive effects of nicotinic treatments can be discerned by the circumstances under which improvements are and are not seen. Improvement of working memory by nicotine is clearly seen, while reference memory is relatively unaffected by either acute[29] or chronic nicotine administration.[30] Tasks with a large component of proactive interference are also not improved by nicotine[36] or can actually be impaired[22] because there is a proactive effect of nicotine enhancing the partial reinforcement extinction effect.[26] Also, tasks that require animals to pay close attention (sustained-attention or vigilance tasks) are only moderately[12,23] or not at all facilitated by nicotine[65] in animals, although this is typically not the case in humans. However, failures to show enhanced performance following nicotine administration may be due to differences between studies regarding factors such as dose, strain, the nature of the task, and level of training (see Levin[31] for a more complete review).

Taken together, these results suggest that the nature of the task can profoundly influence nicotinic effects in animals. Performance on tasks with any component of proactive interference likely will not be enhanced by nicotinic agonists,[22] but may be facilitated by nicotinic antagonists.[36] Proactive interference is a phenomenon whereby information learned in the past interferes with the learning or memory of more recently presented material. For example, spatial alternation tasks have a high degree of proactive interference because animals must remember previously executed responses and switch to make a different response. Certain "tracking" tasks[23] also have a high component of proactive interference because correct choices have a probability of occurring in a particular position according to the position of the last choice. In these tasks, then, animals get reinforced for remembering where the last response occurred, and then "forgetting" that information in order to make an alternative choice. What is likely happening with nicotinic agonists, is that memory is facilitated to such a degree that animals fail to "forget" their previous responses appropriately. Conversely, nicotinic antagonists such as mecamylamine can facilitate performance in these tasks by impairing memory of the last response choice.

7.6 NICOTINIC RECEPTOR SUBTYPE INVOLVEMENT IN MEMORY FUNCTION

Considerable progress has been made concerning the identity and structure of nicotinic receptor subtypes. The $\beta 2$ nicotinic subunit appears to be widely expressed in the central nervous system, although other prominent subunit combinations including the α4β2, α3β4, and α7 exist as well. The functional roles played by these subtypes are being discovered. Hippocampal α4β2, α3β4, and α7 receptors appear to be important for working memory functions. Studies have shown that most hippocampal neurons have nicotinic receptors that respond to nicotine with a Type IA nicotinic current characterized by rapid desensitization and blockade by MLA,[2,13] a selective α7 nicotinic receptor antagonist.[59] Fewer neurons have receptors that respond with Type II currents and are blocked by DHβE and Type III currents, which are blocked by mecamylamine and are more slowly desensitized.[2] Some neurons show a mixed response termed Type IB which is partially blocked by either MLA or DHβE, but completely blocked by both antagonists.[2] The pharmacological and kinetic properties of these currents support their correspondence to α7 (Type 1A), α4β2 (Type II), and α3β4 (Type III) nicotinic receptor subtypes.[1] However, it is important to note that one hippocampal neuron can express more than one nicotinic receptor subtype.[2] ACh-evoked electrical response in hippocampal neurons in an *in vitro* preparation are blocked by MLA[8] and ACh and anatoxin elicited currents blocked in hippocampal neurons by either MLA or DHβE.[3] As described previously and shown in Figure 7.2, local infusions of either MLA or DHβE into the ventral hippocampus have been found to impair working memory performance on the radial-arm maze signiticantly.[24]

Found in high concentrations in the rat hippocampus,[15,25,60] the α7 type of nicotinic receptor appears to be important in cognitive functioning. α7 receptors are found on hippocampal interneurons and seem to regulate neurite outgrowth and survival. The functional roles of α7 type nicotinic receptors in the brain are still being determined; they may be important for hippocampal processing of information. Central blockade of α7 nicotinic receptors with α-bungarotoxin or (+)-tubocurarine disrupted habituation of evoked responses in the hippocampus to repeated auditory stimuli, while blockade of α4β2 nicotinic receptors or muscarinic receptors was ineffective.[52] Both α4β2 and α7 nicotinic receptors in the hippocampus may be important for aspects of cognitive function.

β2 units are important in cognitive functioning as well. For example, it has been demonstrated that mice homozygous for a β2-subunit mutation (β2-/β2-) do not demonstrate high-affinity binding sites for nicotine; behaviorally, these mice fail to show facilitation of a passive avoidance task following nicotine as normal controls do, suggesting that β2 sites are important for nicotinic memory enhancement.[56] It has been found that nicotinic antagonists with differential blockade of α3β2 (mecamylamine), α4β2 (DHβE), and α7 (MLA) nicotinic receptors impair working memory. Local infusion of the nicotinic antagonists mecamylamine, DHβE, and MLA into the hippocampus impairs working memory in rats.[7,24,28,33,55]

7.7 NICOTINIC-DOPAMINERGIC INTERACTIONS

The brain is an organ of communication. No receptor system acts in isolation from others in providing the functional output of the brain. Nicotinic receptor stimulation induces the release of a variety of neurotransmitters including dopamine (DA). Interactions between nicotinic and DA receptor systems are important for nicotinic

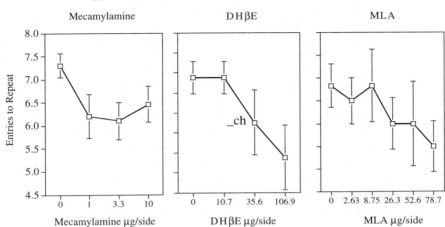

FIGURE 7.2 Mecamylamine, MLA, and DHβE effects when infused into the ventral hippocampus (entries to repeat mean±sem).

involvement with cognitive function. Nicotinic receptors are heavily concentrated on midbrain dopaminergic nuclei. Nicotine administration promotes DA efflux in the striatum and nucleus accumbens,[50] and increases DA concentration in the cortex.[62] Conversely, D_1 antagonist treatment inhibits ACh release, and D_1 agonist treatment enhances ACh release. Curiously, both stimulation and blockade of D_2 receptors have been found to reduce ACh release.[21] Nicotine-induced DA release has been thought to be important for nicotine dependence. Nicotinic-DA interactions have also been found to be important for cognitive function. Mecamylamine-induced memory deficits are potentiated by D_2 antagonist cotreatment and are reversed by D_2 agonist cotreatment.[41,53] Infusion of nicotinic antagonist mecamylamine into the midbrain DA nuclei significantly impairs working memory performance in the radial-arm maze.[34] Other studies have shown that nicotine disrupts latent inhibition learning, an effect that can be blocked by pretreatment with the DA antagonist haloperidol.[27] Studies in the laboratory have also shown that the memory-enhancing effects of both acute and chronic nicotine[44] interact with DA systems. Chronic (5 mg/kg/day) nicotine increased the memory impairment caused by the D_1 agonist dihydrexidine and decreased the improvement caused by the D_1 antagonist SCH 23390.[30] These findings suggest that nicotinic functions are important in some cognitive disorders traditionally believed to involve primarily DA mechanisms, including attention deficit/hyperactivity disorder and schizophrenia.

7.8 CONCLUSIONS

There is convincing evidence that CNS nicotinic systems are clearly important for memory function. The behavioral nature of the effect has helped to define its specificity. Working memory is improved, but reference memory is relatively unaffected. Tasks with a pronounced component of proactive interference are not improved by nicotine or can actually be impaired. While several studies have failed to find significant facilitation of vigilance task performance with nicotine, these failures may be the consequence of the nature of the task, baseline rates of behavior, and/or dose. Nicotinic-induced improvements in a variety of cognitive functions including learning, attention, and memory have been documented in several different species, including rats, monkeys, and humans performing a range of tasks.

Promising areas for nicotine research in the future include the determining the role of different nicotinic receptor subtypes in the neural substrates of cognitive function, the particular parts of the brain in which nicotinic receptors are important for cognition, the non-nicotinic systems that interact with nicotinic systems with regard to cognitive function, and the utility of nicotinic ligands for the treatment of cognitive disorders. One important area for investigation is the relationship of nicotinic involvement in cognitive function to nicotinic involvement in other types of function. Resolving these issues will aid in understanding the role of nicotinic systems in memory function and development of nicotinic treatments for cognitive dysfunction.

REFERENCES

1. Albuquerque, E.X., Alkondon, M., Pereira, E.F.R., Castro, N.G., Schrattenholz, A., Barbosa, C.T.F., Bonfante-Cabarcas, R., Aracava, Y., Eisenberg, H.M., and Maelicke, A., Properties of neuronal nicotinic acetylcholine receptors: pharmacological characterization and modulation of synaptic function, *J. Pharmacol. Exp. Ther.,* 280, 1117–1136, 1997.
2. Alkondon, M. and Albuquerque, E., Diversity of nicotinic acetylcholine receptors in rat hippocampal neurons. I. Pharmacological and functional evidence for distinct structural subtypes, *J. Pharmacol. Exp. Ther.,* 265, 1455–1473, 1993.
3. Alkondon, M. and Albuquerque, E.X., alpha-Cobratoxin blocks the nicotinic acetylcholine receptor in rat hippocampal neurons, *Eur. J. Pharmacol.,* 191, 505–506, 1990.
4. Andrews, J.S., Jansen, J.H.M., Linders, S., and Princen, A., Effects of disrupting the cholinergic system on short-term spatial memory in rats, *Psychopharmacology,* 115, 485–494, 1994.
5. Arendash, G.W., Sanberg, P.R., and Sengstock, G.J., Nicotine enhances the learning and memory of aged rats, *Pharmacol. Biochem. Behav.,* 52, 517–523, 1995.
6. Arendash, G.W., Sengstock, G.J., Sanberg, P.R., and Kem, W.R., Improved learning and memory in aged rats with chronic administration of the nicotinic receptor agonist GTS-21, *Brain Res.,* 674, 252–259, 1995.
7. Bancroft, A. and Levin, E.D., Ventral hippocampal $\alpha 4\beta 2$ nicotinic receptors and chronic nicotine effects on memory, *Neuropharmacology,* 39, 2770–2778, 2000.
8. Barbosa, C.T.F., Alkondon, M., Aracava, Y., Maelicke, A., and Albuquerque, E.X., Ligand-gated ion channels in acutely dissociated rat hippocampal neurons with long dendrites, *Neurosci. Lett.,* 210, 177–180, 1996.
9. Benchierif, M., Lovette, M.E., Fowler, K.W., Arrington, S., Reeves, L., Caldwell, W.S., and Lippiello, P.M., RJR-2403: a nicotinic agonist with CNS selectivity I. *In vitro* characterization, *J. Pharmacol. Exp. Ther.,* 279, 1413–1421, 1996.

10. Brioni, J.D. and Arneric, S.P., Nicotinic receptor agonists facilitate retention of avoidance training — participation of dopaminergic mechanisms, *Behav. Neural Biol.*, 59, 57–62, 1993.

11. Brioni, J.D., Decker, M.W., Sullivan, J.P., and Arneric, S.P., The pharmacology of (-)-nicotine and novel cholinergic channel modulators, *Adv. Pharmacol.*, 37, 153–214, 1997.

12. Bushnell, P.J., Oshiro, W.M., and Padnos, B.K., Detection of visual signals by rats: effects of chlordiazepoxide and cholinergic and adrenergic drugs on sustained attention, *Psychopharmacology*, 134, 230–241, 1997.

13. Castro, N.G. and Albuquerque, E.X., Brief-lifetime, fast-inactivating ion channels account for the alpha-bungarotoxin-sensitive nicotinic response in hippocampal neurons, *Neurosci. Lett.*, 164, 137–140, 1993.

14. Clarke, P.B.S. and Fibiger, H.C., Reinforced alternation performance is impaired by muscarinic but not by nicotinic receptor blockade in rats, *Behav. Brain Res.*, 36, 203–207, 1990.

15. Clarke, P.B.S., Pert, C.B., and Pert, A., Autoradiographic distribution of nicotine receptors in rat brain, *Brain Res.*, 323, 390–395, 1984.

16. Cregan, E., Ordy, J.M., Palmer, E., Blosser, J., Wengenack, T., and Thomas, G., Spatial working memory enhancement by nicotine of aged Long-Evans rats in the T-maze, *Soc. Neurosci. Abs.*, 15, 731, 1989.

17. Decker, M.W., Brioni, J.D., Bannon, A.W., and Arneric, S.P., Diversity of neuronal nicotinic acetylcholine receptors: lessons from behavior and implications for CNS therapeutics — minireview, *Life Sci*, 56, 545–570, 1995.

18. Decker, M.W., Curzon, P., Brioni, J.D., and Arneric, S.P., Effects of ABT-418, a novel cholinergic channel ligand, on place learning in septal-lesioned rats, *Eur. J. Pharmacol.*, 261, 217–222, 1994.

19. Decker, M.W. and Majchrzak, M.J., Nicotine-induced attenuation of spatial memory deficits in rats with septal lesions, *Abs. Soc. Neurosci.*, 17, 1235, 1991.

20. Decker, M.W., Majchrzak, M.J., and Arneric, S.P., Effects of lobeline, a nicotinic receptor agonist, on learning and memory, *Pharmacol. Biochem. Behav.*, 45, 571–576, 1993.

21. DiChiara, B., Morelli, M., and Consolo, S., Modulatory functions of neurotransmitters in the striatum: ACh/dopamine/NMDA interactions, *Trends Neurosci.*, 17, 228–233, 1994.

22. Dunnett, S.B. and Martel, F.L., Proactive interference effects on short-term memory in rats: 1. Basic parameters and drug effects, *Behav. Neurosci.*, 104, 655–665, 1990.

23. Evenden, J.L., Turpin, M., Oliver, L., and Jennings, C., Caffeine and nicotine improve visual tracking by rats — a comparison with amphetamine, cocaine and apomorphine, *Psychopharmacology*, 110, 169–176, 1993.

24. Felix, R. and Levin, E.D., Nicotinic antagonist administration into the ventral hippocampus and spatial working memory in rats, *Neuroscience*, 81, 1009–1017, 1997.

25. Freedman, R., Wetmore, C., Stromberg, I., Leonard, S., and Olson, L., alpha-bungarotoxin binding to hippocampal interneurons — immunocytochemical characterization and effects on growth factor expression, *J. Neurosci.*, 13, 1965–1975, 1993.

26. Grigoryan, G. and Gray, J.A., A single dose of nicotine proactively enhances the partial reinforcement extinction effect in the rat, *Psychobiology*, 24, 136–146, 1996.

27. Joseph, M.H., Peters, S.L., and Gray, J.A., Nicotine blocks latent inhibition in rats: evidence for a critical role of increased functional activity of dopamine in the mesolimbic system at conditioning rather than pre-Exposure, *Psychopharmacology*, 110, 187–192, 1993.

28. Kim, J. and Levin, E., Nicotinic, muscarinic and dopaminergic actions in the ventral hippocampus and the nucleus accumbens: effects on spatial working memory in rats, *Brain Res.*, 725, 231–240, 1996.

29. Levin, E., Kaplan, S., and Boardman, A., Acute nicotine interactions with nicotinic and muscarinic antagonists: working and reference memory effects in the 16-arm radial maze, *Behav. Pharmacol.*, 8, 236–242, 1997.

30. Levin, E., Kim, P., and Meray, R., Chronic nicotine effects on working and reference memory in the 16-arm radial maze: Interactions with D1 agonist and antagonist drugs, *Psychopharmacology*, 127, 25–30, 1996.

31. Levin, E.D., Nicotinic systems and cognitive function, *Psychopharmacology*, 108, 417–431, 1992.

32. Levin, E.D., Bettegowda, C., Blosser, J., and Gordon, J., AR-R17779, an alpha 7 nicotinic agonist, improves learning and memory in rats, *Behav. Pharmacol.*, 10, 675–680, 1999.

33. Levin, E.D., Bradley, A., Addy, N., and Sigurani, N., Hippocampal $\alpha7$ and $\alpha4\beta2$ nicotinic receptors and working memory, *Neuroscience*, in press.

34. Levin, E.D., Briggs, S.J., Christopher, N.C., and Auman, J.T., Working memory performance and cholinergic effects in the ventral tegmental area and substantia nigra, *Brain Res.*, 657, 165–170, 1994.

35. Levin, E.D., Briggs, S.J., Christopher, N.C., and Rose, J.E., Chronic nicotinic stimulation and blockade effects on working memory, *Behav. Pharmacol.*, 4, 179–182, 1993.

36. Levin, E.D., Christopher, N.C., and Briggs, S.J., Chronic nicotinic agonist and antagonist effects on T-maze alternation, *Physiol. Behav.*, 61, 863–866, 1997.

37. Levin, E.D., Christopher, N.C., Briggs, S.J., and Auman, J.T., Chronic nicotine-induced improvement of spatial working memory and D2 dopamine effects in rats, *Drug Dev. Res.*, 39, 29–35, 1996.

38. Levin, E.D., Christopher, N.C., Briggs, S.J., and Rose, J.E., Chronic nicotine reverses working memory deficits caused by lesions of the fimbria or medial basalocortical projection, *Cog. Brain Res.*, 1, 137–143, 1993.

39. Levin, E.D., Damaj, M.I., Glassco, W., May, E.L., and Martin, B.R., Bridged nicotine, isonicotine, and norisonicotine effects on working memory performance of rats in the radial-arm maze, *Drug Develop. Res.*, 46, 107–111, 1999.

40. Levin, E.D., Lee, C., Rose, J.E., Reyes, A., Ellison, G., Jaravik, M., and Gritz, E., Chronic nicotine and withdrawal effects on radial-arm maze performance in rats, *Behav. Neural. Biol.*, 53, 269–276. 1990.

41. Levin, E.D., McGurk, S.R., Rose, J.E., and Butcher, L.L., Reversal of a mecamylamine-induced cognitive deficit with the D2 agonist, LY 171555, *Pharmacol. Biochem. Behav.*, 33, 919–922, 1989.

42. Levin, E.D. and Rose, J.E., Anticholinergic sensitivity following chronic nicotine administration as measured by radial-arm maze performance in rats, *Behav. Pharmacol.*, 1, 511–520, 1990.

43. Levin, E.D. and Rose, J.E., Nicotinic and muscarinic interactions and choice accuracy in the radial-arm maze, *Brain Res. Bull.*, 27, 125–128, 1991.

44. Levin, E.D. and Rose, J.E., Acute and chronic nicotinic interactions with dopamine systems and working memory performance, in A. Lajtha and L. Abood, Eds., *Functional Diversity of Interacting Receptors*, The New York Academy of Sciences, New York, 1995, 218–221.

45. Levin, E.D., Rose, J.E., and Abood, L., Effects of nicotinic dimethylaminoethyl esters on working memory performance of rats in the radial-arm maze, *Pharmacol. Biochem. Behav.*, 51, 369–373, 1995.

46. Levin, E.D. and Simon, B.B., Nicotinic acetylcholine involvement in cognitive function in animals, *Psychopharmacology*, 138, 217–230, 1998.

47. Levin, E.D., Toll, K., Chang, G., Christopher, N.C., and Briggs, S.J., Epibatidine, a potent nicotinic agonist: Effects on learning and memory in the radial-arm maze, *Med. Chem. Res.*, 6, 543–554, 1996.

48. Levin, E.D. and Torry, D., Nicotine effects on memory performance. In P.B.S. Clarke, M. Quik, K. Thurau, and F. Adlkofer, Eds., *International Symposium on Nicotine: The Effects of Nicotine on Biological Systems II*, Birkhäuser, Boston, 1995, 329–336.

49. Levin, E.D. and Torry, D., Acute and chronic nicotine effects on working memory in aged rats, *Psychopharmacology*, 123, 88–97, 1996.

50. Lichtensteiger, W., Hefti, F., Felix, D., Huwyler, T., Melamed, E., and Schlumpf, M., Stimulation of nigrostriatal dopamine neurons by nicotine, *Neuropharmacology*, 21, 963–968, 1982.

51. Lippiello, P.M., Benchierif, M., Gray, J.A., Peters, S., Grigoryan, G., Hodges, H., and Collins, A.C., RJR-2403: A nicotinic agonist with CNS selectivity II. *In vivo characterization, J. Pharmacol. Exp. Ther.*, 279, 1422–1429, 1996.

52. Luntz-Leybman, V., Bickford, P.C., and Freedman, R., Cholinergic gating of response to auditory stimuli in rat hippocampus, *Brain Res.*, 587, 130–136, 1992.

53. McGurk, S.R., Levin, E.D., and Butcher, L.L., Nicotinic-dopaminergic relationships and radial-arm maze performance in rats, *Behav. Neural. Biol.*, 52, 78–86, 1989.

54. Moran, P.M., Differential effects of scopolamine and mecamylamine on working and reference memory in the rat, *Pharmacol. Biochem. Behav.*, 45, 533–538, 1993.

55. Ohno, M., Yamamoto, T., and Watanabe, S., Blockade of hippocampal nicotinic receptors impairs working memory but not reference memory in rats, *Pharmacol. Biochem. Behav.*, 45, 89–93, 1993.

56. Picciotto, M.R., Zoll, M., Lena, C., Bessis, A., Lallemand, Y., Le Novère, N., Vincent, P., Pich, E.M., Brulet, P., and Changeux, J.P., Abnormal avoidance learning in mice lacking functional high-affinity nicotine receptor in the brain, *Nature*, 374, 65–67, 1995.

57. Rochford, J., Sen, A.P., and Quirion, R., Effect of nicotine and nicotinic receptor agonists on latent inhibition in the rat, *J. Pharmacol. Exp. Ther.*, 277, 1267–1275, 1996.

58. Sasaki, H., Yanai, M., Meguro, K., Sekizawa, K., Ikarashi, Y., Maruyama, Y., Yamamoto, M., Matsuzaki, Y., and Takishima, T., Nicotine improves cognitive disturbance in rodents fed with a choline-deficient diet, *Pharmacol. Biochem. Behav.*, 38, 921–925, 1991.

59. Schoepfer, R., Conroy, W.G., Whiting, P., Gore, M., and Lindstrom, J., Brain alpha-bungarotoxin binding protein cDNAs and MAbs reveal subtypes of this branch of the ligand-gated ion channel gene superfamily, *Neuron*, 5, 35–48, 1990.

60. Segal, M., Dudai, Y., and Amsterdam, A., Distribution of an alpha-bungarotoxin-binding cholinergic nicotinic receptor in rat brain, *Brain Res.*, 148, 105–119, 1978.

61. Socci, D.J., Sanberg, P.R., and Arendash, G.W., Nicotine enhances Morris water maze performance of young and aged rats, *Neurobiol. Aging*, 16, 857–860, 1995.

62. Summers, K.L. and Giacobini, E., Effects of local and repeated systemic administration of (-)nicotine on extracellular levels of acetylcholine, norepinephrine, dopamine, and serotonin in rat cortex, *Neurochem. Res.*, 20, 753–759, 1995.

63. Terry, A.V., Buccafusco, J.J., and Decker, M.W., Cholinergic channel activator, ABT-418, enhances delayed- response accuracy in rats, *Drug Develop. Res.*, 40, 304–312, 1997.

64. Terry, A.V., Buccafusco, J.J., Jackson, W.J., Zagrodnik, S., Evansmartin, F.F., and Decker, M.W., Effects of stimulation or blockade of central nicotinic-cholinergic receptors on performance of a novel version of the rat stimulus discrimination task, *Psychopharmacology*, 123, 172–181, 1996.

65. Turchi, J., Holley, L.A., and Sarter, M., Effects of nicotinic acetylcholine receptor ligands on behavioral vigilance in rats, *Psychopharmacology, 118, 195–205, 1995.*

66. Widzowski, D.V., Cregan, E., and Bialobok, P., Effects of nicotinic agonists and antagonists on spatial working memory in normal adult and aged rats, *Drug Dev. Res.*, 31, 24–31, 1994.

67. Zarrindast, M.R., Sadegh, M., and Shafaghi, B., Effects of nicotine on memory retrieval in mice, *Eur. J. Pharmacol.*, 295, 1–6, 1996.

8 Nicotine and Cognition in Young and Aged Nonhuman Primates

Jerry J. Buccafusco and Alvin V. Terry, Jr.

CONTENTS

8.1 NICOTINE AS A COGNITIVE ENHANCING AGENT IN MONKEYS

The earliest studies that sought to examine the ability of nicotine to enhance cognitive performance were performed in rodents, and many of these experiments were predicated on the known alerting or nootropic effect of certain CNS stimulant drugs. We entered this area in 1988[1] and chose to study the potential cognitive-enhancing actions of nicotine in nonhuman primates with the premise that nicotine's actions involved more complex mechanisms than a nonspecific sharpening of attention or arousal. Interest was based largely on two findings. In studying the effects of nicotine in humans 2 years earlier, Wesnes and Warburton[2] concluded that nicotine facilitates state-dependent learning and does not affect associative processes. Secondly, the strong relationship between the ability of the centrally acting nicotinic receptor antagonist mecamylamine to impair performance of the retention trial of an inhibitory avoidance task in rats, and its ability to inhibit the biosynthesis of acetylcholine in cortical and limbic structures was impressive.[3] These findings, along with earlier and concurrent work showing that nicotine enhances the release of brain acetylcholine motivated continuing study of the effect of nicotine on a complex memory task in monkeys. In the first study, five young adult macaques were well trained in the performance of a computer-assisted delayed matching-to-sample (DMTS) task.[1] The subjects were tested individually in a sound-attenuated chamber (subsequent to this initial study, subjects were, as they are now, routinely tested within their home cages). It was during this study that each animal was normalized for his performance in the task so that delay intervals were adjusted on an individual basis to provide for a standard level of performance efficiency (forgetting curve). A near-zero-second delay also was imposed to ascertain task motivation and any nonmnemonic effect of drug administration. This normalization approach is currently recommended for most delayed response tasks.[4] This first study also demonstrated that mecamylamine was indeed amnestic in the primate subjects, and that pretreatment with low doses of mecamylamine (that did not affect task performance) completely blocked the task-enhancing effect of nicotine. A curious finding was that the lowest dose of mecamylamine (0.25 mg/kg) seemed to enhance task performance efficiency slightly. This unexpected effect of mecamylamine is discussed in more detail later.

The first study with nicotine also provided additional aspects of the drug's pharmacology not previously known. For example, the doses of nicotine used to enhance memory in our subjects were very low, ranging from 1 to 20 µg/kg, i.m. This range appeared to be much lower than those used in rodent studies for memory enhancement. Although plasma levels of nicotine were not measured, it is unlikely that they would have exceeded 100 nM. However, this concentration represents the threshold range used to achieve significant neurotransmitter release (e.g., Wilkie et al. and Reuben and Clarke[5,6]). This apparent discrepancy in the potency of nicotine for the two processes does not necessarily rule out a role for evoked transmitter release in nicotine's positive mnemonic actions. For example, it is possible that small, but important levels of transmitter release might occur below the sensitivity of most biochemical measures. In support of this possibility, Wesnes and colleagues[7,8] reported that the muscarinic antagonist scopolamine blocked the effect of nicotine on information processing. They initially interpreted this finding to indicate that the effect of nicotine on encoding was pharmacologically nonspecific. This finding has essentially been replicated in monkeys,[9] and the preferred interpretation of the blocking effect of scopolamine is that it indicates nicotine does indeed enhance the release of acetylcholine, which in turn activates muscarinic receptors as a component of nicotine's beneficial effects on cognition. Part of this enthusiasm for the concept also is derived from similar experiments in rodents, along with Levin and colleagues' interesting findings of synergism in the amnestic actions of combined central nicotinic and muscarinic receptor blockade.[10] Moreover, some years ago, it was noted that the hypertensive response to central injection of nicotine was blocked by pretreatment with central injection of atropine. The cardiovascular response to central injection of nicotine also was blocked by prior depletion of brain acetylcholine with hemicholinium-3, again suggesting that nicotine's central cardiovascular response was mediated through release of endogenous acetylcholine.[11] Thus, there appears to be a common mechanistic feature mediating these two rather disparate (memory and blood pressure) pharmacological actions of nicotine. It is interesting to note that the reverse experiment was not successful; that is, nicotine failed to reverse scopolamine-induced impairment of a visual recognition memory task in monkeys.[12]

Again returning to the initial study with nicotine in monkeys, one other feature of nicotine's actions is exemplified in those experiments. This is the very narrow range of potential therapeutic action of nicotine. Generally, responsiveness is relegated to one or two doses in a series of less than two log units. This is not a feature *specific* to nicotine; indeed, it is the case for most drugs in various pharmacological classes that have been demonstrated to improve cognitive performance in behavioral tasks. The highly individualized response to memory-enhancing drugs led Bartus[13] to suggest that the effectiveness of a drug could mainly be determined by performing a dose-response series, and then selecting the individualized optimum dose or "best dose." Bartus used this approach to help identify nonresponders and it has also been used as a means of comparison of drug effectiveness.[14] However, the question remains as to the mechanism(s) contributing to the inverted-U-dose-response relationship. For nicotine, it has been suggested that this effect is related to nicotine's ability to produce a desensitization type of receptor blockade at higher doses. Alternatively, higher doses of nicotine may be associated with side effects that could interfere with task motivation. The observation that drugs from other pharmacological classes exhibit a similar dose-response relationship profile[15–18] speaks against the depolarization hypothesis, although it cannot be completely ruled out (e.g., blockade of nicotinic receptors does result in an amnestic response). However, in the initial study with mecamylamine,[1] the quaternary nicotinic antagonist hexamethonium was used to control the potential peripheral actions (mainly ganglionic blockade) of mecamylamine on subjects performing the DMTS task. When monkeys were pretreated with hexamethonium, the nicotine-induced improvement in average DMTS efficiency was enhanced across all delays (although the effect was not statistically significant). Thus, it may be possible to widen the therapeutic window of certain agents like nicotine by preventing peripheral side effects with low levels of peripheral nicotinic receptor blockade.

Although a significant body of evidence in rodents largely supported these findings in monkeys, there was very little similar work in nonhuman primates from other labs with which to compare the early results. Rupniak and coworkers[12] examined a wide range of doses of nicotine in young rhesus monkeys trained to perform a visual recognition memory task. In their study the lowest dose of nicotine tested (1 µg/kg) improved the group's task performance by about 16% of baseline levels. The authors admit, however, that the effect would have been more dramatic had they considered each subject's most effective dose. Hironaka and colleagues[19] reported that nicotine improved the performance efficiency of four young adult rhesus monkeys trained to perform a version of the DMTS task. In their case, the reward consisted of a sweetened drink (banana-flavored reinforcement pellets can be used). There were other differences between the two paradigms: in the Hironaka report the subjects were not as well trained (30 to 63 sessions), and the animals performed 4 fixed-retention intervals of up to only 8 seconds in duration. Also, these animals performed the task in special test chambers (although it was not clear whether the subjects were restrained in a chair) and nicotine was administered by the subcutaneous route. Although this group reported that nicotine did enhance task performance in their subjects, there were some differences with respect to those described for the experiments above: for example, the Hironaka group observed significant task improvement with a rather high dose of nicotine (500 µg/kg). In contrast, in the experiments described earlier, the descending limb of the dose-response relationship occurred with doses as low as 20 µg/kg. Part of this discrepancy could be due to the differences in the respective routes of administration. The majority of task improvement observed by the Hironaka group was relegated to trials associated with short delay interval; these authors concluded that nicotine's actions might not be specific to memory function, but instead the drug might improve attention or ease of response. This possibility was supported by a later study in rhesus monkeys in which levels of arousal and orientation to peripheral targets were measured.[20] Low doses of nicotine (3 to 12 µg/kg) reduced mean reaction times to onset of the target. The other studies with nicotine in a primate behavioral model of attention, on the other hand, support the ability of nicotine to enhance attention (see below); however, it is more than likely that effects attributed to low doses of nicotine also have relevance to other aspects of memory such as consolidation and recall. Perhaps the lower level of DMTS training by the subjects in the Hironaka study allowed nicotine to enhance attention selectively to the task, or perhaps to enhance aspects of reference memory over working memory.

Hudzik and Wenger[21] examined the effect of nicotine on squirrel monkeys trained to perform a titrating (incrementing or decrementing retention intervals depending upon the previous trial's outcome) DMTS task. The dose range they used was rather high (10 to 1000 µg/kg), and they observed no statistically significant effect of the drug on titrating DMTS performance. However, on average, the group did exhibit improvement in terms of achieving longer durations of maximal and average retention intervals for sessions following administration of the lowest dose of nicotine, although, again, marked individual variability precluded statistical significance.

More recently, Schneider and his colleagues[22] studied the effects of SIB-1508Y in a model of chronic low-dose MPTP (1-methyl-4-phenyl-1,2,3,6-tetrahydropyridine) in 4 cynomolgus macaques. The model was designed to permit Parkinsonian-like cognitive deficits to emerge, without confounding effects on motor function. SIB-1508Y, a new centrally acting nicotinic receptor agonist, had been characterized as a drug with striatal dopamine and hippocampal acetylcholine release properties. This combination of neuronal targets could prove useful for treating both motor and cognitive symptoms associated with Parkinson's disease. In the study, the authors used a small test battery (which included a version of DMTS task) to better ascertain the drug's potential actions on cognitive function. The chronic MPTP regimen resulted in a shift from mnemonic to nonmnemonic (delay-independent responding) strategies in the delayed response tasks. Treatment with SIB-1508Y (0.5 to 2.5 mg/kg, i.m.) significantly improved task performance and tended to resolve the MPTP-induced change in mnemonic strategy. In this case, performance for trials associated with Short delay intervals was improved to levels above those for longer delay intervals, in effect reconstituting a delay-dependent (mnemonic) approach to the task. A similar ability of nicotine to improve shorter delay interval-associated performance in aged subjects, who tend to have flatter delay-performance curves than younger subjects, (see below) has been noted. Curiously, in the Schneider study, nicotine administration (100 to 500 µg/kg) to two subjects failed to improve delayed response performance significantly. Again, this dose-range is considerably higher than that found to be effective in either young monkeys or in age-impaired subjects (see below).

8.2 THE PROTRACTED MNEMONIC RESPONSE TO NICOTINE

As more experience working with nicotine in primates developed, it was noted that, when animals were tested on the day following nicotine pretreatment (in the absence of further drug or vehicle treatment), significant enhancement of performance efficiency continued to be maintained.[23] Performance levels returned to prenicotine levels on the following day (i.e., within 36 hours after injection). This protracted feature of nicotine's beneficial mnemonic actions was unexpected, particularly in view of the short plasma half-life of the drug in rhesus monkeys.[24] Levin and colleagues[10] reported similar findings in rats. In fact, they demonstrated that this protracted effect of nicotine was not dependent upon the presence of the drug at the time of behavioral training. The improvement in task efficiency measured on the day after nicotine administration generally occurred for trials associated with long delay intervals, as they had for the previous day's session. This pattern (of retention interval receiving the greatest improvement on the day of testing being the same retention interval receiving the greatest improvement upon testing 24 hours later) generally was maintained for other compounds evaluated under similar conditions. Table 8.1 presents the increase in DMTS task performance efficiency by macaques obtained from the best dose for eight nicotinic compounds of differing chemical structure and nicotinic receptor selectivity. Each of these compounds was administered by the intramuscular route, and DMTS testing was initiated within 30 min after drug injection. The animals were also tested without drug or vehicle treatment on the following day, approximately 24 h after drug administration. Each of these compounds was first administered as an increasing dose-response regimen, and each compound produced an inverted-U-type relationship between effect and dose. The exception was for a study with isoarecolone in which the highest dose tested represented the maximal response for the dose-response.[25] Therefore, with the possible exception of isoarecolone, the individualized best dose determined for each animal represents the theoretical maximal potential level of improvement in task performance efficiency for each compound.

On the day of administration, the degree of task improvement ranged from about 9 to 19% of the sessions' 96 trials answered correctly, with SIB-1553A, ABT-418, and nicotine representing the compounds with the greatest level of apparent effectiveness (Table 8.1). ABT-089 and isoarecolone represented an intermediate level of improvement and the two GTS-21 analogs represented the lowest degree of effectiveness. The nicotinic receptor antagonist mecamylamine provided a level of mnemonic effectiveness equivalent to that of nicotine. The mechanism for the positive cognitive actions of low doses of mecamylamine is not completely understood[26] (see below), but since the drug is classified as nicotinic it did enhance task performance,

TABLE 8.1
Rank Order of Eight Nicotinic Drugs in Their Ability to Improve DMTS Performance Efficiency by Macaques

Rank Order Initial Effect	Change in % Trials Correct	Delay Interval (Initial Effect)	Rank Order Protracted Effect	Change in % Trials Correct
SIB-1553A	18.5	Short	GTS-21	15.2
ABT-418	17.3	Long	4OH-GTS-21	11.6
Nicotine	15.1	Long	Nicotine	10.9
Mecamylamine[a]	14.2	Long	Mecamylamine	10.9
ABT-089	12.7	Long	ABT-089	7.1
Isoarecolone	11.0	Average[b]	Isoarecolone	4.4 (n.s.)
4OH-GTS-21	9.3	Average[b]	ABT-418	3.7 (n.s.)
GTS-21	9.1	Long	SIB-1553A	3.1 (n.s.)

The data represent average increases in DMTS performance efficiency derived from estimation of best dose. The best doses for each subject (N = 5–6) were obtained from dose-response data.

[a] Mecamylamine, a nicotinic receptor antagonist, also exhibits positive mnemonic actions at low doses (see text).

[b] Task performance improvement was not specific to a particular delay interval; the average for the effect across all four delay intervals is presented. (n.s.) not significant

and was included for purposes of comparison. Table 8.1 also lists the rank order of the same 8 compounds in their ability to improve DMTS task performance efficiency 24 h after drug administration (also see Figures 8.1 and 8.3). Individual data were derived from those presented in the table for the initial effect, i.e., the data from the sessions run 24 h after drug administration were compared to vehicle levels of task performance in the same manner as were the initial (same-day) sessions. Although nicotine, mecamylamine, ABT-089, and isoarecolone maintained the same relative order in terms of mnemonic effectiveness as they had for their initial effects, SIB-1553A and ABT-418, essentially exchanged places in the ranking order with the two analogs of GTS-21. In fact, SIB-1553A and ABT-418, which were the most effective compounds in the initial sessions, evoked no significant level of task improvement during the 24-h sessions. Perhaps even more surprisingly, for GTS-21 analogs, the degree of task improvement measured during 24-h sessions was, on average, greater than that for the initial sessions.

The mechanism underlying the protracted positive mnemonic actions produced by nicotine has not been determined; however, certain points may be relevant. For example, compounds listed in Table 8.1 exhibit rather short plasma half-lives, and it is unlikely that significant compound remains in the brain or circulation 24 h after administration. These drugs also do not give rise to more active or long-lived metabolites. In fact, the active metabolite of GTS-21, 4OH-GTS-21,[27] produced responses almost indistinguishable from those produced by the parent compound. Another observation that may seem heretical is that the agonist properties of these compounds may not be required for eliciting a positive mnemonic response. The antagonist mecamylamine produced a profile of enhanced task performance efficiency virtually identical to that for nicotine; mecamylamine is only ten-fold less potent than nicotine in this regard.[26] Moreover, the anabaseine analog GTS-21, which appears to offer some selectivity for the α7 subtype,[28] has been characterized as a weak (12% of the nicotine effect) partial agonist in a functional assay using expressed human α7 receptors.[29] In fact, GTS-21 exhibited no functional agonism (ion flux) in cell systems expressing predominantly α4β2 and ganglionic (α3) nicotinic receptors despite very good *in vitro* binding potency (e.g., Ki = 20 nM for α4β2 receptors).[29] However, GTS-21 blocked the ion flux response to nicotine with an IC_{50} of 2.5 μM. Memory and cognition are not the only physiological substrates for mecamylamine's actions, however. Low concentrations (10 to 100 nM) of the antagonist also produce the nicotine-like action of enhancing the expression of cell surface NGF receptors on differentiated PC-12 cells.[30,31] This response was not mimicked by the α4β2-preferring dihydro-β-erythroidine (DHBE) (unpublished data). This antagonist differs in two important ways from mecamylamine: it is not lipid soluble and it is a competitive antagonist for the agonist recognition site. Mecamylamine is a lipid-soluble, noncompetitive, nicotinc channel blocking agent.

Which, if any, of these properties contributes to either the initial or the protracted phases of nicotinic-mediated positive cognitive effects in animals and humans is unknown. In terms of the protracted effect, one additional lead may be derived from Table 8.1. Since the analogs of GTS-21 are α7-preferring agents, and since both ABT-418 and SIB-1553A are compounds that prefer subtypes other than α7, it may be reasonable to implicate the α7 subtype in mediating the expression of the protracted mnemonic action produced by nicotinic drugs. Isoarecolone also evoked no significant protracted improvement in task performance. Although the subtype binding characteristics of isoarecolone have not been ascertained, the compound exhibits several characteristics in behavioral assays that are unlike nicotine's. For example, the drug does not substitute for nicotine in the nicotine discrimination task in rats.[25] As such, it appears that chemical synthetic approaches to dissociate nicotine from its toxicity (cardiovascular, gastrointestinal, addictive, etc.) by creating more subtype selective agents also may have limited the ability of these newer drugs to evoke the protracted positive mnemonic response. However, it is not yet clear whether this ability is a necessary property for a useful memory-enhancing agent.

8.3 THE MNEMONIC EFFECTS OF MECAMYLAMINE IN MONKEYS

Administration of nicotine in experimental animals has been reported to facilitate neurotransmitter release and synaptic transmission in thalamo-cortical, ascending mesolimbic-dopaminergic, noradrenergic, and cholinergic pathways.[32,33] To what extent any of these effects contributes to nicotine's cognitive-enhancing action is not known. However, the concentrations of nicotine required to elicit receptor up-regulation[34-36] or increased neurotransmitter release[32,37] require μM concentrations. Alternatively, nicotine can serve functionally as an antagonist by desensitizing[38,39] or by inactivating[40] nicotinic receptors in a dose-dependent fashion. Based on these observations, it may not be so obvious which effects of nicotine (i.e., the agonist or the functional antagonist effects) truly underlie the positive mnemonic effects of the compound. Furthermore, in certain circumstances, as discussed previously, mecamylamine paradoxically has been shown to enhance performance of some memory-related tasks in animals. It should be noted that mecamylamine has been used extensively to induce memory deficits in lower animals (i.e., rodents, lagomorphs, etc.[41,42]), as well as occasionally in nonhuman primates[1] and humans;[43] this effect is by far the most common observation with the use of mecamylamine. In the majority of these studies, however, doses significantly above 1.0 mg/kg have been used, which may produce confounding pharmacologic effects such as NMDA receptor interactions[44] and changes in acetylcholine turnover.[3]

Recently we tested the hypothesis that low (i.e., μg/kg) doses could improve memory in aged rhesus monkeys.[26] As indicated in Figure 8.1, some doses of mecamylamine appear to have the potential to improve working memory both at 10 min and 24 h after parenteral administration thus appearing to mimic certain memory-enhancing effects produced by nicotinic receptor agonists. The mechanism for this effect is not clear from this initial investigation. One might speculate that mecamylamine is in some manner acting as a partial agonist, or producing a low-level nicotinic antagonism (at these low doses), which produces a cellular response that may be analogous to nicotine-induced receptor desensitization. Several previous observations may have relevance to the memory enhancement induced by mecamylamine and they may relate to the excitatory electrophysiological effects of nicotine and (paradoxically) some nicotinic antagonists. For example, Freund and coworkers[45] reported that two nicotinic antagonists, d-tubocurarine and α-bungarotoxin, produced excitatory effects (i.e., increased population spikes in mouse hippocampal slices) quantitatively similar to those evoked by nicotine. Mecamylamine, d-tubocurarine, and α-bungarotoxin (at relatively high concentrations) released iontophoretically into the hippocampus (CA1 region) of anesthetized rats produced excitation (increased population spikes) rather than inhibition of electrical discharges.[46] Intrahypothalamic injection of d-tubocurarine produced excitatory behavioral responses (i.e., fear and escape reactions) in rats[47,48] in a similar fashion to carbachol. In reference to the underlying physiological basis for these observations, Freund and colleagues[45] suggested that the excitatory effects produced by nicotine might be a consequence of diminished GABAergic transmission, i.e., nicotine desensitized nicotinic receptors controlling GABA release. The excitatory effects on mouse hippocampal slices in their study also were elicited by certain nicotinic receptor antagonists (presumably also through GABA disinhibiton). This action was not shared by mecamylamine and, in fact, the compound inhibited population spikes. However, mM concentrations of mecamylamine were used as opposed to the μM concentrations used for other antagonists. It would have been interesting to know what the effects of low μM or even nM concentrations of mecamylamine (more relevant to this study) would have been in this assay. It certainly appears that further investigation of these paradoxical actions of mecamylamine (and potentially other nicotinic antagonists) is warranted.

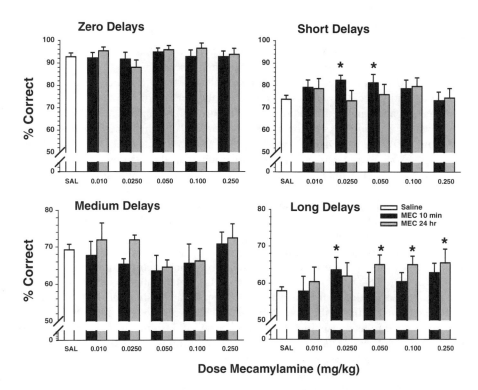

FIGURE 8.1 Dose-effect relationships for each delay in the DMTS task by 8 aged-adult rhesus macaques, 10 min and 24 h following the intramuscular administration of mecamylamine or saline. Data represented are mean percentage correct ± SEM over 96 trials per session. * = significantly different from saline associated baseline (P<0.05, repeated measures ANOVA).

8.4 REVERSAL OF DISTRACTOR-INDUCED PERFORMANCE DEFICITS BY NICOTINIC DRUGS

Attention deficit hyperactivity disorder (ADHD) is among the most prevalent of childhood and adolescent disorders, accounting for up to 50% of clinic visits in these populations. The predominant characteristics of ADHD, inattention and distractibility, are also among the symptoms associated with prenatal and early postnatal exposure to a variety of toxicants, as well as with a wide variety of neurologic and psychiatric disorders including AD, Parkinson's disease, Huntington's disease, Tourettes syndrome, and schizophrenia. The primary therapeutic agents utilized for ADHD and other disorders where distractibility and inattention are prominent features include methylphenidate, dextroamphetamine, mixed amphetamine isomers, and pemoline. While the efficacy of these agents has been demonstrated, they are also associated with a wide variety of adverse effects including insomnia, decreased appetite and weight loss, irritability, elevated heart rate and blood pressure, etc.[49] Furthermore, the most common pharmacologic agent used, methylphenidate, appears to lack efficacy in neurodegenerative conditions such as AD.[50] A number of secondary treatments are available (e.g., antidepressants, clonidine, etc.); however, it is clear that a need for alternative therapeutic agents exists.

Nicotine has been shown in a number of studies to enhance arousal, visual attention, and perception and may ameliorate the deficits in vigilance and memory task performance induced by fatigue.[51,52] The drug also improves reaction time and shortens information processing time.[53] Accordingly, it has been evaluated as a potential therapy for ADHD as well as for a variety of other cognitive disorders. Due to the social stigma associated with nicotine as well as its untoward physiological effects (e.g., elevated blood pressure and heart rate, gastrointestinal side effects, etc.) and undesirable pharmacokinetic profile, a search for structurally related compounds which do not exhibit these limitations has been underway for several years. Both nicotine and an isoxazole isostere of nicotine, ABT-418, have shown efficacy in small clinical trials in human adult ADHD patients.[54,55]

For the purposes of novel drug development in this area, a new version of the DMTS task in monkeys was recently introduced which incorporates a task relevant distractor (DMTS-D task) that appears on a limited number of trials in order to evaluate test compounds for their ability to ameliorate distractibility.[56] The task is conducted in a similar fashion to standard DMTS except that on 18 of the 96 trials a distractor stimulus consisting of a random array of flashing colored lights appears on the test panels and lasts for 3 seconds. Distractor lights are generated by the same 3 colored diodes as those used for sample and choice stimuli. The 3-second duration of the distractor was chosen based on observation that distractors of lesser duration were not effective in disrupting DMTS performance. The remaining trials are presented as standard DMTS trials distributed across all delay intervals. To reduce the extent to which habituation to the distractors may develop during repeated testing, DMTS-D sessions are conducted a maximum of 3 times per 2-week interval, with a minimum of 3 days of standard DMTS testing conducted in between. Furthermore, saline is administered in a random fashion (i.e., on days before and after test compound doses) to minimize any potential artifact of the order of drug administration.

The distractors most notably reduce performance associated with short delay intervals in young monkeys, an effect which is sensitive to methylphenidate associated improvement.[56] The nicotinic agonists, nicotine, ABT-418, ABT-089,[57] and SIB 1553A, an arylalkyl pyrrolidine[58] have also been evaluated, with positive results. A comparison of the effects of these compounds in young-adult macaques performing the DMTS-D appears in Figure 8.2. Each of the compounds improved performance of the DMTS-D at one or more of the doses evaluated and, in fact, at least two of the compounds (i.e., ABT-089 and SIB 1553A) had a superior dose-effect profile when compared to methylphenidate (i.e., more doses were effective). SIB 1553A appeared to be a particularly effective agent as indicated by the fact that several (quite low) doses were effective. None of the nicotinic agents were associated with untoward effects at the doses tested.

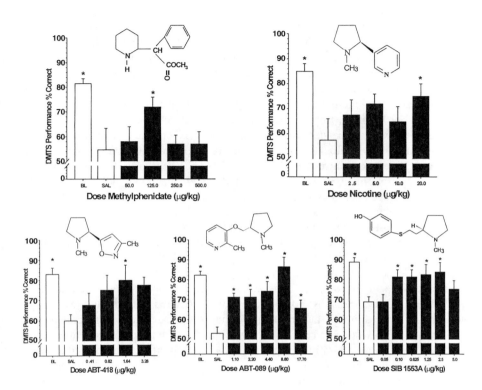

FIGURE 8.2 Dose-effect relationship for several nicotinic receptor agonists compared to methylphenidate. Short-delay distractor trial performances (i.e., mean percentage correct $\pm$ SEM) during interference sessions by young adult monkeys, 10 min following IM administration of test compound are presented. * Indicates a significant ($p<0.05$, repeated measures ANOVA) difference between the test compound and the saline-associated performance. Medium- and long-delay trials were not significantly affected by the distractor or the compounds illustrated. BL = saline baseline for standard DMTS performance; SAL = saline baseline performance associated with 3-sec distractor trials. N = 8 for each test compound, with the exception of SIB-1553A, N = 5.

8.5 EFFECT OF NICOTINE IN YOUNG AND AGED MONKEYS

One particularly notable and reproducible finding encountered over the years is improvement in DMTS task performance by both young adult and aged monkeys administered nicotine, as well as a variety of drugs with nicotinic properties. Enhanced DMTS performance in older monkeys is certainly relevant for the therapeutics of neurodegenerative conditions such as AD, Parkinson's disease, dementia with Lewy bodies and other age-related conditions in which subtypes of nicotinic receptors may be deficient. Aged monkeys (particularly aged rhesus) have been characterized as one of the most useful animal models for the CNS and behavioral abnormalities that occur in aged humans and AD patients. This assertion is based on observations that they begin to encounter learning and memory deficits during the second decade of life, with more substantial deficits apparent by the mid- to late 20s. With age, nonhuman primates have been shown to develop abnormal neurites, amyloid deposition, altered levels of neurotransmitters, and reductions in synaptic densities and pyramidal neurons with age. They also express the apoenzyme E isoform that is analogous to the human apoE4 isoform implicated as a risk factor in AD.[59,60] Furthermore, improvements in DMTS performance by nicotine may be especially relevant since this working memory task engages many of the same neuronal substrates in monkeys as in humans (e.g., prefrontal cortex, hippocampus, etc.). A modified version of the DMTS paradigm has been used to study cognitive ability and impairment in AD patients, and delay dependent deficits in DMTS performance have been documented in both aged humans and those with AD.[61,62]

Evaluation and comparison of nicotine and other nicotinic receptor agonists in both young adult and aged monkeys trained to perform the DMTS task has shown several age-related differences in response profile. For example, the improvement in task efficiency elicited by nicotinic drugs (particularly nicotine and ABT-418) appears to be more delay-specific in younger animals as compared with aged animals. As indicated in Figure 8.3, in young animals both compounds enhanced accuracy of DMTS performance, most notably for trials associated with the longest delays, whereas in older animals, enhanced performance appeared across several delays. The lack of a delay-specific response to nicotinic drugs in aged subjects may reflect the (generally) shorter durations required for each retention interval in aged animals, thus compressing the delay intervals relative to those used in young monkeys. Alternatively, aged monkeys are more prone to become distracted than are younger subjects, and it is more difficult to reverse the distractor-induced performance deficit with drug therapy in these animals relative to their younger cohorts.[56] It is possible that the longer delay intervals offer more opportunity for distraction than do short delay intervals, thereby having a more profound role in the process of forgetting. If this were the case, nicotine would be less effective in reversing distraction-mediated reductions in task accuracy that might occur in aged subjects.

As discussed earlier, an important component of the ability of nicotine to improve DMTS performance efficiency is its protracted effect on task performance. This nicotinic property was evident in both younger and older subjects performing the task. As indicated in Figure 8.3 (and in the previous discussion), ABT-418 did not exhibit this property even after stimulating quite robust improvements on the day of administration.

8.6 SEXUALLY DIMORPHIC MNEMONIC RESPONSES TO NICOTINE IN AGED MONKEYS

Nicotinic receptor agonists have been shown to exert sexually dimorphic actions in laboratory animals. These nicotinic effects are as broadly physiological as sensory gating phenomenon,[63] tuberoinfundibular dopaminergic neuron activity,[64] active avoidance learning,[65] and analgesia.[66] The same may be said for nicotine's action in

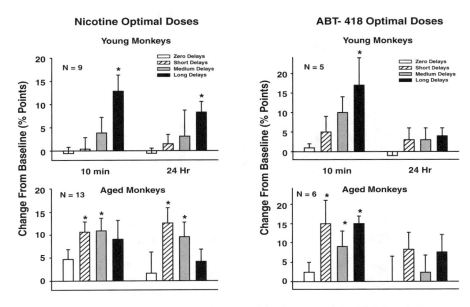

FIGURE 8.3 Comparisons in accuracy for each delay interval of the DMTS task by young adult and aged monkeys 10 min and 24 h following the IM administration of nicotine and ABT-418. Data represented are the mean change in percentage correct ± SEM. (96 trials/session) for performances associated with the individualized optimal doses of each compound. * = significantly different from saline associated baseline (P<0.05, repeated measures ANOVA).

the clinic. For example, male/female differences have been noted regarding nicotine withdrawal in smokers,[67] smoking cessation,[68] nicotine reinforcement,[69] body weight loss,[70] and the hemodynamic response to stress.[71] In these examples, the differential responsiveness attributed to sex often was manifest as an altered sensitivity to nicotine by females — reflected in a differential dose-response relationship. During early studies of the effect of nicotine and its analogs in aged macaques, the first impression was of no sex difference in response to nicotine. However, in a recent head-to-head comparison of the effect of nicotine in aged male and female rhesus monkeys,[72] the most pertinent finding was the marked difference between males and females in the profile of their responses to nicotine. Aged male monkeys showed improvement at lower doses of nicotine, and exhibited much less variance in the response compared with the females; this, despite the fact that response to the averaged (of all four delay intervals) best dose was similar for both sexes (Figure 8.4). It is not known whether these differences in response to nicotine are related to estrogen deprivation in the females, or whether they occur in younger animals. It is also not known whether the response to other potential AD drugs will exhibit these apparent sex-related differences. From an optimistic viewpoint, there is every reason to consider that, with proper dose titration, the nicotinic class of cognitive enhancing agents now under development will prove as useful in females as in males.

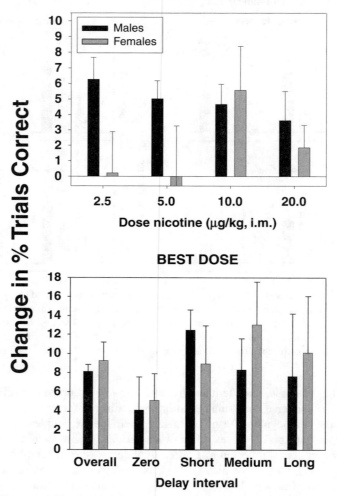

FIGURE 8.4 Upper panel: The nicotine-induced change in overall (all four delay intervals combined) DMTS performance by aged male (filled circles) and female (open circles) rhesus monkeys plotted as a function of dose during sessions initiated 10 min after nicotine injection (not depicted: data derived from sessions run on the day after nicotine injection). For males, there was a significant overall treatment (nicotine) effect, $F_{(2,10)}$ = 11.8, P=0.002), but there was no significant interaction between treatment and dose, $F_{(6,29)}$ = 0.55, P = 0.76. For females, there was no significant overall treatment effect, $F_{(3,18)}$ = 2.46, P = 0.096 and no significant interaction between treatment and drug, $F_{(6,36)}$ = 0.83, P = 0.55. Also, the male subjects exhibited a significantly greater response to nicotine than did the females, $F_{(1,38)}$ = 4.63, P = 0.0378. Lower panel: The nicotine-induced change in DMTS performance by aged male and female rhesus monkeys plotted as a function of delay interval. The data are presented as the average best dose for each group. The best dose was determined for each subject as that dose which provided the greatest overall response (all delays averaged) to nicotine. The data are derived from sessions initiated 10 min after nicotine injection (not depicted: data derived from sessions run on the day after nicotine injection). In males and females, nicotine improved task performance overall, $F_{(2,10)}$ = 12.8, P = 0.002, and $F_{(2,12)}$ = 10.7, P = 0.002, respectively, but there were no significant interactions between treatment effect and delay for either group (P>0.80).

Alternatively, the differential response to the cognitive enhancing properties to nicotine observed in females may suggest accelerated neural deterioration with developing loss of responsiveness to drug therapy.

REFERENCES

1. Elrod, K., Buccafusco, J. J., and Jackson, W. J., Nicotine enhances delayed matching-to-sample performance by primates, *Life Sci.*, 43, 277, 1988.
2. Warburton, D. M., Wesnes, K., Shergold, K., and James, M., Facilitation of learning and state dependency with nicotine, *Psychopharmacology*, 89, 55, 1986.
3. Elrod, K. and Buccafusco, J. J., Correlation of the amnestic effects of nicotinic antagonists with inhibition of regional brain acetylcholine synthesis in rats, *J. Pharmacol. Exp. Ther.*, 258, 403, 1991.
4. Paule, M. G., Bushnell, P. J., Maurissen, J. P. J., Wenger, G. R., Buccafusco, J. J., Chelonis, J. J., and Elliott, R., Symposium overview: the use of delayed matching-to-sample procedures in studies of short-term memory in animals and humans, *Neurotoxicol. Teratol.*, 20, 493, 1998.
5. Wilkie, G. I., Hutson, P., Sullivan, J. P., and Wonnacott, S., Pharmacological characterization of a nicotinic autoreceptor in rat hippocampal synaptosomes, *Neurochemi. Res.*, 21, 1141, 1996.
6. Reuben, M. and Clarke, P. B., Nicotine-evoked [3H]5-hydroxytryptamine release from rat striatal synaptosomes, *Neuropharmacology*, 39, 290, 2000.
7. Wesnes, K. and Warburton, D. M., Effects of scopolamine and nicotine on human rapid information processing performance, *Psychopharmacology*, 82, 147, 1984.
8. Wesnes, K. and Revell, A., The separate and combined effects of scopolamine and nicotine on human information processing, *Psychopharmacology*, 84, 5, 1984.
9. Terry, A. V., Jr., Buccafusco, J. J., and Jackson, W. J., Scopolamine reversal of nicotine enhanced delayed matching-to-sample performance by monkeys, *Pharmacol. Biochem. Beh.*, 45, 925, 1993.
10. Levin, E. D., Nicotinic systems and cognitive function, *Psychopharmacology*, 108, 417, 1992.
11. Buccafusco, J. J., and Yang, X-H., Mechanism of the hypertensive response to central injection of nicotine in conscious rats, *Brain Res. Bull.*, 32, 35, 1993.
12. Rupniak, N. M., Steventon, M. J., Field, M. J., Jennings, C. A., and Iverson, S. D., Comparison of the effects of four cholinomimetic agents on cognition in primates following disruption by scopolamine or by lists of objects, *Psychopharmacology*, 99, 189, 1989.
13. Bartus, R.T., On neurodegenerative diseases, models, and treatment strategies: lessons learned and lessons forgotten a generation following the cholinergic hypothesis, *Exp. Neurol.*, 163, 495, 2000.
14. Buccafusco, J. J. and Terry, A. V., Jr., Multiple CNS targets for eliciting beneficial effects on memory and cognition, *J. Pharmacol. Exp. Ther.*, (in press).
15. Jackson, W. J. and Buccafusco, J. J., Clonidine enhances delayed matching-to-sample performance by young and aged monkeys, *Pharmacol. Biochem., Beh.*, 39, 79, 1991.
16. Terry, A. V., Jr., Jackson, W. J., and Buccafusco, J. J., Effects of concomitant cholinergic and adrenergic stimulation on learning and memory performance by young and aged monkeys, *Cerebral Cortex*, 3, 304, 1993.

17. Terry, Jr., A. V., Buccafusco, J. J., Prendergast, M. A., Jackson, W. J., Fontana, D. L., Wong., E. H. F., Whiting, R. L., and Eglen, R. M., The 5-HT$_3$ receptor antagonist, RS-56812, enhances delayed matching performance in monkeys, *NeuroReport,* 8, 49, 1996.

18. Bartolomeo, A. C., Morris, H., Buccafusco, J. J., Kille, N., Rosenzweig-Lipson, S., Husbands, M. G., Sabb, A. L., Abou-Gharbia, M., Moyer, J. A., and Boast, C. A., The preclinical pharmacological profile of WAY-132983, a potent M1 Preferring Agonist, *J. Pharmacol. Exp. Ther.,* 292, 584, 2000.

19. Hironaka, N., Miyata, H., and Ando, K., Effects of psychoactive drugs on short-term memory in rats and rhesus monkeys, *Japan. J. Pharmacol.,* 59, 113, 1992.

20. Witte, E. A., Davidson, M. C., and Marrocco, R. T., Effects of altering brain cholinergic activity on covert orienting of attention: comparison of monkey and human performance, *Psychopharmacology,* 132, 324, 1997.

21. Hudzik, T. J. and Wenger, G. R., Effects of drugs of abuse and cholinergic agents on delayed matching-to-sample responding in the squirrel monkey, *J. Pharmacol. Exp. Ther.,* 265, 120, 1993.

22. Schneider, J. S., Tinker, J. P., Van Velson, M., Menzaghi, F., and Lloyd, G. K., Nicotinic acetylcholine receptor agonist SIB-1508Y improves cognitive functioning in chronic low-dose MPTP-treated monkeys, *J. Pharmacol. Exp. Ther.,* 290, 731, 1999.

23. Buccafusco, J. J. and Jackson, W. J., Beneficial effects of nicotine administered prior to a delayed matching-to-sample task in young and aged monkeys, *Neurobiol. Aging,* 12, 233, 1991.

24. Buccafusco, J. J., Jackson, W. J., Terry, Jr., A. V., Marsh, K. C., Decker, M. W., and Arneric, S. P., Improvement in performance of a delayed matching-to-sample task by monkeys following ABT-418: a novel cholinergic channel activator for memory enhancement, *Psychopharmacology,* 120, 256, 1995.

25. Buccafusco, J. J., Jackson, W. J., Gattu, M., and Terry, A. V., Jr., Isoarecolone-induced memory enhancement in monkeys: Role of nicotinic receptors, *NeuroReport,* 6,1223, 1995.

26. Terry, A. V., Jr., Buccafusco, J. J., and Prendergast, M. A., Dose-specific improvements in memory-related task performance by rats and aged monkeys administered the nicotinic-cholinergic antagonist mecamylamine, *Drug Develop. Res.,* 47, 127, 1999.

27. Azuma, R., Komuro, M., Korsch, B. H., Andre, J. C., Onnagawa, O., Black, S. R., and Mathews, J. M., Metabolism and disposition of GTS-21, a novel drug for Alzheimer's disease, *Xenobiotica,* 29, 747, 1999.

28. van Haaren, R., Anderson, K. G., Haworth, S. C., and Kern, W. R., GTS-21, a mixed nicotinic receptor agonist/antagonist, does not affect the nicotine cue, *Pharmacol. Biochem. Beh.,* 64, 439, 1999.

29. Briggs, C. A., Anderson, D. J., Brioni, J. D., Buccafusco, J. J., Buckley, M. J., Campbell, J. E., Decker, M. W., Donnelly-Roberts, D., Elliott, R. L., Holladay, M. W., Hui, Y-H., Jackson, W. J., Kim, D. J. B., Marsh, K. C., O-Neill, A., Prendergast, M. A., Ryther, K. B., Sullivan, J. P., and Arneric, S. P., Functional characterization of the novel neuronal nicotinic acetylcholine receptor ligand GTS-21. *In vitro* and *in vivo, Pharmacol. Biochem. Beh.,* 57, 231, 1997.

30. Terry, A. V., Jr., and Clarke, M. S. F., Nicotine stimulation of nerve growth factor receptor expression, *Life Sci.,* 55, PL 91, 1994.

31. Jonnala, R. R. and Buccafusco, J. J., Nicotine increases the expression of TrkA receptors on differentiated PC12 cells, *Soc. Neurosci. Abs.,* 24, 1340, 1998.

32. Grady, S. R., Marks, M. J., Wonnacott, S., and Collins, A. C., Characterization of nicotinic receptor-mediated [3H]dopamine release from synaptosomes prepared from mouse striatum, *J. Neurochem.,* 59, 848, 1992.

33. Granon, S., Poucet, B., Thinus-Blanc, C., Changeux, J. P., and Vidal, C., Nicotinic and muscarinic receptors in the rat prefrontal cortex: differential roles in working memory, response selection and effortful processing, *Psychopharmacology,* 119,139, 1995.

34. Rowell, P. P. and Wonnacott, S., Evidence for functional activity of up-regulated nicotine binding sites in rat striatal synaptosomes, *J. Neurochem.,* 55, 2105, 1990.

35. Collins, A. C., Romm, E., and Wehner, J. M., Dissociation of the apparent relationship between nicotine tolerance and up-regulation of nicotinic receptors, *Brain Res. Bull.,* 25, 373, 1990.

36. Yang, X. H. and Buccafusco, J. J., Comparison of the ability of direct and indirect-acting nicotinic agonists to induce nicotinic receptor up-regulation in rat cerebral cortex, *Drug Devel. Res.,* 31, 95, 1994.

37. Clarke, P. B. S. and Reuben, M., Release of [3H]-noradrenaline from rat hippocampal synaptosomes by nicotine: mediation by different nicotinic receptor subtypes from striatal [3H]-dopamine release, *Br. J. Pharmacol.,* 117, 607, 1996.

38. James, J. R., Villanueva, H. F., Johnson, J. H., Arezo, S., and Rosecrans, J. A., Evidence that nicotine can acutely desensitize central nicotinic acetylcholinergic receptors, *Psychopharmacology,* 114, 456, 1994.

39. Damaj, M. I., Welch, S. P., and Martin, B. R., Characterization and modulation of acute tolerance to nicotine in mice, *J. Pharmacol. Exp. Ther.,* 277, 454, 1996.

40. Rowell, P. P. and Duggan, D. S., Long-lasting inactivation of nicotinic receptor function *in vitro* by treatment with high concentrations of nicotine, *Neuropharmacology,* 37,103, 1998.

41. Levin, E. D. and Rose, J. E., Anticholinergic sensitivity following chronic nicotine administration as measured by radial-arm maze performance in rats, *Behav. Pharmacol.,* 1, 511, 1990.

42. Woodruff-Pak, D. S. and Hinchliffe, R. M., Mecamylamine-or scopolamine-induced learning impairment: ameliorated by nefiracetam, *Psychopharmacology,* 131,130, 1997.

43. Newhouse, P. A., Potter, A., Corwin, J., and Lenox, R., Age-related effects of the nicotinic antagonist mecamylamine on cognition and behavior, *Neuropsychopharmacology,* 10, 93, 1994.

44. O'Dell, T. J. and Christensen, B. N., Mecamylamine is a selective noncompetitive antagonist of n-methyl-d-aspartate and aspartate-induced currents in horizontal cells dissociated from catfish retina, *Neuroscience Lett.,* 94, 93, 1988.

45. Freund, R. K., Jungschaffer, D. A., and Collins, A. C., Nicotine effects in mouse hippocampus are blocked by mecamylamine, but not other nicotinic antagonists, *Brain Res.,* 511, 187, 1990

46. Ropert, N. and Krnjevic, K., Pharmacological characteristics of facilitation of hippocampal population spikes by cholinomimetics, *Neuroscience,* 7, 1963, 1982.

47. Decsi, L. and Karmos-Varszegi, M., Fear and escape reaction evoked by the intrahypothalamic injection of d-tubocurarine in unrestrained cats, *Acta Physiologica. Acad. Sci. Hungary,* 36, 95, 1969.

48. Buccafusco, J. J. and Brezenoff, H. E. ,Opposing influences on behavior mediated by muscarinic and nicotinic receptors in the rat posterior hypothalamic nucleus, *Psychopharmacology,* 67, 249, 1980.

49. Elia, J., Ambrosini, P. J., and Rapoport, J. L., Treatment of attention-deficit–hyperactivity disorder, *New Eng. J. Med.,* 340, 780, 1999.

50. Elia, J., Drug treatment for hyperactive children. Therapeutic guidelines, *Drugs,* 46, 863, 1993.

51. Jones, G. M., Sahakian, B. J., Levy, R., Warburton, D. M., and Gray, J., A. Effects of acute subcutaneous nicotine on attention, information processing and short-term memory in Alzheimer's disease, *Psychopharmacology,* 108, 485, 1992.

52. Newhouse, P. A., Potter, A., Corwin, J., and Lenox, R., Acute nicotinic blockade produces cognitive impairment in normal humans, *Psychopharmacology,* 108, 480, 1992.

53. Le Houezec, J., Halliday. R., Benowitz, N. L., Callaway, E., and Naylor, H., A low dose of subcutaneous nicotine improves information processing in nonsmokers, *Psychopharmacology,* 114, 628, 1994.

54. Levin, E. D. and Simon, B. B., Nicotinic acetylcholine involvement in cognitive function in animals, *Psychopharmacology,* 138, 217, 1998.

55. Wilens, T. E., Biederman, J., Spencer, T. J., Bostic, J., Prince, J., Monuteaux, M. C., Soriano, J., Fine, C., Abrams, A., Rater, M. and Polisner, D. A., pilot controlled clinical trial of ABT-418, a cholinergic agonist, in the treatment of adults with attention deficit hyperactivity disorder, *Am. J. Psych.,* 156, 1931, 1999.

56. Prendergast, M. A., Jackson, W. J., Terry, A. V. Jr., Kille, N. J., Arneric, S. P., Decker, M. W., and Buccafusco, J. J., Age-related differences in distractibility and response to methylphenidate in monkeys, *Cerebral Cortex,* 8, 164, 1998.

57. Prendergast, M. .A, Jackson, W. J., Terry, A. V. Jr., Decker, M. W., Arneric, S. P., and Buccafusco, J. J., Central nicotinic receptor agonists ABT-418, ABT-089, and (-)-nicotine reduce distractibility in adult monkeys, *Psychopharmacology,* 136, 50, 1998.

58. Terry, A. V. Jr, Buccafusco, J. J., Menzaghi, F., and Lloyd, G. K., Nicotinic-cholinergic receptor (nAChR) agonist SIB 1553A reduces distractibility in adult macaques, *Soc. Neurosci. Abs.,* 25, 1981, 1999.

59. Price, D. L. and Sisodia, S. S., Cellular and molecular biology of Alzheimer's disease and animal models, *Ann. Rev. Med.,* 45, 435, 1994.

60. Voytko, M. L., Nonhuman primates as models for aging and Alzheimer's disease, *Lab. Animal Sci.,* 48, 611, 1998.

61. Oscar-Berman, M. and Bonner, R. T., Matching-and delayed matching-to-sample performance as measures of visual processing, selective attention, and memory in aging and alcoholic individuals, *Neuropsychologia,* 23, 639, 1985.

62. Perryman, K. M. and Fitten, L. J., Delayed matching-to-sample performance during a double blind trial of tacrine (THA) and lecithin in patients with Alzheimer's disease, *Life Sci.,* 53, 479, 1993.

63. Faraday, M. M., Rahman, M. A., Scheufele, P. M., and Grunberg, N. E., Nicotine administration impairs sensory gating in Long-Evans rats, *Pharmacol. Biochem. Behav.,* 61, 281, 1998.

64. Shieh, K.R. and Pan, J. T. ,Sexual differences in the diurnal changes of tuberinfundibular dopaminergic neuron activity in the rat — role of cholinergic control, *Biol. Reproduc.,* 54, 987, 1996.

65. Yilmaz, O., Kanit, L., Okur, B. E., and Pogun, S., Effects of nicotine on active-avoidance learning in rats - sex-differences, *Behav. Pharmacol.,* 8, 253, 1997.

66. Chiari, A. and Eisenach, J. C., Effects of RJR 2403, a selective nicotinic agonist, in male and female rats, *FASEB J.,* 12, A158, 1998.

67. Rojas, N. L., Killen, J. D., Haydel, K. F., and Robinson, T. N., Nicotine dependence among adolescent smokers, *Arch. Pediat. Adoles. Med.,* 152, 151, 1998.

68. Gritz, E. R., Nielsen, I. R., and Brooks, L. A., Smoking cessation and gender: the influence of physiological and psychological, and behavioral factors, *J. Am. Med. Womens Assoc.,* 51, 35, 1996.
69. Perkins, K. A., Sanders, M., D-Amico, D., and Wilson, A., Nicotine discrimination and self-administration in humans as a function of smoking status, *Psychopharmacology,* 131, 361, 1997.
70. Perkins, K. A., Sexton, J. E., and DiMarco, A., Acute thermogenic effects of nicotine and alcohol in healthy male and female smokers, *Physiol. Behav.,* 60, 305, 1996.
71. Girdler, S. S., Jamner, L. D., Jarvik, M., Soles, J. R., and Shapiro, D., Smoking status and nicotine administration differentially modify hemodynamic stress reactivity in men and women, *Psychosom. Med.,* 59, 294, 1997.
72. Buccafusco, J. J., Jackson, W. J., Jonnala, R. R., and Terry, A. V., Jr., Differential improvement in memory-related task performance to nicotine by aged male and female rhesus monkeys, *Behav. Pharmacol.,* 10, 681, 1999.

9 Neuroscience Research with Nicotine Self-Administration

William A. Corrigall

CONTENTS

9.1 INTRODUCTION

Other chapters in this book have described methods to study important features of neuronal nicotine receptors (nAChRs), including their structure, response to ligand binding, and ability to influence other neurochemical systems. These receptors obviously play a number of roles in normal brain function. In addition, they are co-opted by the nicotine delivered by tobacco products which acts through these receptors to produce addiction or dependence.[1] In animals, one of the principal ways to study dependence phenomena is with the self-administration paradigm.[2–4] Early studies with primates and other species demonstrated that nicotine delivered intravenously could maintain self-administration behavior as other drugs do.[5,6] We capitalized on early attempts by others to demonstrate intravenous (IV) nicotine self-administration in rodents,[7-11] and developed a model with rats in which self-administration is entrained by a limited-access schedule with rapid delivery of the drug.[12] The basic procedures used in our model have since been used by several research groups to obtain nicotine self-administration.[13-15] Thus, despite some difficulties experienced in early studies of nicotine self-administration in rats reviewed previously,[16] it is now well accepted that IV nicotine will maintain self-administration behavior in this species.

The availability of any animal model in which nicotine is self-administered provides the means to study the mechanisms of dependence at the cellular and receptor level. This is especially true for rodent models, which lend well to manipulations of the central nervous system. The objective of this chapter is to provide a summary of one set of methods used to elaborate the neuroscience of nicotine reinforcement, that is, the ones used in our research and, to that end, the explicitly "how to" parts of the chapter are based on the specific methods we currently use. However, our methods are by no means the only way to obtain nicotine self-administration, or to study CNS mechanisms. To give more context to the description of our self-administration methods, a subsequent section articulates issues related to our model and outlines different approaches used by other researchers since the time we published our method. One caveat: although nicotine has been shown to be voluntarily consumed orally by rodents,[17,18] the focus of this chapter is on the intravenous model which mimics the pharmacokinetics of nicotine delivery by the inhalation of tobacco smoke. Following a description of these methods for self-administration, the chapter concludes with an account of the application of particular neuroscience techniques to study CNS mechanisms of reinforcement of nicotine self-administration. As with the account of self-administration methods, this description focuses on our own work on mechanisms that modify the mesolimbic dopamine system. This description is complemented by recent reviews which discuss self-administration more in the context of the mesolimbic dopamine system.[19,44]

9.2 METHODS FOR NICOTINE SELF-ADMINISTRATION

9.2.1 Surgical Techniques

One of the main issues in carrying out drug self-administration studies is the surgical implantation of catheters to permit drug delivery into the circulatory system. In our laboratory, catheters are prepared from a combination of silicone rubber and polyethylene tubing; full details about design and construction have been reported.[16] Figure 9.1 shows the structure of the catheter schematically. The silicone tubing is the end that is implanted into the circulatory system, in this case, the jugular vein. Polyethylene tubing, stronger than silicone, forms the middle and the external end of the catheter. The actual external terminus of the catheter is made by enclosing the end of the polyethylene tubing in a length of polyolefin heat-shrink tubing, and fastening this with cyanoacrylate adhesive into a nylon bolt which has been drilled axially. A piece of polypropylene surgical mesh is cemented to the head of the bolt

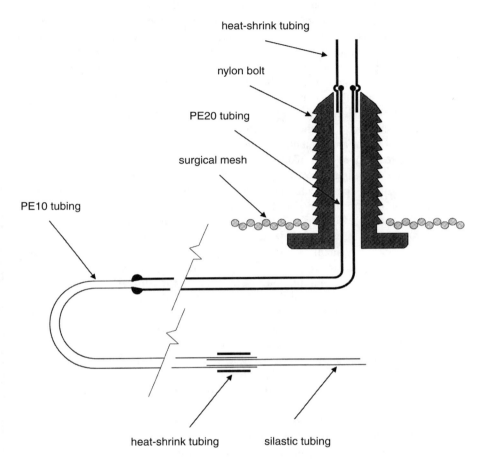

FIGURE 9.1 Schematic of the intravenous catheter. The connection between the two sizes of polyethylene tubing (PE10 and PE20; Intramedic tubing; Clay Adams) is made by melting the ends in a jet of hot air; during this process a mandrel maintains the lumen. With a similar approach, a small bubble of polyethylene is raised at the external end of the PE20 tubing to assist in anchoring the heat-shrink tubing in place. The heat-shrink tubing at the external end of the PE20 tubing is cemented within the nylon bolt with cyanoacrylate adhesive. Silastic tubing (Dow Corning) is stretched over the PE10 tubing and secured with heat-shrink tubing. The PE10 tubing is prepared with a full return bend (i.e., 180°) as shown to facilitate implantation and suturing. The schematic is not to scale, but drawn to show salient features of the catheter (tubing lengths are as follows: PE20, 170 mm; PE10, 65 mm; Silastic, 37 mm; the nylon bolt is 6-32 × 3/4 inch).

with dental acrylic. During surgery, this mesh is implanted subcutaneously between the scapulae, where tissue growth engages it and anchors the external end of the catheter.

There are several advantages to this catheter system. First, construction of the external end allows easy connection to and disconnection from the drug delivery system. This is accomplished by pressure fit: the drug delivery line in the experimental chamber terminates in a piece of stainless steel needle tubing, which can be pushed into the polyolefin heat-shrink end of the catheter. Second, this heat shrink end is easily rejuvenated several times if it fatigues; the polyolefin end is simply heated locally while the patency of the bore is maintained with the appropriately sized mandrel. Third, and most important for neuroscience research, this method of externalizing and anchoring the catheter leaves access to the skull unimpeded, unlike catheter systems that are cemented to the skull.

Surgical techniques to implant intravenous catheters (and the CNS techniques described later) are highly invasive, and require that particular attention be given to sepsis. In addition, these techniques obviously need to be done by well-trained individuals, with care and respect for the experimental animal. Preparation of the animal for surgery is done in an area separate from the surgical one. Animals are anesthetized with a ketamine/xylazine cocktail, administered intraperitoneally (IP; ketamine 75 mk/kg; xylazine 10 mg/kg). Animals are shaved dorsally between and around the scapulae, in an area sufficient to allow the exit-anchor part of the catheter to be installed without contamination by contact with fur. Similarly animals are also shaved ventrally over the area of the jugular vein in an area large enough to accommodate the incision and the implantation of the catheter. Incision sites are infused with bupivacaine and the shaved areas are disinfected with a surgical scrub, rinsed with 70% ethyl alcohol, and painted with iodine-based solution (Betadine; Purdue Frederick, Inc.). Ophthalmic lubricant is applied to prevent ocular drying during surgery. At this point, the level of anesthesia is monitored and, if still acceptable, the animal is moved to the surgical area. Supplemental boosts of anesthetic are provided, as needed, during surgery.

Surgical implantation of intravenous catheters is done in a clean operating area, with an approach designed to mimic aseptic technique, even though the work is not done in an aseptic environment per se. This approach includes disinfection of the surgical area, and the sterilization of all instruments, implants, and other material that will contact the open incision. Surgical packs are prepared in advance for each animal, and sterilized by autoclave. Each pack contains all the drapes, sutures, gauze, and swabs to be used in the surgery of one animal. Catheters will not withstand high-temperature sterilization, and are instead sterilized by immersion in and flushing with zephiran chloride. Prior to implantation in the vein, catheters are filled with sterile saline.

The actual implantation of the catheter system has been well described;[16] the essentials of this technique are summarized here. Again, respecting aseptic technique, the surgeon wears sterile operating gloves. The right jugular vein of the rat is exposed with ventral dissection, and presterilized 4-0 silk sutures are placed under it. A second incision is made dorsally, at midline between the scapulae. At this latter location, blunt dissection is used to open a small subcutaneous area that will be used to contain the surgical mesh on the nylon bolt of the catheter. Blunt dissection is also used to open a subcutaneous path from the ventral to dorsal incision, caudal to the animal's right front leg. A length of large-bore polyethylene tubing is drawn though this subcutaneous path, and the catheter passed through this tubing. The large-bore tubing is then carefully withdrawn, leaving the main length of the catheter in place subcutaneously. (From this position, the catheter can be adjusted to permit a greater or lesser amount at the ventral or dorsal location; placement with the large-bore tubing is to obviate the need to pull on the small, fragile silastic end of the catheter to implant the catheter subcutaneously.) The silastic end of the catheter, prefilled with sterile saline, is then implanted in the jugular vein as previously described.[16] At the exit point on the dorsal surface, the surgical mesh and the head of the nylon bolt are set in the previously prepared subcutaneous pocket so that the bolt projects vertically from the plane of the animal's back at midline, and the overlying skin and tissue are sutured. Once closed, all incisions are dressed with antiseptic wound spray. Finally, a nylon nut is threaded onto the bolt, to reside above the skin, and secured with a droplet of cyanoacrylate adhesive. This provides a stop against which a threaded spring, which protects the drug delivery line in the experimental environment, can be tightened.

Following surgery, animals receive buprenorphine for post-operative anesthesia (0.01 mg/kg SC), and penicillin (30,000U IM), as well as fluid replacement (isotonic saline, IP). Animals recover from anesthesia on a thermostat-controlled heating pad, and are turned regularly to prevent hypostasis. Because of experience with this surgical procedure, recovery is typically uneventful. For the occasional animal in which respiration rate falls, doxapram hydrochloride (5.6 to 11 mg/kg IV or IP) can be administered and oxygen ventilation provided for a short time.

Animals recover from the overall surgical intervention for a period of 1 week before self-administration procedures begin. During this week, catheters are flushed several times with sterile saline containing heparin (0.1ml volume, 30 units/ml heparin).

9.2.2 EXPERIMENTAL ENVIRONMENT

Prior to surgical implantation of the intravenous catheters described, animals are acclimatized to the housing facilities for a minimum period of one week. Next they are deprived of food overnight, and trained to respond to receive 45 mg food pellets. Training continues until each animal responds to obtain 100 pellets. This training takes from 1 day to several days.

The experimental environment used for self-administration is typical of that used in other laboratories. The experimental chamber presents two levers to each animal; one of these delivers drug when the response criterion is reached, and the other provides a control for the specificity of lever-pressing behavior. (The specificity of drug-taking behavior is also confirmed by extinction studies in which the nicotine solution is replaced by saline vehicle or by systemic treatments with nicotine antagonists during self-administration, both of which lead to reductions in responding on the active lever.[12-15]) Signal lights over each lever, a tone, and the house light in the chamber itself can all be programmed to indicate the availability of drug, or to serve as secondary reinforcers. The drug solution reaches the animal by way of a fluid swivel to prevent tangling the delivery line, which is covered by a metal spring that threads onto the nylon bolt at the end of the catheter, as described earlier, and protects the line from damage that might be caused by the animal.

Control equipment is needed to record the animals' responses on the levers, and to deliver drug infusions when criteria are reached. Like the experimental chambers themselves, control and data acquisition hardware and software can be "home made" or commercially obtained. It is important that this equipment maintain a temporal record of responding and drug delivery so that the data can be analyzed for response patterns during the session as well as for total events, as required.

The pH of nicotine solutions is adjusted to 7 with NaOH, and solutions are sterilized prior to use by filtration through 0.22 micrometer syringe filters. Control equipment must be able to deliver nicotine rapidly to the experimental animal when response criteria have been met. To achieve this rapid delivery the pneumatically-driven microsyringe developed by Weeks [20] is used in some experimental chambers, and the rapid syringe pump system sold by Med Associates (St Albans, VT) in others.

9.2.3 TRAINING AND SCHEDULES OF REINFORCEMENT

Self-administration training begins with a simple schedule of reinforcement in which a single press on the appropriate lever in the experimental chamber results in the delivery of nicotine (fixed-ratio 1 or FR1). Over several weeks, the response require-ment is increased in steps to its final value. After each infusion of nicotine that is obtained, there is a time-out (TO) period during which presses do not accrue toward the next infusion. Other parameters include a final response requirement of fixed-ratio 5 (FR5), which the animals reach over a 3-week period, and a session duration of 60 minutes.

During the period of self-administration, catheters are flushed daily with hep-arinized saline (0.1 ml) both before and after the session. The presession flush allows assessment of resistance to flow, increases in which are an indication of catheter blockade. The postsession flush allows replacement of nicotine solution in the dead-space of the catheter with sterile saline.

9.3 ISSUES AND MODIFICATIONS USED BY OTHER RESEARCH GROUPS

9.3.1 PRIOR TRAINING

In all the drug self-administration studies, including those with drugs other than nicotine, animals are regularly trained to respond for food prior to surgery to implant IV catheters. This was originally done so that training time after catheter implantation would be shortened, and for consistency the procedure has continued. Recently, several studies have shown that this is not necessary to obtain nicotine self-admin-istration,[21–24] although weight-restricted/food-deprived animals have been found to have higher rates of self-administration.[24] This is consistent with well-established observations that food deprivation increases the rate of responding for other drugs.[25] For nicotine self-administration, it has also been shown that animals trained to respond for food subsequently maintain active-lever responding only when delivery of nicotine is contingent on this responding.[24] Hence the prior training to respond for food and deprivation may affect rates, but does not appear to account for the self-administration behavior *per se.*

9.3.2 Schedule Requirements

The TO ensures that drug deliveries are spaced apart, originally believed to be important, particularly during acquisition of self-administration behavior. However, others have used smaller TOs successfully in limited access paradigms,[14,15,21,22] and reducing the TO after acquisition has been found to make little difference to the extent of self-administration.[12] Thus there may be no need to limit access this way. Like the TO, the duration of the session may not be critical over comparably short periods.[15,21,22]

9.3.3 Doses of Nicotine and Dose-Effect Curve

In all of the studies of nicotine self-administration to date using the rat as the experimental animal, there is general agreement that the dose-effect curve is relatively flat over the range of doses that support behavior.[11,12,14,15] However, one study in which nicotine was available in an unlimited access fashion reported that the number of self-administered infusions decreased consistently as the dose increased.[23] In this study, very low doses, approximately three- to tenfold lower than those found to maintain self-administration in other studies, were found to be self-administered. However, as noted elsewhere,[26] it is not known what plasma or tissue dose is achieved by the temporally distributed intake of these low doses. A dose of 30 μg/kg for acquisition is routinely used; the unit dose is subsequently changed as required for the experimental design, over a range of 10 to 60 μg/kg. This is the dose range generally found to be self-administered in other studies, with some differences in the reported efficacy of doses at the low-dose end of the range (e.g., compare Donny et al.[24] and Corrigall et al.[27]). The presence of these small differences at threshold doses across different studies is not surprising. Otherwise, the doses that support IV nicotine self-administration, and the shape of the dose-response curve, are remarkably similar across rat studies, and similar to the doses and dose-response curve for nicotine self-administration by primates.[5,28]

9.3.4 Strain Differences

Nicotine self-administration has been obtained successfully with a range of rat strains, including Long-Evans,[11,27,29–32] Sprague-Dawley,[12,13,21,33] Lister,[22] Wistar,[7,14] and Holtzmann.[23] One report has compared the acquisition of nicotine self-administration in both Long-Evans and Sprague-Dawley rats, with and without chronic nicotine treatment, in a paradigm in which the operant was a nose-poke.[21] The data suggest that there may be differences between these two strains, but both appear to self-administer the drug well. In contrast, the Fisher 344 and Lewis inbred strains do not appear to self-administer nicotine.[21,34]

9.4 CNS STUDIES WITH NICOTINE SELF-ADMINISTRATION

From a neuroscience perspective, establishing a self-administration system can be rewarded by the ability to investigate drug dependence at the cellular level. Obviously a number of techniques could be used to further this strategy — examples include using knock-out animals in a mouse model of nicotine self-administration,[35] and using fos-reactivity to identify active neuronal populations in a rat model.[36,37] The approach that we have taken with nicotine self-administration has been to identify the sites of action for the drug in the brain, and to follow with an examination of the neurochemical mechanisms involved. This research has relied extensively on the placement of microcannulae into the brain to allow the delivery of small amounts of neurochemicals, and on the generation of selective neurotoxin lesions. It is these techniques and their application to nicotine self-administration that are reviewed here. However, both of these methods, which rely on stereotaxic surgery, are commonly used in neuroscience research. Hence, description of the techniques is brief and focused on their application to animals trained for drug self-administration.

9.4.1 MICROCANNULAE AND LESION METHODS

Microcannulae guides are prepared in-house from 22-gauge stainless steel needle tubing obtained from disposable hypodermic needles. The plastic hub of the needle can be easily pulled from the stainless steel tubing if the former is heated. To do this, a small jet of hot air is used. The tubing from the hypodermic needle has a small cross bar fixed to it that normally secures the plastic hub; the cross bar provides an excellent anchor for the guide cannulae within the dental acrylic substrate used to attach the guide to the skull. Each guide cannula is prepared with an obturator made of stainless steel suture wire to prevent foreign material from entering it. Guides and obturators are ground to the required length depending on the target brain region. The length is chosen so that the guide/obturator combination will terminate above the brain region of interest, leaving it undamaged. The microinjection cannulae (28 gauge) used on test days are cut to reach beyond the guides to the target area. The guide cannulae/obturator combination that we use has the advantage of presenting a small profile on the skull surface, so that the possibility of mechanical damage and/or injury to the animal is minimized.

Neurochemical lesions are made by attaching a microsyringe to the stereotaxic carrier, and lowering this assembly into the brain, although on one occasion previously implanted cannulae have been used to allow toxin delivery with minimal respiratory depression.[32] In this instance, the guide cannulae were implanted under full anesthesia, but toxin treatment was done in lightly anesthetized animals, taking advantage of the previously implanted guides. (Microinfusions are normally done in nonanesthetized animals; the use of light anesthesia in this case was due to the longer duration of toxin treatment, and the need to keep the animal still and comfortable over this period).

To effect either cannula implants or lesions, stereotaxic surgery is usually carried out once the animals have learned to self-administer nicotine and their responding is stable at the FR5 schedule, generally at three weeks after first exposure to the experimental chambers. The approach to preoperative preparation, anesthesia, and postoperative handling for these surgeries is identical to the approach used to implant intravenous catheters, except that the area involved is the skull and the underlying brain. However, since the time required to implant microcannulae bilaterally, or to deliver a toxin, can be longer than the time needed for catheter surgery, careful attention needs to be given to the depth of anesthesia throughout the procedure.

Interaural zero, rather than skull landmarks, is used to locate brain regions. Once the stereotaxic carriers, with the guide cannulae to be implanted or the microsyringe to deliver the toxin, have been set to zero, the animal is fixed in the frame, and skull locations to be drilled are identified with the aid of the tips of the guides or injection needles. Cannulae are fixed to the skull with dental acrylic, which moulds around the cannulae and around several small screws partially threaded into the skull before the stereotaxic placement. For lesions, the burr holes are filled with bone wax after toxin delivery and the skin similarly closed over the skull. Recovery from either procedure follows the regimen used for intravenous catheter surgery. Details of stereotaxic placements or dose of toxins delivered are all specific to the target site and the particular experiment, and can be found in various publications.[27,30,32]

After experiments have been completed, it is usually necessary to examine brain tissue, at the very least to confirm cannulae placements, if not for more quantitative histological analysis. Animals are deeply anesthetized with pentobarbital, and perfused transcardially with saline followed by 10% formalin. Brains are removed and fixed for sectioning. Other fixatives and processing may obviously be used depending on the histological assessment to be done.

9.4.2 Self-Administered Nicotine: Sites and Mechanisms of Action

The self-administration of nicotine intravenously by rodents is only a partial model of the inhalation of tobacco smoke by humans.[2] That said, similarities between the self-administration model and smoking behavior by humans[26,38] suggest that data derived from the former are relevant to understanding the etiology and treatment of the latter, at least by providing signposts critical to mechanisms. In this section, research from the laboratory, using techniques described earlier, will be used to show how basic neuroscience data about the mechanisms of nicotine dependence can be obtained. Figure 9.2 summarizes schematically the brain systems that will be discussed.

Given the prominent role ascribed to dopamine systems in drug-rewarded behavior, one of the first questions asked with this model was whether dopamine mechanisms were involved in nicotine self-administration. Several studies contributed to answering this question. An early study found that selective dopamine antagonists, administered systemically before nicotine self-administration sessions, reduced drug-taking behavior, and did so in a fashion that suggested an effect on reinforcement rather that on motor performance. In achieving this result, the average patterns of responding for nicotine throughout the time of the sessions were assessed, an example of the methodological advantage of being able to examine and analyze the pattern of responding throughout the sessions.[29] This result with systemic administration of dopaminergic compounds led to examination of the mesolimbic dopamine system itself by means of a selective neurotoxin lesion made with 6-hydroxydopamine

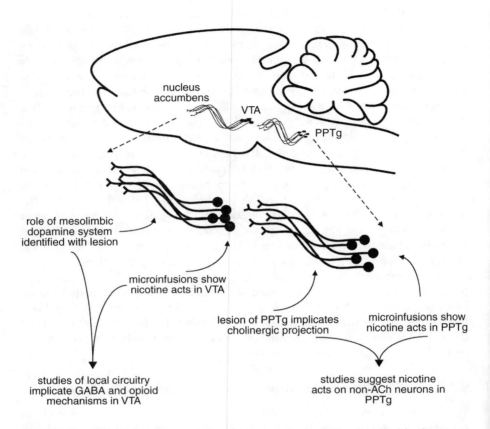

FIGURE 9.2 This schematic summarizes the neuroscience studies described in the text to identify sites at which nicotine acts, and the neurochemical processes involved.

(6-OHDA).[31] This manipulation, carried out by infusion of 6-OHDA into the nucleus accumbens in the manner described previously, confirmed an involvement of the mesolimbic system. In addition, post-mortem measurement of dopamine levels showed that the lesion had depleted dopamine in the nucleus accumbens, but depleted striatal dopamine only marginally — less than the depletion usually associated with the production of motor deficits. A comparison of responding for food by these same animals showed a difference in this behavior as well, with 6-OHDA-lesioned animals responding less that the sham-lesioned animals. However, the lesioned animals responded at a higher rate for food than for nicotine, suggesting that their response for the drug was not motorically limited. Nonetheless, data from CNS lesion experiments need to be interpreted carefully; therefore, the role of the mesolimbic dopamine system in nicotine self-administration was examined with a manipulation that did not alter dopamine function.[30] In this study, animals were trained to self-administer nicotine, and prepared with brain cannulae directed to the nucleus accumbens or ventral tegmental area (VTA). The extent of nicotine self-administration was then measured after microinfusions of the nicotinic antagonist dihydro-β-erythroidine

(DHβE) delivered before certain test sessions. In this experiment, nicotine self-administration was reduced when the antagonist was infused into the VTA, but not after infusions into the nucleus accumbens. In contrast to its effect on nicotine, DHβE delivered into the VTA had no effect on cocaine self-administration. As a result of this set of studies, it was concluded that the mesolimbic dopamine system is involved in the self-administration of nicotine, and that nicotine targets the mesolimbic system by acting at nicotine receptors on or near dopamine cells in the VTA.

One of the consequences of this conclusion has been an examination of mechanisms in the VTA that might modify the action of nicotine. In particular, mu opioid receptors are located within the VTA, probably on interneurons and projection terminals that are largely GABAergic in nature. These two neurochemical systems constitute one focus of our studies of mechanism in the VTA carried out with microinfusions of receptor selective agonists and antagonists, using the methods described earlier in this chapter. This work is ongoing; at this stage differential sensitivity between nicotine self-administration and cocaine self-administration (which is used as a comparison) appears to exist.[27,39] The selective mu receptor agonist DAMGO infused into the VTA has a lesser effect on the self-administration of nicotine than of cocaine, but the opposite order of efficacy occurs with GABA agonists of both A and B selectivity; that is, nicotine self-administration is more sensitive than cocaine to intra-VTA microinfusions of these compounds. These observations suggest that nicotine and cocaine activate VTA circuitry differently, which is consistent with the idea that nicotine targets the mesolimbic system through the VTA, while cocaine acts in the terminal field. The data also demonstrate how neuroscience manipulations can expand the understanding of the mechanisms of nicotine dependence.

A second consequence of the demonstration that nicotine acts in the VTA is the implication of a natural cholinergic input to the midbrain dopamine cell population that might be a part of the neuronal circuitry in drug-rewarded behavior. The cholinergic input to the VTA arises from the laterodorsal tegmental (LDTg) and pedunculopontine tegmental (PPTg) nuclei located in the brain stem. Because the PPTg itself has been implicated in aspects of drug-rewarded behavior for substances other than nicotine, the role of its cholinergic projection in nicotine self-administration was explored.[32] Using a toxin selective for cholinergic neurons, lesions of the PPTg in animals trained to self-administer nicotine were made. Toxin treatment produced a sustained attenuation of nicotine self-administration compared to vehicle-treated animals. Quantitative analysis of histology after the completion of behavioral testing showed that the toxin produced a loss of cholinergic neurons, but largely spared noncholinergic cells. In addition, cholinergic and noncholinergic neurons in the adjacent LDTg were comparable in vehicle- and toxin-treated animals. In a separate experiment, DHβE infused into the PPTg also attenuated nicotine self-administration. As a result of these experiments, it has been concluded that nicotine acts within the PPTg, and that the cholinergic output of the PPTg to the VTA is a part of the reward circuitry.

The next step in this research has been to identify the mechanisms within the PPTg recruited between the action of nicotine at receptors within the nucleus and the cholinergic output. This question has been addressed with experimenter-administered nicotine, using Fos expression to identify neurons activated by nicotine and NADPH-diaphorase labelling to identify the cholinergic neurons in the PPTg.[32,40] Examination of the histology from this experiment has shown that the population of fos-positive neuronal nuclei is typically not located in cholinergic neurons, strongly suggesting that nicotine acts on a noncholinergic population of neurons within the PPTg, and that this population in turn ultimately activates the cholinergic projection to the VTA.

9.5 FUTURE DIRECTIONS

The research summarized earlier has been intended as an example to illuminate how the application of neuroscience techniques to the self-administration paradigm can contribute to the discovery of areas of the brain relevant in nicotine dependence, and to the identification of neurochemical mechanisms in these areas that are important elements.

These manipulations are not without limitations. It is evident that various manipulations can alter nicotine self-administration, and the specificity of these for nicotine self-administration behavior can be assessed by performing the same manipulations on animals trained to self-administer other drugs, or to obtain nondrug reinforcers. However, given the nature of the regulation of nicotine self-administration, it has been difficult to know whether these manipulations are increasing or decreasing the *reinforcing value* of the drug. Often, inferences must be drawn indirectly, by comparison to manipulations with a logical conclusion, for example, the reduction of the reinforcing efficacy of nicotine by a nicotinic antagonist. Nonetheless, one can recognize the risk of even this extrapolation. Recently, it has been shown that nicotine will also support self-administration with a progressive-ratio (PR) schedule.[33] In a schedule of this kind, the response requirement increases according to a particular algorithm after each infusion is obtained. This schedule is believed to be less sensitive to the effect of the self-administered drug on response rate, and may therefore permit a more direct measure of reinforcing efficacy. The successful application of this schedule to nicotine is welcome; self-administration of nicotine on a PR schedule may permit conclusions about central nervous system mechanisms not accessible with FR-based responding. Thus, one future direction is likely to be further development of PR schedules for nicotine self-administration and their use to examine the mechanisms of nicotine reinforcement per se.

Along with the further development of these models, use of a wider range of neuroscience tools in conjunction with them could contribute to the discovery of processes fundamental to nicotine dependence. For example, the increasing availability of selective agents to target the subtypes of the nicotine receptor will be welcome. Another is the development of IV nicotine self-administration techniques in the mouse, which appear to be growing in sophistication.[35, 41–43] Stolerman has recently reviewed these.[26] Nicotine self-administration in the mouse represents more than a simple extension of the model. Given the development of knock-out mice, the ability to establish self-administration in the mouse will contribute to insights about the role of particular receptors and neurochemical systems.[35] In addition, the mouse model may yield a better appreciation of the role of genetics in nicotine dependence,[43] research that has been initiated with oral consumption models.[18]

ACKNOWLEDGMENTS

A large portion of the author's research described in this chapter has been supported by a grant from the National Institute on Drug Abuse (DA 09577). The success of the research described here has been due to the contributions of a number of people, including Laurie Adamson, Betty Chow, Jianhua Zhang, Antonio Lança, and Teresa Sanelli. The outstanding contributions made over the years by Kathy Coen deserve particular mention.

REFERENCES

1. U.S. Department of Health and Human Services, *The Health Consequences of Smoking: Nicotine Addiction*. A report of the Surgeon General. Public Health Service, Centers for Disease Control, Office on Smoking and Health, Rockville, MD, 1998.
2. Corrigall, W.A., Nicotine self-administration in animals as a dependence model, *Nicotine Tobacco Res.*, 1, 11, 1999.
3. Collins, R.J. et al., Prediction of abuse liability of drugs using i.v. self-administration by rats, *Psychopharmacology*, 82, 6, 1984.
4. Griffiths, R.R., Bigelow, G.E., and Henningfield, J.E., Similarities in animal and human drug taking behavior, in *Advances in Substance Abuse*, Mello, N.K., Ed., JAI Press, Greenwich, CT, 1980, 1–90.
5. Goldberg, S.R., Spealman, R.D., and Goldberg, D.M., Persistent behavior at high rates maintained by intravenous self-administration of nicotine, *Science*, 214, 573, 1981.
6. Risner, M.E. and Goldberg, S.R., A comparison of nicotine and cocaine self-administration in the dog: fixed-ratio and progressive-ratio schedules of intravenous drug infusion, *J. Pharmacol. Exp. Ther.*, 224, 319, 1983.
7. Cox, B.M., Goldstein, A., and, Nelson W.T. Nicotine self-administration in rats, *Br. J. Pharmacology*, 83, 49, 1984.
8. Lang, W.J. et al., Self administration of nicotine with and without a food delivery schedule, *Pharmacol. Biochem. Behav.*, 7, 65, 1977.
9. Singer, G. and Simpson, F., Schedule induced self-injections of nicotine with recovered body weight, *Pharmacol. Biochem. Behav.*, 9, 387, 1978.
10. Latiff, A.A., Smith, L.A., and Lang, W.J., Effects of changing dosage and urinary pH in rats self-administering nicotine on a food delivery schedule, *Pharmacol. Biochem. Behav.*, 13, 209, 1980.
11. Smith, L.A. and Lang, W.J., Changes occurring in self-administration of nicotine by rats over a 28-day period, *Pharmacol. Biochem. Behav.*, 13, 215, 1980.
12. Corrigall, W.A. and Coen, K.M., Nicotine maintains robust self-administration in rats on a limited-access schedule, *Psychopharmacology*, 99, 473, 1989.
13. Donny, E.C. et al., Nicotine self-administration in rats, *Psychopharmacology*, 122, 390, 1995.
14. Tessari, M. et al., Nicotine reinforcement in rats with histories of cocaine self-administration, *Psychopharmacology*, 121, 282, 1995.

15. Watkins, S.S. et al., Blockade of nicotine self-administration with nicotinic antago-
    nists in rats, *Pharmacol. Biochem. Behav.*, 62, 743, 1999.
16. Corrigall, W.A., A rodent model for nicotine self-administration, in *Neuromethods:
    Animal Models of Drug Addiction*, Boulton, A., Baker, G., and Wu, P., Eds., Humana
    Press, Clifton, NJ, 1992, 315–344.
17. Smith, A. and Roberts, D.C.S., Oral self-administration of sweetened nicotine solu-
    tions by rats, *Psychopharmacology*, 120, 341, 1995.
18. Robinson, S.F., Marks, M.J., and Collins, A.C., Inbred mouse strains vary in oral
    self-selection of nicotine, *Psychopharmacology*, 124, 332–339, 1996
19. Merlo Pich, E., Chiamulera, C., and Carboni, L., Molecular mechanisms of the posi-
    tive reinforcing effect of nicotine, *Behav. Pharmacol.*, 10, 587, 1999.
20. Weeks, J.R., An improved pneumatic syringe for self-administration of drugs by rats,
    Pharmacol. Biochem. Behav., 14, 573, 1981.
21. Shoaib, M., Schindler, C.W., and Goldberg, S.R., Nicotine self-administration in rats:
    strain and nicotine pre-exposure effects on acquisition, *Psychopharmacology*, 129,
    35, 1997.
22. Shoaib, M. and Stolerman, I.P., Plasma nicotine and cotinine levels following intra-
    venous nicotine self-administration in rats, *Psychopharmacology*, 143, 318, 1999.
23. Valentine, J.D. et al., Self-administration in rats allowed unlimited access to nicotine.
    Psychopharmacology, 133, 300, 1997.
24. Donny, E.C. et al., Acquisition of nicotine self-administration in rats: the effects of
    dose, feeding schedule, and drug contingency, *Psychopharmacology*, 136, 83, 1998.
25. Carroll, M. E. et al., Effects of naltrexone on intravenous cocaine self-administration
    in rats during food satiation and deprivation, *J. Pharmacol. Exp. Ther.*, 238, 1, 1986.
26. Stolerman, I.P., Inter-species consistency in the behavioral pharmacology of nicotine
    dependence, *Behav. Pharmacol.*, 10, 559, 1999.
27. Corrigall, W.A. et al., Response of nicotine self-administration in the rat to manipu-
    lations of mu-opioid and gamma-aminobutyric acid receptors in the ventral tegmental
    area, *Psychopharmacology*, 149, 107, 2000.
28. Sannerud, C.A. et al., The effects of sertraline on nicotine self-administration and
    food-maintained responding in squirrel monkeys, *Eur. J. Pharmacol.*, 271, 461, 1994.
29. Corrigall, W.A. and Coen, K.M., Selective dopamine antagonists reduce nicotine self-
    administration, *Psychopharmacology*, 104, 171, 1991.
30. Corrigall, W.A., Coen, K.M., and Adamson, K.L., Self-administered nicotine activates
    the mesolimbic dopamine system through the ventral tegmental area, *Brain Res.*, 653,
    278, 1994.
31. Corrigall, W.A. et al., The mesolimbic dopamine system is implicated in the rein-
    forcing effects of nicotine, *Psychopharmacology*, 107, 285, 1992.
32. Lança et al., The pedunculopontine tegmental nucleus and the role of cholinergic
    neurons in nicotine self-administration in the rat: a correlative neuroanatomical and
    behavioral study, *Neuroscience*, 96, 735, 2000.
33. Donny, E.C. et al., Nicotine self-administration in rats on a progressive ratio schedule
    of reinforcement, *Psychopharmacology*, 147, 135, 1999.
34. Dworkin, S.I. et al., Comparing the reinforcing effects of nicotine, caffeine, meth-
    ylphenidate and cocaine, *Med. Chem. Res.*, 2, 593, 1993.
35. Picciotto, M.R. et al., Acetylcholine receptors containing the beta2 subunit are
    involved in the reinforcing properties of nicotine, *Nature*, 391, 173, 1998.
36. Pagliusi, S.R. et al, The reinforcing properties of nicotine are associated with a specific
    patterning of c-fos expression in the rat brain, *Eur. J. Neurosci.*, 8, 2247, 1996.

37. Pich, E.M. et al., Common neural substrates for the addictive properties of nicotine and cocaine, *Science*, 275, 83, 1997.
38. Rose, J.E. and Corrigall, W.A., Nicotine self-administration in animals and humans: similarities and differences, *Psychopharmacology*, 130, 28, 1997.
39. Corrigall, W.A. et al., Manipulations of mu-opioid and nicotinic cholinergic receptors in the pontine tegmental region alter cocaine self-administration in rats, *Psychopharmacology*, 145, 412, 1999.
40. Lança, A.J., Sanelli, T.R., and Corrigall, W.A., Nicotine-induced fos expression in the pedunculopontine mesencephalic tegmentum in the rat, *Neuropharmacology*, in press.
41. Martellotta, M.C. et al., Isradipine inhibits nicotine intravenous self-administration in drug-naive mice, *Pharmacol. Biochem. Behav.*, 52, 271, 1995.
42. Rasmussen, T. and Swedberg, M.D.B., Reinforcing effects of nicotinic compounds: intravenous self-administration in drug-naive mice, *Pharmacol. Biochem. Behav.*, 60, 567, 1998.
43. Stolerman, I.P, et al., Discrimination and self-administration of nicotine by inbred strains of mice, *Psychopharmacology*, 141, 297, 1999.
44. Di Chiapa, G., Behavioural pharmacology and neurobiology of nicotine reward and dependence, in *Handbook of Experimental Pharmacology, Vol. 144 Neuronal Nicotinic Receptors,* Clementi, F., Fornasari, D., Gotti, C., Eds., Springer-Verlag, Berlin, 2000, 603–750.

Section 3

10 Evaluation of Pharmacologic Treatments for Smoking Cessation

Jed E. Rose

CONTENTS

10.1 INTRODUCTION

In the last decade there has been an upsurge in the development of new smoking cessation treatments. Although in 1990 the only pharmacologic aid approved by the U.S. Food and Drug administration was nicotine chewing gum, as of the year 2001, four modes of nicotine replacement therapy (NRT) are available (nicotine gum, skin patch, nasal spray, and inhaler) as well as a nonnicotine pharmaceutical, bupropion (Fiore et al. 2000; Hajek et al. 1995). Moreover, there are several other promising treatments at various stages of development, including nicotine antagonist treatment (Covey et al. 2000; Rose et al. 1998; Rose et al. 1994b; Rose et al. 1996b). With a multiplicity of therapeutic techniques and potential combination therapies, it is inevitable that comparisons between the efficacy are sought, and this raises the question of how to compare treatments on a level playing field, using similar methodology and criteria for success. The presentation here reviews some of the most important decisions to be made regarding evaluation of pharmacotherapies for

0-8493-2386-X/02/$0.00+$1.50
© 2002 by CRC Press LLC

smoking cessation, including sample selection, timing of treatment, outcome measures, and rationale for evaluation of combination treatments. Awareness of the difference in outcome that may result from different decisions made by research teams studying these varied methods, may make it possible to achieve a more accurate evaluation of new treatments and treatment combinations.

10.2 SAMPLE SELECTION

Samples of cigarette smokers can be recruited by a variety of means, including newspaper, radio or television advertisements, press releases, word of mouth, posted flyers, clinic referrals, or other means. Usually subjects must report smoking a certain minimum number of cigarettes — frequently a pack a day or more. Often medical exclusion criteria must be followed that are unique to the study medication being evaluated. A variable not often discussed explicitly in articles is the effort required on the part of a potential research participant to be enrolled in a clinical trial. Some centers place several hurdles in the path of potential participants, requiring them to report for several visits prior to formal enrollment in the study, in order to select the most highly motivated subjects. This likely affects overall success rates following treatment, although it is not clear that the ratio of success of active over placebo treatment is affected. Nonetheless, it would be helpful if investigators reported the number of subjects contacted at each stage of the recruitment process, and the proportion of initial respondents actually enrolled in the trial. This would enable readers to better evaluate the generalizability of the findings reported.

To characterize the study sample, a number of dependent measures have generally been of interest to researchers in the field. For purposes of comparison, it would be desirable to collect these measures in every treatment trial. They include basic demographic variables including age, sex, and race, and smoking history variables including number of cigarettes smoked daily, nicotine yield of cigarettes, number of years smoking, number of prior quit attempts, and nicotine dependence (e.g., Fagerstrom test for nicotine dependence questionnaire, Heatherton et al., 1991). Biochemical measures such as plasma nicotine and cotinine levels and expired air carbon monoxide (CO) are also useful in characterizing habitual nicotine intake. Plasma nicotine concentrations fluctuate considerably, based on the time since the last cigarette, because nicotine follows a rapid distributional half-life of 5 min and a metabolic half-life of typically 2 h (Benowitz et al. 1990). Cotinine, the main metabolite, has a half-life of approximately 20 h and hence is more stable. Expired air CO, which can be measured quickly and inexpensively, has a 4 to 6 h half-life and provides a reliable measure of chronic smoke intake (Horan et al. 1978). For measures of smoking behavior or smoking withdrawal symptoms, a baseline (pre-quit) measure using the same instrument is helpful for assessing changes.

10.3 TIMING OF TREATMENT

One important experimental design issue involves whether to begin treatment prior to the target quit-smoking date. There are at least two reasons why this may be desirable, First, it may be important that therapeutic levels of drug be in the system on the quit-smoking day. For example, with bupropion treatment it is believed that approximately one week is required to achieve steady-state blood levels (Holm and Spencer, 2000; *Physician's Desk Reference*, 2000); this is the rationale usually provided for recommending that patients begin treatment several days before quitting smoking. Second, there may be benefits of having a period of time (e.g., 2 weeks) during which participants continue to smoke while receiving treatment. The rationale behind nicotinic blockade treatment is to provide nonreinforced trials of smoking prior to the quit-date (i.e., behavioral extinction). Indeed, some evidence suggests that efficacy of a nicotine antagonist (mecamylamine) is enhanced by extinction resulting from smoking while receiving treatment for 2 to 4 weeks prior to the target quit-date (Rose et al., 1998).

It is interesting that bupropion has recently been found to act as an antagonist at nicotinic receptors (Fryer and Lukas, 1999); thus, although the mechanism of action is not currently known, it is conceivable that nicotinic blockade contributes to its therapeutic efficacy. If so, there may be an additional reason for initiating bupropion treatment prior to the quit-smoking date, aside from attaining steady-state blood levels. Further research is needed to resolve this issue and methodologically, it is difficult to tease apart the two mechanisms just described (achieving therapeutic blood levels vs. extinction) because administering a medication prior to the quit-smoking date inherently involves its presence when subjects are smoking. However, one method for distinguishing between attainment of sufficient blood levels vs. extinction would be to evaluate a pharmacotherapy in the context of different cigarette brand-switching manipulations. For example, subjects might be instructed to switch to cigarettes of similar nicotine yield but having different sensory cues (e.g., mentholated vs. nonmentholated) for 2 weeks prior to the quit-smoking date. Because the cues would be novel, participants would not be learning that the familiar cues of their usual brand of cigarette are no longer predictive of reward. Thus, extinction resulting from nicotinic blockade should be weakened. A second informative brand-switching manipulation is switching to denicotinized tobacco cigarettes (Hasenfratz et al. 1993; Westman et al. 1996). A denicotinized cigarette provides an extinction manipulation even in the absence of pharmacologic treatment. Thus, one would expect little additional benefit of antagonist administration, although it could potentially improve compliance with smoking the denicotinized cigarettes by making nicotine-containing cigarettes unappealing. However, if studies with a known antagonist such as mecamylamine demonstrate efficacy only in the context of patients continuing to smoke their habitual brands of cigarette, but not when switched to denicotinized cigarettes or to cigarettes with different sensory cues, then extinction as a mechanism may be implicated.

Aside from the use of brand-switching to explore mechanisms underlying the efficacy of potential nicotine blockade therapies, other prequit brand-switching manipulations are of interest. For instance, brand fading, i.e., switching to cigarette brands of successively lower nicotine delivery during the weeks leading up to the quit-smoking date, has often been recommended as a smoking cessation treatment (Prue et al. 1983), although it has not been proven to be efficacious (Fiore et al. 2000). Nonetheless, if this strategy is used, and subjects reduce dependence on nicotine prior to the quit-smoking date, there may be influences on the efficacy of pharmacologic treatment being evaluated. Moreover, the dose of NRT may need to be adjusted based on the decreased tolerance to nicotine after completion of the brand fading regimen (Shipley and Rose, 2000).

10.4 OUTCOME MEASURES

The selection of primary outcome measures is, of course, critical in assessing a potential treatment and there are several factors which can greatly affect the success rate obtained. Key factors include whether a successful outcome is defined as continuous vs. point abstinence, whether self-reported abstinence is biochemically verified by biochemical measures, and how drop-outs are classified.

Continuous abstinence from the target quit-smoking date to the observation point is the most rigorous criterion for abstinence; however, abstinence criteria have often allowed for slips within the first 2 weeks of the target quit-smoking date. Although any smoking during this period is highly predictive of eventual relapse (Fiore et al. 1992; Westman et al. 1997), most relapses occur within the first few weeks and thus a high concordance exists between the classification of patients based on continuous abstinence from weeks 2 to 6 and weeks 1 to 4 post-quit. Four weeks of continuous abstinence, but not necessarily beginning with the quit-smoking date, has been the basis of FDA approval of NRT products. However, this is only a fair indicator of long-term success as roughly 50% of patients relapse between 1 month and 12 months post-quit. After 12 months, abstinence rates seem to be relatively constant up to 3 years (Richmond et al. 1997).

In addition to continuous abstinence, point measures of abstinence are often reported. Point abstinence is usually defined as abstinence for the preceding seven days (Fiore et al. 1994; Holm and Spencer, 2000). Abstinence rates can vary considerably depending on whether point or continuous abstinence is used as a measure of success. For example, in a recent study of bupropion the rate of abstinence using a 7-day point abstinence criterion was 26.9% as opposed to 12.2% using a continuous abstinence criterion (Hurt et al. 1997). The main message is not that one particular measure has been shown to be the correct one, but rather that the rates of success depend on the criteria employed; this fact should be borne in mind in making comparisons across various treatments.

A self-report of abstinence, whether point or continuous, is generally viewed as an inadequate criterion without some biochemical verification, such as measures of expired air CO or plasma or saliva measures of nicotine, cotinine, or thiocyanate. In studies that entail NRT, use of nicotine and cotinine measures becomes problematic and expired air CO is frequently used. One other potential solution to this problem involves assay of tobacco-specific alkaloids such as anabasine and anatabine, which are not contained in NRT products and are not metabolically derived from nicotine (Jacob et al. 1999).

The issue of study drop-outs has been most often addressed within the context of an intent-to-treat approach, according to which drop-outs are classified as treatment failures. However, some researchers have questioned whether this is the best strategy (Society for Research on Nicotine and Tobacco, 2000). Ideally, every effort to obtain follow-up data is made.

The intent-to-treat analysis may require modification when used to analyze the effects of treatments geared to different stages of the cessation process. For example, consider a study in which a precessation pharmacotherapy is administered, followed by a postquit treatment such as NRT. Dropouts during the prequit phase should probably be counted as failures of the prequit treatment, but in evaluating the efficacy of the post-quit treatment, the denominator should only include those subjects who have reached their quit-smoking dates.

Depending on decisions regarding these criteria, different success rates may be computed in a given treatment trial. However, the absolute rate will still be highly dependent on other factors, such as subject selection, as discussed earlier, and type of behavioral support provided, if any. Thus, in comparing across treatments, absolute success rates are not very informative. One should also examine the odds ratio of active vs. placebo treatment and, in particular, focus on the success rate of the placebo therapy group in making comparisons across different treatments evaluated at different study centers. If one site obtains results of 20% vs. 10% for placebo in evaluating a form of NRT, for example, but another study reports success rates of 40% vs. 20% (placebo) in evaluating a new treatment, it would not constitute convincing evidence that the new treatment is superior; it is more likely that both treatments are equivalent in terms of doubling success over placebo.

10.4.1 OTHER OUTCOME MEASURES

It should be acknowledged that there has been increasing attention given to other outcome measures besides complete smoking abstinence. For example, it has been argued that a sustained reduction in smoking might be a legitimate therapeutic goal for those smokers unable to quit smoking entirely (Hughes, 2000). Other outcome measures may be informative in terms of mechanisms underlying treatment efficacy. Thus, for a treatment designed to alleviate withdrawal symptoms, such as NRT, assessing withdrawal at several time points during treatment may be useful. Peak intensity of withdrawal symptoms occurs within the first few days, and a method that has been applied in several clinical trials is to distribute take-home withdrawal symptoms questionnaires to capture data from the quit-smoking day (Rose et al. 1994b). By 1 week post-quitting, symptoms are often much lower in intensity and the power to detect treatment effects will be attenuated. Other dependent measures may be informative in elucidating mechanisms underlying treatment efficacy — measures of smoking satisfaction and other rewarding and/or aversive aspects of smoking may shed light on mechanisms relating to pharmacologic blockade when nicotinic antagonist therapy is being tested (Rose et al. 1994a; Shiffman et al. 2000).

10.5 CONTEXT IN WHICH TO EVALUATE PHARMACOTHERAPIES

The social and behavioral context in which to evaluate treatments also varies widely, from individual behavioral therapy to group support to minimal contact. Although there may be value to a variety of contexts, the most accurate appraisal of real-world success rates obtained with a given treatment is probably a minimal behavioral intervention setting, as only a small percentage of smokers seek time-intensive behavioral therapy (Lichtenstein and Hollis, 1992). The initial NRT trials often employed behavioral therapy (e.g. [Transdermal Nicotine Study Group, 1991]) and long-term success rates (6 months continuous abstinence) with behavioral support have often been obtained in the range of 20 to 30%. However, trials in a minimal intervention setting as well as experience with over-the-counter availability of NRT products have more typically yielded long-term abstinence rates of 10 to 20% (Hughes, 1999; Russell et al. 1993; Westman et al. 1993).

10.6 COMBINATION PHARMACOTHERAPIES

As the number of smoking cessation treatments increases, the number of potential combinations escalates geometrically. Including the five existing approved pharmacotherapies (four modes of NRT and bupropion), there are ten potential pair-wise combinations. Methods of focusing on the more promising combinations are needed. Some obvious factors to consider are the acceptability and efficacy of each method, as combining two methods with relatively low appeal will yield a treatment possibility that will be even more limited in its applicability. Thus, combining the two modes of NRT that are associated with the most frequent adverse effects, nasal spray and chewing gum (Hajek et al. 1999), would not seem to present as convincing a rationale as combining nicotine patch plus inhaler. However, pharmacokinetic considerations would also affect the strength of the rationale for evaluating combinations of NRT methods. Combining a slow steady delivery of nicotine via a skin patch with a more discrete rapid delivery system such as nasal spray would be one example. Combining bupropion with NRT has already been reported in one trial in which there was a suggestion of addictive efficacy (Jorenby et al. 1999). Drugs that work through different mechanisms would be more likely to yield increments in efficacy. One nonobvious combination treatment that has received attention involves combining nicotinic agonist treatment (as with NRT) with nicotinic blockade (Rose and Levin, 1991a). The rationale is that a nicotinic antagonist such as mecamylamine attenuates the rewarding effects of cigarette smoking; however, when used alone, it does not ameliorate tobacco withdrawal symptoms. By coadministering nicotine, withdrawal symptoms can be alleviated; even though mecamylamine shifts the dose-effect curve for nicotine to the right, relief of craving for cigarettes can still be achieved with NRT (Rose et al. 1996a). Use of mecamylamine in conjunction with NRT raises more possibilities in terms of dose and timing of treatment. Several options that need to be compared include mecamylamine prequit plus nicotine postquit, nicotine plus mecamylamine prequit followed by nicotine postquit, etc. Additionally, if nicotine is administered with mecamylamine, exploration of higher than usual doses of nicotine may be warranted.

An important methodologic issue arises when comparing, for example, nicotine skin patch plus gum to nicotine skin patch alone (Fagerström, 1994). If a superior outcome is obtained with the combined treatment, one does not know if it is due to the combination of two different modes of delivery per se, as opposed simply to delivering a higher total dose of nicotine. A useful control comparison in this case would be to provide the same average total daily dose of nicotine from the skin patches. Then, any enhancement in outcome could accurately be ascribed to the use of the two NRT modalities.

Another combination approach involves addressing the distinct components of smoking aside from the role of nicotine's pharmacologic effects. At least two other components missing from NRT may be critical. One component consists of the sensory and behavioral cues associated with inhalation upon which smokers have become dependent (Rose, 1988; Rose and Behm, 1995; Rose and Levin, 1991b). Although the nicotine inhaler provides some of these cues, it does not deliver tobacco taste or replace smokers' enjoyable sensations of inhaling cigarette smoke. Similarly, some promising results have been reported using inhalers that deliver citric or ascorbic acid to mimic some of the sensations accompanying inhalation of cigarette smoke (Behm et al. 1993; Levin et al. 1993; Westman et al. 1995). However, these too fall short of providing the full constellation of cues to which smokers are accustomed.

A second component that may also be important entails nonnicotine constituents in tobacco that inhibit an enzyme (monoamine oxidase) important in the breakdown of neurotransmitters in the brain (e.g., dopamine), which, in turn, may mediate the chemical reward of nicotine (Fowler et al. 1996). Methods of replacing these missing components are being developed and may yield further improvements in treatment efficacy.

10.7   USE OF OPEN-LABEL PILOT EVALUATIONS

The large number of potential treatments and treatment combinations highlights the need to develop strategies to evaluate these possible therapies efficiently. It is clear that conducting large randomized controlled trials as an initial step in the evaluation of new smoking cessation treatments is likely to be an inefficient strategy. An alternative approach is to conduct pilot open-label investigations of new treatments; the comparison of abstinence rates and tolerability to the large historical database of NRT of a given research center (using consistent methodology across studies) can then be used to decide whether to pursue the evaluation further. For example, if one exposes 50 volunteers to a new treatment and observes a success rate of 10%, then one can be 95% confident that the population success lies between 0 to 20% (Sheskin, 1997); one can then probably conclude that it is not worth expending the substantial resources needed to test efficacy in a large controlled trial involving hundreds of volunteers. On the other hand, a 6-month continuous abstinence rate of 50% would be highly suggestive of a real enhancement beyond the typical 10 to 20% success rate of NRT (with minimal supportive counseling). Of course, there is a trade-off in terms of sample size and statistical power: if the actual success rate were 35%, for example, a sample size of 50 would yield an approximate probability of 80% of detecting an improvement of this magnitude over an assumed 20% success rate of standard treatment, using a (one-sided) alpha criterion of .05 (Cohen, 1988). In contrast, it would require approximately 135 subjects per group, or a total of 270 subjects, to conduct a randomized controlled trial with sufficient power to detect the same difference between 20% and 35% abstinence, using a two-tailed criterion of .05. Clearly, a preliminary open-label trial with 50 participants would provide a reasonable chance of detecting a promising treatment, and prevent expending large resources on treatments not likely to provide significant improvements over current practice.

10.8 CONCLUSION

The evaluation of pharmacotherapies for smoking cessation offers an increasingly challenging area for creative endeavor to develop and test improved treatments. The rate at which new treatments can be devised and treatment combinations envisioned offers many opportunities, but also challenges, in the development and evaluation of new treatments. Some of the most important decisions and strategic approaches to this problem have been described here. Through recognition of the explicit and implicit decisions made regarding experimental design, subject selection, outcome measures, timing of treatment, and related methodologic issues, it may be possible not only to develop improved smoking cessation treatments but also to gain an accurate appraisal of their relative effectiveness when judged within a consistent framework.

REFERENCES

Behm, F.M., Schur, C., Levin, E.D., Tashkin, D.P., and Rose, J.E. (1993), Clinical evaluation of a citric acid inhaler for smoking cessation, *Drug Alc. Depend.*, 31:131–138.
Benowitz, N.L., Porchet, H., and Jacob, P.I. (1990), Pharmacokinetics, metabolism, and pharmacodynamics of nicotine, in Wonnacott, S., Russell, M.A.H., and Stolerman, I.P., Eds., *Nicotine Psychopharmacology*, Oxford University Press, Oxford, 112–157.
Cohen, J. (1988), *Statistical Power Analysis for the Behavioral Sciences*, Lawrence Erlbaum Associates, Hillsdale, NJ.
Covey, L.S., Sullivan, M.A., Johnston, J.A., Glassman, A.H., Robinson, M.D., and Adams, D.P. (2000), Advances in nonnicotine pharmacotherapy for smoking cessation, *Drugs*, 59:17–31.
Fagerström, K. (1994), Combined use of nicotine replacement products, *Health Values*, 18:15–20.
Fiore, M., Jorenby, D., Baker, T., and Kenford, S. (1992), Tobacco dependence and the nicotine patch, Clinical guidelines for effective use, *JAMA*, 268:2687–2694.
Fiore, M.C., Bailey, W.C., Cohen, S.J., Dorfman, S.F., Goldstein, M.G., Gritz, E.R., Heyman, R.B., Jaen, C.R., Kottke, T.E., Lando, H.A., Mecklenburg, R., Mullen, P.D., Nett, L.M., Robinson, L., Stitzer, M.L., Tommasello, A.C., Villejo, L., and Wewers, M.E. (2000), Treating tobacco use and dependence, Clinical Practice Guidelines, Rockville, MD.
Fiore, M.C., Smith, S.S., Jorenby, D.E., and Baker, T.B. (1994), The effectiveness of the nicotine patch in smoking cessation, *JAMA*, 271:1940–1947.
Fowler, J.S., Volkow, N.D., Wang, G-J., Pappas, J., Logan, J., MacGregor, R.R., Alexoff, D., Shea, Cl, Schlyer, D.J., Wolf, A.P., Warner, D., Zezulkova, I., and Cilento, R. (1996), Inhibition of monoamine oxidase B in the brains of smokers, *Nature*, 379:732–736.
Fryer, J.D. and Lukas, R.J. (1999), Noncompetitive functional inhibition at diverse, human nicotinic acetylcholine receptor subtypes by bupropion, phencyclidine, and ibogaine, *JPET*, 288:88–92.

Hajek, P., West, R., Foulds, J., Nilsson, F., Burrown, S., and Meadow, A. (1999), Randomized comparative trial of nicotine polacrilex, a transdermal patch, nasal spray, and an inhaler, *Arch. Int. Med.*, 159:2033–2036.

Hasenfratz, M., Balding, B., and Battig, K. (1993), Nicotine or tar titration in cigarette smoking behavior? *Psychopharmacology*, 112:253–258.

Heatherton, T.F., Kozlowski, L.T., Frecker, R.C., and Fagerstrom, K.L. (1991), the Fagerstrom test for nicotine dependence: a revision of the Fagerstrom tolerance questionnaire, *Br. J. Addict.*, 86:1119–1127.

Holm, K.J. and Spencer, C.M. (2000), Bupropion: a review of its use in the management of smoking cessation, *Drugs*, 59:1007–1024.

Horan, J.J., Hackett, G., and Linberg, S.E. (1978), Factors to consider when using expired air carbon monoxide in smoking assessment, *Addict. Behav.*, 3:25–28.

Hughes, J. (1999), Four beliefs that may impede progress in the treatment of smoking, *Tobacco Control*, 8:323–326.

Hughes, J. (2000), Reduced smoking: an introduction and review of the evidence, *Addiction*, 95:S3–S7.

Hurt, R.D., Sachs, D.P.L., Glover, E.D., Sullivan, C.R., Croghan, I.T., and Sullivan, P.M. (1997), A comparison of sustained-release bupropion and placebo for smoking cessation, *N. Eng. J. Med.*, 337:1195–1202.

Jacob, P.I., Yu, L., Shulgin, A.T., and Benowitz, N.L. (1999), Minor tobacco alkaloids as biomarkers for tobacco use: comparison of users of cigarettes, smokeless tobacco, cigars, and pipes, 89:731–736.

Jorenby, D.E., Leischow, S.J., Nides, M.A., Rennard, S.I., Johnston, J.A., Hughes, A.R., Smith, S.S., Muramoto, M.L., Daughton, D.M., Doan, K., Fiore, M.C., and Baker, T.B. (1999), A controlled trial of sustained-release bupripion, a nicotine patch, or both for smoking cessation, *N. Eng. J. Med.*, 340:685–691.

Levin, E.D., Behm, F.M., Carnahan, E., LeClair, R., Shipley, R., and Rose, J.E. (1993), Clinical trials using ascorbic acid aerosol to aid smoking cessation, *Drug Alc. Depend.*, 32:211–223.

Lichtenstein, E. and Hollis, J.F. (1992), Patient referral to a smoking cessation program: who follows through? *J. Fam. Pract.*, 34:739–744.

Physicians' Desk Reference (2000), Medical Economics Company, Inc., Montvale, NJ.

Prue, D.M., Davis, C.J., Martin, J.E., and Moss, R.A. (1983), An investigation of a minimal contact brand fading program for smoking treatment, *Addict. Behav.*, 8:307–310.

Richmond, R.L., Kehoe, L., de Almeida, C., and Neto, A. (1997), Three year continuous abstinence in a smoking cessation study using the nicotine transdermal patch, *Heart*, 78:617–618.

Rose, J.E. (1988), The role of upper airway stimulation in smoking, in Pomerleau, O.F. and Pomerleau, C.S., Eds., *Nicotine Replacement: A Critical Evaluation*, Alan R. Liss, Inc., New York, 95–106.

Rose, J.E. and Behm, F.M. (1995), There is more to smoking than the CNS effects of nicotine, in Eds., *Effects of Nicotine on Biological Systems II*, Burkhäuser Verlag, Basel, 9–16.

Rose J.E., Behm, F.M., and Westman, E.C. (1996a), Interactive effects of nicotine and mecamylamine, presented at the Soc. Res. Nicotine Tobacco, March 15–17, Washington, DC.

Rose, J.E., Behm, F.M., and Westman, E.C. (1998), Nicotine/mecamylamine treatment for smoking cessation: the role of precessation therapy, *Exp. Clin. Psychopharmacol.*, 6:331–343.

Rose, J.E., Behm, F.M., Westman, E.C., Levin, E.D., Stein, R.M., Lane, J.D., and Ripka, G.V. (1994a), Combined effects of nicotine and mecamylamine in attenuating smoking satisfaction, *Exp. Clin. Psychopharmacol.*, 2:1–17.

Rose, J.E., Behm, F.M., Westman, E.C., Levin, E.D., Stein, R.M., Lane, J.D., and Ripka, G.V. (1994b), Mecamylamine combined with nicotine skin patch facilitates smoking cessation beyond nicotine patch treatment alone, *Clin. Pharmacol. Ther.*, 56:86–99.

Rose, J.E. and Levin, E.D. (1991a), Concurrent agonist-antagonist administration for the analysis and treatment of drug dependence, 41:219–226.

Rose, J.E. and Levin, E.D. (1991b), Interrelationships between conditioned and primary reinforcement in the maintenance of cigarette smoking, in West, R. and Grunberg, N.E., Eds., *Br. J. Addic.*, Carfax Publishing Company, London, 605–610.

Rose, J.E., Westman, E.C., and Behm, F.M. (1996b), Nicotine/mecamylamine combination treatment for smoking cessation, *Drug Devel. Res.*, 38:243–256.

Russell, M.A.H., Stapleton, J.A., Feyerabend, C., Wiseman, S.M., Gustavsson, G., Sawe, U., and Connor, P. (1993), Targeting heavy smokers in general practice: randomized controlled trial of transdermal nicotine patches, *Br. Med. J.*, 306:1308–1312.

Sheskin, D.J. (1997), *Handbook of Parametric and Nonparametric Statistical Procedures*, CRC Press, Boca Raton, FL.

Shiffman, S., Johnston, J.A., Khayrallah, M., Elash, C.A., Gwaltney, C.J., Paty, J.A., Gnys, M., Evoniuk, G., DeVeaugh-Geiss, J. (2000), The effect of bupropion on nicotine craving and withdrawal, *Psychopharmacology*, 148:33–40.

Shipley, R.H. and Rose, J.E. (2000), *Quit Smart: Stop Smoking Guidebook*, QuitSmart Stop Smoking Resources, Inc., Durham, NC.

Society for Research on Nicotine and Tobacco (2000), Treatment methodology meeting, presented at the meeting of the Soc. Res. Nicotine and Tobacco, Feb. 18–20, Arlington, VA.

Transdermal Nicotine Study Group (1991), Transdermal nicotine for smoking cessation, *JAMA*, 266:3133–3138.

Westman, E.C., Behm, F.M., and Rose, J.E. (1995), Airway sensory replacement combined with nicotine replacement for smoking cessation: a randomized, placebo controlled trial using a citric acid inhaler, *Chest*, 107:1358–1364.

Westman, E.C., Behm, F.M., and Rose, J.E. (1996), Dissociating the nicotine and airway sensory effects of smoking, *Pharmacol. Biochem. Behav.*, 53:309–315.

Westman, E.C., Behm, F.M., Simel, D.L., and Rose, J.E. (1997), Smoking behavior on the first day of a quit-attempt predicts long-term abstinence, *Arch. Int. Med.*, 157:335–340.

Westman, E.C., Levin, E.D., and Rose, J.E. (1993), The nicotine patch in smoking cessation: a randomized trial with telephone counseling, *Arch. Int. Med.* 153:1917–1923.

11 Nicotine and Schizophrenia

Joseph P. McEvoy

CONTENTS

11.1 WHY FOCUS ON SMOKING IN PATIENTS WITH SCHIZOPHRENIA?

Patients with schizophrenia smoke at much higher prevalence rates (70–80%) than the general population (25 to 30%).[1–5] This is true even when patients are identified in their first psychotic episodes, before they have been institutionalized or treated with antipsychotic medications.[6] Smokers with schizophrenia also smoke more heavily than smokers in the general population, or smokers with other psychiatric illnesses.[7,8] The increased prevalence of smoking, and the heavy smoking, suggest that nicotinic mechanisms are involved in the pathophysiology of schizophrenia.

Smoking takes its cost in this population. Mortality rates for patients with schizophrenia are two to four times that of the general population, and patients with schizophrenia die, on average, ten years earlier than would otherwise be expected. The prevalence rates for respiratory and cardiovascular disease are significantly elevated among patients with schizophrenia — in some studies twice as high as seen in age-matched controls from the general population.[9]

11.2 NICOTINE-RESPONSIVE ELEMENTARY PHENOTYPES IN SCHIZOPHRENIA

Schizophrenia is a complex genetic disorder; i.e., the illness does not have a pattern of inheritance resulting from a single genetic abnormality.[10] Two nicotine-responsive neurophysiological abnormalities, one in auditory sensory gating and the other in smooth pursuit eye movements, are currently under investigation as potential elementary phenotypes representing gene effects that, in combination with other specific gene effects, may result in the development of schizophrenic illness.[11,12] These neurophysiological abnormalities appear to be transmitted as autosomal dominant characteristics in some families with high occurrence rates for schizophrenia, and they are both normalized by nicotine administration. The subjective experience of this normalization may contribute to the drive to smoke among patients with schizophrenia.

11.2.1 AUDITORY SENSORY GATING ABNORMALITIES

In nearly all neuronal systems, when stimuli are repeated, the electroencephalographic response to the second stimulus is less than that to the first. The first stimulus activates not only the initial response, but also inhibitory circuits that reduce the response to the second stimulus. When presented with two-click stimuli 500 msec apart, patients with schizophrenia do not inhibit one aspect of their response (the P50 wave of the auditory evoked potential) to the second click as much as normal individuals do. After these patients smoke as much as they wish, this deficit in auditory sensory gating disappears and the responses to the second stimulus are decreased, as occurs in normal individuals, for 15 to 30 minutes before again returning to the abnormal pattern as nicotine effects dissipate. Fifty percent of the nonafflicted, first-degree relatives of patients with schizophrenia who have this auditory sensory gating deficit also demonstrate the deficit. Chewing nicotine gum corrects the deficit in these nonafflicted, nonsmoking relatives as well.

Linkage analysis of P50 auditory sensory gating abnormalities in families that display this trait has provided evidence for linkage in the chromosome 15 q 14 region, a site later shown to code for the alpha-7 nicotine receptor subunit.[13]

11.2.2 SMOOTH PURSUIT EYE MOVEMENT ABNORMALITIES

Patients with schizophrenia have significantly more frequent intrusions of small anticipatory saccades into their smooth pursuit eye movements than do normal individuals. Like the auditory sensory gating abnormalities, these intrusions can be conceptualized as deficits in inhibition. The eye movement abnormalities can be dispelled temporarily by high doses of nicotine, such as those delivered by the heavy smoking of patients with schizophrenia. The eye-tracking abnormalities also appear to be transmitted as an autosomal dominant characteristic, and can be demonstrated in 50% of the offspring of a parent with schizophrenia. Linkage studies suggest a locus on chromosome 6.[12]

Childhood onset schizophrenia is a rare, severe form of the illness with early onset and poor prognosis. A study of ten proband/both parent trios revealed bilineal transmission of sensory gating and/or eye movement abnormalities to these probands, in contrast to the more common transmission of one of these traits down one parental line that characterizes most patients with adult-onset schizophrenia.[12]

Currently, ongoing research is exploring whether individuals with these potential elementary phenotypes can be consistently discriminated from individuals without the phenotypes, and whether these potential elementary phenotypes reflect single genetic phenomena. Comorbidity appears to be the rule rather than the exception in psychiatric disorders, and potential nicotinic mechanisms involved in highly comorbid disorders (e.g., substance use disorders) must be distinguished from those involved specifically in schizophrenia. Several genetic disorders involving the long arm of chromosome 15 are associated with increased incidence of schizophrenia-like psychosis, lending validation to the auditory sensory gating/alpha-7 nicotine receptor linkage with schizophrenia. The ultimate goal of such work is to understand how genotypes associated with elementary phenotypes contribute to risk for schizophrenia.[11]

11.3 ANTIPSYCHOTIC DRUGS, SMOKING, AND SCHIZOPHRENIA

Antipsychotic drugs ameliorate the psychopathology and course of schizophrenia. Recent research has begun to examine how antipsychotic drugs affect smoking among patients with schizophrenia, how they affect nicotine-responsive phenotypes, and how these effects relate to the drugs' effects on psychopathology and cognitive psychomotor performance.

An initial caveat in designing and evaluating studies in this area is that smoking approximately doubles the metabolic clearance rate of many antipsychotic drugs; excipients in tobacco smoke stimulate hepatic microsomal enzymes that degrade many antipsychotic drugs.[14]

Studies of how much patients smoke usually collect patients' self-reports of their smoking. When patients are actively psychotic it can be helpful to administer questionnaires twice, separated by a brief interval, to identify patients who did not adequately understand what was being asked. Biological measures of smoking (e.g., expired carbon monoxide levels, plasma nicotine and cotinine levels) are highly desirable because these bypass whatever difficulties patients may have in recall or communication.

There are substantial advantages to studying smoking among hospitalized inpatients with schizophrenia: 1) patients with schizophrenia have high rates of noncompliance with clinical or investigational procedures without the close supervision that the inpatient setting can provide; 2) the inpatient environment is stable, and patients are largely protected from stressful events or concurrent substance abuse that may alter their response to nicotine; and, 3) the protective containment of the inpatient environment permits rapid detection of and attention to any adverse reactions developing in relation to novel interventions under study.

The advantages of studying outpatients include: 1) outpatients are free to smoke as much as they choose in a natural environment, unfettered by limited available times for smoking that most inpatient units have established (for example, inpatients at our hospital can smoke one cigarette per hour between 7 a.m. and 10 p.m. daily), and 2) the psychiatric condition of outpatients is usually relatively stable. It is important to recognize that any changes in the background rates of patients' smoking (e.g., moving from the outpatient to the inpatient setting) can have substantial effects on measures of smoking.

Two-hour free-smoking sessions are utilized as a measure of drive-to-smoke in outpatient studies. These sessions begin after lunch (1 p.m.) and run until 3 p.m. Expired carbon monoxide (CO) levels are measured at the beginning of each session (approximately 1 hour after the last time cigarettes were available on the ward). Patients then have free access to cigarettes, snacks, and sodas over the ensuing 120 minutes, and watch a movie or regular TV programming, according to their preference. Research staff keep count of the number of cigarettes each patient smokes, and obtain repeated measures of expired CO after 30, 60, 90, and 120 minutes of free smoking. Intraclass correlations for eight patients who had three free-smoking sessions over consecutive days while on fixed doses of antipsychotic medication were beginning CO, .80; end CO, 93; difference in CO, .90; and number of cigarettes smoked, .81.

11.3.1 CONVENTIONAL NEUROLEPTICS (E.G., HALOPERIDOL)

These older drugs do not correct the nicotine-sensitive failure of inhibition in auditory sensory gating in patients with schizophrenia, even among those patients showing clear declines in psychopathology. Measures of smoking taken in acutely psychotic medication-free patients (admitted to hospital because they had stopped their prescribed medications and relapsed) at the time of admission, and then repeated after treatment with conventional neuroleptics, show an increase in smoking during treatment with conventional neuroleptics, despite resolving psychopathology.[15]

Not only does haloperidol fail to correct nicotine-responsive deficits, it also impairs performance on certain cognitive psychomotor tests in a dose-dependent manner; these impairments in performance can be reversed in a dose-dependent manner by nicotine.[16] Thus, both nicotine-sensitive deficits, and treatment with conventional neuroleptics may contribute to prevalent and heavy smoking among patients with schizophrenia.

11.3.2 ATYPICAL ANTIPSYCHOTICS (E.G., CLOZAPINE)

The most atypical antipsychotic, clozapine, offers therapeutic benefit to approximately 50% of patients with schizophrenia who fail to respond to conventional neuroleptics. This suggests that clozapine brings therapeutic pharmacologic mechanisms to bear beyond those offered by the conventional neuroleptics (which primarily block D_2-dopamine receptors). Clozapine does correct auditory sensory gating abnormalities in those patients who show a therapeutic response to treatment.[17] Being a smoker is a powerful predictor of therapeutic response to clozapine. When switched from conventional neuroleptics to clozapine, smokers with schizophrenia smoke significantly less; these findings suggest that some of the therapeutic advantages of clozapine may be explainable via actions on nicotinic mechanisms.[18] Those patients whose pathophysiologies include defective nicotinic mechanisms find that smoking is helpful to them. As smokers, they signal that they are potentially good responders to clozapine. As they reap the benefits of clozapine on nicotinic mechanisms, which appear to be more long-lasting and consistent than those of nicotine, they find less need to smoke and their smoking declines. Clozapine is known to bind to nicotine receptors in the central nervous system and research is ongoing to explicate its actions at these receptors. Research is also currently examining the effects on smoking and potential elementary phenotypes of other, newer atypical antipsychotics (e.g., risperidone, olanzapine).[19]

11.4 NICOTINIC AGONISTS, ANTAGONISTS, AND ALLOSTERIC MODULATORS

A small number of studies have examined the effects of nicotine, the nicotine antagonist, mecamylamine, or allosteric modulators of the nicotine receptor in patients with schizophrenia.

11.4.1 NICOTINIC AGONISTS

The effects of nicotine itself, delivered via transdermal patch or chewing gum, have been examined in a number of different study designs. As mentioned earlier, nicotine gum has been shown to correct abnormalities in sensory gating and smooth-pursuit eye tracking in patients with schizophrenia and in their nonafflicted relatives. Of note, high doses (6 mg, i.e., 3 pieces) are necessary for the effect; the alpha-7 nicotine receptor is a low-affinity receptor requiring high nicotine concentrations for activation.[20,21]

Several studies have included nicotine transdermal patches in smoking cessation trials. Addington et al.[22] offered a smoking cessation program, including nicotine patches and weekly group therapy, to outpatients with schizophrenia. Sixty-five patients referred themselves to this program, but 15 dropped out after the first session. Nicotine patches were available at 21 mg daily for 6 weeks, then 14 mg daily for 2 weeks, then 7 mg daily for 2 weeks. Forty of the 50 patients who participated used patches and 20 of these 40 (50%) achieved abstinence for at least 4 weeks. Of the ten patients not using patches, only one (10%) achieved abstinence. The gains achieved were quickly lost after nicotine patches were withdrawn, with only 16% remaining abstinent at 3 months, and 12% at 6 months. Similar results (13% abstinence at 6 months) have been reported by Zeidonis et al.[23] from another program combining group therapy and temporary nicotine patch treatment. These results fall below the expected abstinence rates of 20 to 25% at 6 months seen in smokers from the general population treated initially with group therapy and nicotine patches.

Hartman et al.[24] compared the effects of nicotine and placebo patches in a double-blind crossover design involving 13 psychiatric patients (10 with schizophrenia or schizoaffective disorder) who were **not trying to stop smoking.** These patients smoked significantly less while receiving nicotine patches, raising the possibility of harm reduction through long-term nicotine replacement if complete abstinence cannot be achieved. These investigators report that patients tolerate nicotine patches well, and, in fact, two patches (42 mg total) were applied daily to several patients with heavy smoking histories, without difficulty. However, there is one report of a patient with schizophrenia presenting to an emergency room in a toxic state while wearing a nicotine patch, presumably because of continued smoking.[25] This resolved quickly after the patch was removed.

Ongoing studies are examining the effects of various forms of nicotine delivery on metabolic brain imaging (position emission topography) in patients with schizophrenia.

11.4.2 NICOTINIC ANTAGONISTS

Mecamylamine, a secondary amine that readily penetrates the central nervous system, is a reversible antagonist at human nicotinic acetylcholine receptors, in particular, the $\alpha 2$ $\beta 4$ and $\alpha 4$ $\beta 4$ subtypes. The effects of 0, 5 and 10 mg of mecamylamine on smoking in patients with mania, schizophrenia, or substance use disorders.[8] Mecamylamine produced robust increases, in a dose-dependent manner, on all smoking measures and across all diagnostic groups. No evidence of a differential pattern of response to mecamylamine across these patient groups was found that would suggest diagnosis-specific abnormalities in $\alpha 2$ $\beta 4$ or $\alpha 4$ $\beta 4$ receptors. Currently, the effects of mecamylamine on cognitive psychomotor performance in patients with schizophrenia are being examined.

11.4.3 ALLOSTERIC MODULATORS

The nicotinic acetylcholine receptors (nAChRs) are composed of five subunits arranged pseudosymmetrically around an axis that passes through the gated ion channel, perpendicular to the plane of the neuronal membrane; viewed from the synapse, they appear as rosettes with central depressions. The α-type subunits carry the principal component for the acetylcholine binding site, as well as the complementary component, and are able to form homomeric (e.g., five α-7 subunits) receptors. The non-α-subunits carry only the complementary component for the acetylcholine binding site, and must combine with α-subunits to form receptor binding sites (e.g., α-2 β-4). Differing subunit combinations result in differing binding characteristics across the family of nAChRs. The binding of acetylcholine produces conformational change in the receptor that opens the ion channel to allow for the movement of monovalent and divalent cations along their electrochemical gradients.[26]

The NH_2 hydrophilic extracellular portion of the α subunit bears not only the acetylcholine binding site, but also a newly recognized noncompetitive agonist site. The binding of compounds such as galanthamine to this noncompetitive agonist site induces a conformational change in the nAChR (allosteric modulation) that facilitates the conversion from the resting to the open channel state elicited by acetylcholine, and that which protects the nAChR from desensitization elicited by cholinergic agents.[27]

Because noncompetitive agonists such as galanthamine could enhance nicotinic neurotransmission in any hypofunctioning nicotinic receptors, they could correct nicotine-sensitive neurophysiologic abnormalities in patients with schizophrenia (without the problem of desensitization), and thereby remove a driving force for these patients' heavy smoking. Early studies from the Soviet Union suggest that these drugs are well tolerated, and, when given by themselves, improve negative psychopathology (lack of initiative, blunted emotionality).[28] Studies examining these agents as adjuncts to treatment with atypical antipsychotics are needed.

11.5 SUMMARY

There are compelling reasons to believe that abnormalities in nicotinic neurotransmission contribute to the pathophysiology of schizophrenia. Delineation and understanding of these abnormalities may lead to methods of prevention, and/or better treatments that will improve patients' function and allow them to avoid the destructive effects of smoking.

REFERENCES

1. Masterson, E. and O'Shea, B., Smoking and malignancy in schizophrenia, *Br. J. Psych.*, 145:429–432, 1984.
2. Hughes, J., Hatsukami, D., Mitchell, J., and Dahlgren, L., Prevalence of smoking among psychiatric outpatients, *Am. J. Psych.*, 143:993, 1986.
3. Goff, D., Hendersen, D., and Amico, E., Cigarette smoking in schizophrenia: Relationship to psychopathology and medication side effects, *Am. J. Psych.*, 149:1189–1194, 1992.
4. de Leon, J., Dadvand, M., Canuso, C., White, A., Stanilla, J.K., and Simpson, G., Schizophrenia and smoking: an epidemiological survey in a state hospital, *Am. J. Psych.*, 152;453–455, 1995.
5. Ziedonis, D., Koster, T., Glazer, W.M., and Frances, R.J., Nicotine dependence and schizophrenia, *Hosp. Community Psych.*, 87:204, 1994.
6. McEvoy, J.P. and Brown, S., Smoking in first-episode patients with schizophrenia, *Am. J. Psych.*, 156:1120–1121, 1999.
7. Olincy, A., Young, D.A., and Freedman, R., Increased levels of the nicotine metabolite cotinine in schizophrenic smokers compared to other smokers, *Biol. Psych.*, 45:1–5, 1997.
8. Marx, C.E., McIntosh, E., Wilson, W.H., and McEvoy, J.P., Mecamylamine increases smoking in psychiatric patients, *J. Clin. Psychopharmacol.*, 20:706–707, 2000.
9. Dalack, G.W., Healy, D.J., and Meador-Woodruff, J.H., Nicotine dependence in schizophrenia: clinical phenomena and laboratory findings, *Am. J. Psych.*, 155:1490–1501, 1998.
10. Hyman, S.E., Introduction to the complex genetics of mental disorders. *Biol. Psych.*, 45:518-521, 1999.
11. Freeman, R., Adler, L.E., and Leonard, S., Alternative phenotypes for the complex genetics of schizophrenia, *Biol. Psych.*, 45:551–558,1999.
12. Adler, L.E., Freedman, R., Ross, R.G., Olincy, A., and Waldo, M.C., Elementary phenotypes in the neurobiological and genetic study of schizophrenia, *Biol. Psych.*, 46:8–18, 1999.

13. Freedman, R., Adler, L., Bickford, P., Byerley, W., Coon, H., Cullum, C.M., Griffith, J.M., Harris, J.G., Leonard, S., Miller, C., Myles-Worsley, M., Nagatomo, H.T., Rose G., and Waldo M., 1994. Schizophrenia and nicotine receptors, *Harv. Rev. Psych.,* 2;179–192.

14. Jann, M.W., Saklad, S.R., Ereshefsky, L., Richards, A.L., Harrington, C.A., and Davis, C.M., Effects of smoking on haloperidol and reduced haloperidol plasma concentrations and haloperidol clearance, *Psychopharmacology,* 90:468, 1986.

15. McEvoy, J.P., Freudenreich, O. Levin, E.D., and Rose, J.E., haloperidol increases smoking patients with schizophrenia, *Psychopharmacology,* 119:124, 1995.

16. Levin, E.D., Wilson, W.H., Rose, J.E., and McEvoy, J.P., Nicotine-haloperidol interactions and cognitive performance in schizophrenia, *Neuropsychopharmacology,* 15:429–436, 1996.

17. Nagamoto, H.T., Adler, L.E., Hea, R.A., Griffith, J.M., McRae, K.A., and Freedman, R., Gating of auditory P50 in schizophrenics: Unique effects of clozapine, *Biol. Psych.,* 40:181, 1996.

18. McEvoy, J.P., Freudenreich, O., and Wilson, W.H., Smoking and therapeutic response to clozapine in patients with schizophrenia, *Biol. Psych.,* 46:125–129, 1999.

19. Light, G.A., Geyer, M.A., Clementz, B.A., Cadenhead, K.S., and Braff, D.L., Normal P50 suppression in schizophrenia patients treated with atypical antipsychotic medications, *Am. J. Psych.,* 157:767–771, 2000.

20. Adler, L.E., Hoffer, L.J., Griffith, J., Waldo, M.C., and Freedman, R., Normalization by nicotine of deficient auditory sensory gating in the relatives of schizophrenics, *Biol. Psych.,* 32:607–616, 1992.

21. Adler, L.E., Hoffer, L.J., Griffith, J., Waldo, M.C., and Freedman, R., Normalization of auditory physiology by cigarette smoking schizophrenic patients, *Biol. Psych.,* 150:1856–1861, 1993.

22. Addington, J., el-Guebaly, N., Campbell, W., Hodgins, D.C., and Addington, D., Smoking cessation treatment for patients with schizophrenia, *Am. J. Psych.,* 155:974–976, 1998.

23. Ziedonis, D.M. and George, T.P., Schizophrenia and nicotine use: report of a pilot smoking cessation program and review of neurobiological and clinical issues, *Schizophr. Bull.,* 23:247–254, 1997.

24. Hartman, N., Leong, G., Glyun, S., Wilkins, J., and Jarvik, M., transdermal nicotine and smoking behavior in psychiatric patients, *Am. J. Psych.,* 148:374–375, 1991.

25. Jenkusky, S.M., Use of nicotine patches for schizophrenia patients. *Am. J. Psych.,* 150:1899, 1993.

26. Arias, H.R., Localization of agonist and competitive antagonist binding sites on nicotinic acetylcholine receptors, *Neurochem. Int.,* 36:595–645, 2000.

27. Maelicke, A. and Albuquerque, E.X., Allosteric modulation of nicotinic acetylcholine receptors as a treatment strategy for Alzheimer's disease, *Eur. J. Pharmacol.,* 393:165–170, 2000.

28. Vovin, R.Y., Fakturovich, A.Y., Golenkov, A.V., and Lukin, V.O., Correction of apathoabulic manifestations of process defects with cholinotropic compounds Zh, *Nevropatol. Psikhiatr.,* 91:111–115, 1991.

12 Nicotinic Medications and Tourette's Disorder

R. Douglas Shytle, Archie A. Silver,
Mary Newman, Berney J. Wilkinson,
and Paul R. Sanberg

CONTENTS

12.1 INTRODUCTION

The intent of this chapter is to review the history of research investigating the use of nicotine as a therapeutic agent for the treatment of Tourette's disorder (TD). The first section begins with a brief clinical description of TD and the current pharmacological hypotheses regarding the cause of TD symptoms. This section is followed by a detailed review of animal models used to evaluate possible therapeutic treatment for TD, along with a discussion of the outcome of studies utilizing these animal models. Clinical studies that have examined the effectiveness of nicotine as a treatment of TD are then presented. The second section presents the hypothesis regarding nicotinic receptor inactivation as a mechanism for nicotine's therapeutic effects in TD. The third section introduces the hypothesis of the use of the nicotinic antagonist, mecamylamine, in the treatment of TD. This is followed by a review of current preclinical data and clinical evidence for the use of mecamylamine in the treatment of TD. The last section summarizes conclusions reached thus far and outlines future research directions.

12.2 TOURETTE'S SYNDROME, NICOTINE, AND ANIMAL MODELS

Tourette's disorder (TD) is a neuropsychiatric disorder with childhood onset that is characterized largely by the expression of sudden, rapid and brief, recurrent, non-rhythmic, stereotyped motor movements (motor tics) and sounds (vocal tics) that are experienced as irresistible, but can be suppressed for varying lengths of time.[1] These motoric symptoms range from relatively mild to very severe over the course of a patient's lifetime.[2,3] Most patients with TD also exhibit comorbid neuropsychiatric features including obsessive compulsive symptoms,[4] inattention, hyperactivity, impulsivity,[5,6] emotional liability, anxiety,[7,8] and associated visual-motor deficits.[9] Problems with extreme temper or aggressive behavior are also frequent,[10–12] as are school refusal and learning disabilities.[13,14] While the specific etiology of TD is currently unknown, some believe that the disorder is caused by pathophysiology of cortical-striato-thalamo-cortical circuits in the brain.[15] Several lines of evidence suggest abnormal hyperinnervation of striatal dopamine as a possible cause for the motor symptoms of TD:[16] therapeutic effectiveness of D2 receptor antagonists;[17] reduction of tics by agents that block dopamine synthesis or accumulation;[18] exacerbation of tics by agents, such as amphetamine, that increase central dopaminergic activity; appearance of tics after withdrawal from neuroleptic drugs; and elevated neuronal dopamine uptake sites (37% in caudate and 50% in putamen over controls) in postmortem examination of individuals who had TD.[19]

Historically, neuroleptics have been the first-line treatment for tic symptoms of TD. The neuroleptics, haloperidol (Haldol®) and pimozide (Orap®), are the only two medications currently approved by the FDA for treatment of TD. Although neuroleptics are effective in reducing frequency and severity of the motor and vocal tics,[17] they are not as effective in reducing the behavioral and emotional symptoms of TD. Moreover, neuroleptics often produce adverse side effects including cognitive dulling, sedation, weight gain, acute dystonic reactions, akathisia, parkinsonism, and with long-term treatment, tardive dyskinesia.[20] Clonidine, an α2 agonist, is currently the most widely used drug for treatment of TD, but its efficacy in the disorder remains to be fully established and many patients complain about the accompanying sedation.[21,22] Recent preliminary data suggest that another α2 agonist, guanfacine (Tenex®), may be beneficial in treatment of symptoms of attention-deficit hyperactivity disorder in children with TD while being less sedating than clonidine.[23] However, recent case reports suggest that guanfacine may have the undesirable side effect of precipitating mania in some patients.[24] Due to the side effects associated with current therapies, a number of alternative pharmacological approaches have been investigated in recent years.

12.2.1 FROM LAB TO CLINIC

The preclinical finding that nicotine potentiated the actions of neuroleptics in rats led to clinical studies involving the use of nicotine as an adjunct to neuroleptic treatment of TD.[25] Traditionally, neuroleptic-induced catalepsy in rodents has been used to model the extrapyramidal side effects of neuroleptics.[26] However, the same behavioral effects can be viewed as a therapeutic model to test the potential of drugs for treating hyperkinetic movement disorders. The potential of a drug to produce catalepsy in rodents is measured by the bar test.[27] After a given amount of time, depending on the test drug used, the front feet of the animal are placed on a horizontal bar that is the appropriate height from the bottom of the testing floor for the age of the animal. The degree of catalepsy is measured by how long it takes the animal to remove its front feet from the bar, with a ceiling on the total amount of time the animal is allowed to stay on the bar.

Original studies using this model found that nicotine administered either systemically or by intracaudate infusions augmented fluphenazine-induced catalepsy in rats.[28,29] Later studies also showed that nicotine administered intraperitoneally at 0.1 mg/kg potentiated catalepsy in rats induced by resperine, fluphenazine, and haloperidol. Further animal studies with four experimental conditions (haloperidol/nicotine, nicotine alone, haloperidol alone, and saline control) confirmed the finding that low a dose of nicotine (0.1 mg/kg) did potentiate haloperidol-induced catalepsy in rats, although, nicotine administered alone did not induce catalepsy.[30,31] Since haloperidol blocks both D1 and D2 in the striatum, a study was performed to distinguish whether nicotine's potentiating effect involved D1 or D2 receptors. A D1 antagonist, SCH23390, was used in conjunction with nicotine,[32] but failed to potentiate SCH23390-induced catalepsy. Therefore, the evidence favors the predominate involvement of D2 receptors in nicotine potentiation of neuroleptic-induced catalepsy. It was recently demonstrated that potentiation of haloperidol catalepsy could be achieved by microinjection of nicotine into the striatum or pons in rats.[33]

Tizabi et al. have characterized behavioral sensitization in neonates to the D2/3 receptor agonist, quinpirole (QNP), as an animal model for TD.[34] They found that the sensitized locomotor response to quinpirole in primed neonates is blocked by the acute administration of nicotine. More recent studies by Tizabi et al. have characterized DOI-induced head twitch as a model for testing nicotine's potential to attenuate tics. DOI or (1)-2,5-dimethoxy-4-iodophenyl)-2-aminopropane) is a 5-HT2 receptor agonist that, when administered to rodents, causes head twitches in mice and head and shoulder shakes that appear to resemble motor tics found in patients with TD.[34a] Tizabi et al.[34b] reported that acute nicotine (0.5 and 1.0 mg/kg), when administered 20 minutes prior to DOI (0.5 and 1.0 mg/kg), significantly attenuated DOI-induced head twitches in mice. Furthermore, when nicotine was given chronically (10 days at 1.5 mg/kg) it significantly attenuated DOI-induced head-twitches in mice (0.5 mg/kg). The authors suggested that nicotine might be of therapeutic use in attenuating tics caused by a serotonergic imbalance. The findings from these preclinical studies suggest that nicotine might attenuate motor tics in patients with TD.

12.2.2 CLINICAL STUDIES

Initially, clinical investigations consisted of open-label studies of nicotine gum in TD patients concurrently under treatment with haloperidol. Several open-trial studies conducted from 1987–1991 with nicotine gum and haloperidol reported a decrease in tic frequency and severity,[35–37] and improvements in concentration and attention as reported by parents.[35–37] In a subsequent trial,[38] similar results were reported, with nicotine gum plus haloperidol reducing both tic severity and frequency. Furthermore, nicotine gum alone only reduced tic frequency and placebo gum had no effect on tic frequency or severity.[38] Additional evidence for a therapeutic response to nicotine gum alone was reported in two case reports;[39,40] unfortunately, the effects of nicotine gum were short-lived, lasting 45 minutes to 1 hour after gum chewing. In addition, because of gastrointestinal side effects and the bitter taste of the gum, noncompliance was a significant limitation. Therefore, in an effort to increase compliance, research began with transdermal nicotine patches.

In a study examining the effects of transdermal nicotine patches (TNP) designed to deliver 7 mg of nicotine/24 hr (Nicoderm®), Silver et al. found 47% reduction in tic frequency and 34% reduction in tic severity following TNP application in 11 patients with TD who were not responding well to their neuroleptic treatment.[41] Surprisingly, in two of these patients, the effect of a single nicotine patch persisted for a variable length of time after patch removal. Similar long-term benefits of the TNP were also reported by Dursun et al.,[42–44] who found that applying 2 consecutive 10 mg TNPs, each TNP given for 24 hrs, reduced tic symptoms significantly for 4 but not 16 weeks after TNP removal.

Further evidence for a long-term therapeutic response to the TNP was found when TD patients were followed for various lengths of time following the application of TNP. A retrospective case study found that the application of a single TNP titrated to deliver 7 mg of nicotine in 24 hours resulted in a significant reduction of motor and vocal tics in 17 of 20 patients.[45,46] This reduction in tic symptoms persisted for a mean of 10 days after removal of a single patch applied for 24 hours. In these open trials, side effects included transient itching at the site of application, nausea, and occasional headache and sedation. However, there was no clinical evidence found for nicotine dependence with the TNP. Based on experience with these open trial studies, verification of these findings was sought in a prospective double-blind placebo-controlled trial.

12.2.3 Controlled Study of Transdermal Nicotine in Tourette's Disorder

It was reasoned that one advantage of transdermal nicotine as an adjunct to neuroleptic treatment could be a potential reduction of neuroleptic dose and the associated risks of short- and long-term adverse effects produced by neuroleptics. To test this hypothesis, a controlled trial to investigate the safety and effectiveness of transdermal nicotine as an adjunct to haloperidol for the treatment of TD was conducted. Seventy patients with TD were treated with either transdermal nicotine (7 mg/kg/24hr) or placebo patches for 19 days in a randomized, double-blind study.[47] Each patient received an individually based "optimal" dose of haloperidol for at least 2 weeks prior to randomization to nicotine or placebo treatment. The lowest dosage of nicotine available (7 mg/24hr) was employed and a placebo patch containing a small amount of nicotine, but with a barrier to minimize absorption significantly, was used to ensure that the patches were identical in both appearance and smell. A new patch was worn each day for the first 5 days, following which time the dose of haloperidol was reduced by 50%.

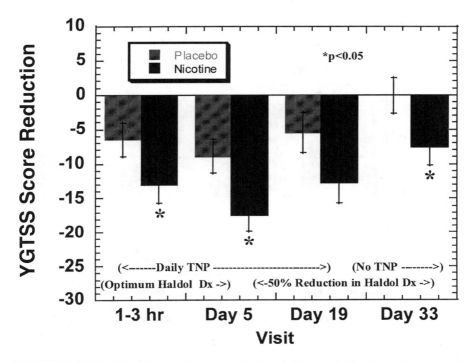

FIGURE 12.1 Yale global tic severity score reductions in Tourette's disorder patients treated with transdermal nicotine (n = 27) or placebo (n = 29).

Daily patch applications were then continued for an additional 2 weeks (day 19), at which time the patch was discontinued; the 50% dose of haloperidol continued for an additional 2 weeks (day 33). As documented by the clinician rated global improvement scale (Figure 12.1) and with the Yale global tic severity scale (Figure 12.2), transdermal nicotine was superior to placebo in reducing the symptoms of Tourette's disorder. The dropout rate was similar for both groups (23% for the mecamylamine group and 28% for the placebo group). Side effects, including nausea and vomiting, were significantly more common in the nicotine group (71% and 40%, respectively) than in the placebo group (17% and 9%, respectively).

These findings are consistent with previous preclinical studies[30,31,35] and, largely, open-label clinical studies[42–45,48,49] suggesting that nicotine potentiates the effects of neuroleptics in the treatment of TD. Moreover, this study substantiated the long-term therapeutic effect of TNP even after an extended period of discontinuation.

Consistent with the hypothesis that temporal absorption determines the addictive nature of nicotine,[50] no evidence was found for withdrawal symptoms following the discontinuation of transdermal nicotine therapy. Despite this desirable characteristic, the high rate of nicotine-related side effects in this study indicates a limitation to the widespread use of transdermal nicotine as a daily adjunctive therapy for the treatment of TD, particularly for children and adolescents.

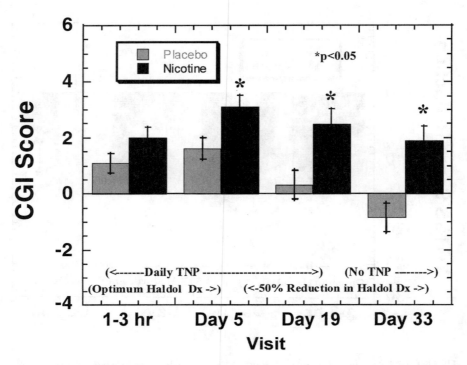

FIGURE 12.2 Clinical global improvement scores in Tourette's disorder patients treated with transdermal nicotine (n = 27) or placebo (n = 29).

12.3 NICOTINIC RECEPTOR INACTIVATION HYPOTHESIS

The rapid proliferation of recent attempts to exploit the therapeutic potential of nicotinic receptor modulation for a variety of neuropsychiatric disorders has resulted in the development and identification of several nicotinic receptor ligands with greater selectivity and substantially improved side-effect profiles.[51] These include novel nicotinic receptor agonists, which are now entering clinical research development for Alzheimer's[52] and Parkinson's diseases.[53]

What remains paradoxical is why nicotine would be therapeutic in certain hyper-dopaminergic disorders (e.g., Tourette's disorder) when nicotine activates the release of dopamine in the striatum, a neuropharmacological effect implicated in the drug's reinforcing properties.[54] The answer might be explained by nicotine's ability to act both as agonist and as antagonist at nicotinic receptors.[55] For example, nicotine acts initially as a rapid agonist at nAChRs, followed by a prolonged inactivation of these receptors shortly thereafter.[56] Hulihan et al.[57] proposed that the predominant effect of nicotine on many nAChR subtypes over time (its time-averaged effect) may be that of an antagonist. Recently, it was demonstrated that nicotine could both activate and inactivate midbrain dopamine neurons.[58]. Thus, nicotinic receptor "modulation" rather than just "receptor activation" may better characterize the pharmacological effects of nicotine. Rate-dependent effects of nicotine are consistent with this hypothesis and were recently reviewed by Perkins.[59]

Agonist effects of nicotine are evident in its ability to produce locomotor stimulant effects, presumably via activation of nicotinic receptors modulating mesolimbic dopamine pathways.[60] However, antagonistic effects of nicotine may explain its ability to potentiate the cataleptic effects of dopamine receptor antagonists.[46,61] While the nicotinic receptor antagonist, mecamylamine was found to attenuate this effect of nicotine partially, mecamylamine alone also potentiated neuroleptic-induced catalepsy.[28] The potentiation of neuroleptic-induced catalepsy by mecamylamine has also been replicated recently.[61]

12.3.1 FROM NICOTINE TO MECAMYLAMINE

Since mecamylamine also potentiates neuroleptic-induced catalepsy, it was reasoned that it may prove beneficial in humans with potentially fewer side effects than obtained with nicotine therapy.[62] Use of mecamylamine in doses ranging from 2.5 to 5 mg per day was initially examined in 13 TD patients (4 adults and 9 children) whose symptoms were poorly controlled by traditional pharmacological treatment.[62] During mecamylamine therapy, the patients continued their previous medication regimen (mostly neuroleptics or sertraline). The results for these patients and an additional 11 patients were reported as a retrospective case series.[63] As measured by the clinical global improvement scale, 22 of 24 patients improved during treatment, and many of those who responded reported improved mood, especially with regard to irritability and aggression. While placebo effects and spontaneous remission of symptoms could explain the reported improvement in this uncontrolled clinical experience, many patients continued to report improvement in symptoms while taking mecamylamine on a daily basis for more than 6 months.

As mentioned earlier, although tic behavior is the key diagnostic feature of TD, it is now recognized that the syndrome comprises several associated pathologies and can be heterogeneous with respect to both clinical presentation and neuropharmacological responses.[3] The effects of mecamylamine on mood and behavior in patients with TD were intriguing, in particular because of the evidence that cholinergic mechanisms may contribute to other neuropsychiatric symptoms, including depression.[64–68]

12.3.2 CONTROLLED STUDY OF MECAMYLAMINE IN TOURETTE'S DISORDER

In order to replicate the clinical experience with mecamylamine therapy under controlled conditions, a double-blind placebo-controlled study was carried out in children and adolescents with a primary diagnosis of TD.[69] While previous preclinical and clinical research suggested that the tic surpressing effects of nicotine and mecamylamine would require concurrent neuroleptic therapy, for several reasons a decision was made to design the initial controlled trial with mecamylamine as a monotherapy for TD. First, for safety reasons, because mecamylamine at high doses can produce significant side effects and had never been studied in a pediatric population before, it was important to establish the drug's safety profile in this population in the absence of concurrent neuroleptic therapy. Second, positive therapeutic response in some patients treated with mecamylamine alone suggested that it may work in the absence of a neuroleptic, particularly for associated behavioral and emotional symptoms. Since neuroleptics are becoming less popular as a first line treatment option for TD, it was important to determine if mecamylamine was efficacious as a monotherapy. Thus, two primary goals of the first controlled study were 1) to establish the mecamylamine's safety profile in this pediatric population and 2) to explore which symptoms of TD were most affected by mecamylamine monotherapy.

Few double-blind, placebo-controlled studies have adequately considered the behavioral and emotional symptoms (BESs) in children and adolescents with TD. However, because clinical experience suggested that mecamylamine may have a therapeutic effect on these symptoms, patients who suffered primarily from BESs as opposed to tics were selected for this study. However, a major obstacle to the initial design of the study was the lack of efficacy measures to address BESs of TD adequately. To resolve this problem, investigators have in the past used scales developed for comorbid disorders in an effort to measure the BESs.[70] However, because these scales do not include measures of tic symptoms, they cannot be used as primary efficacy measures in controlled medication trials seeking FDA approved indication for TD. Moreover, the practice of using different scales with different psychometric properties for such studies limits and further complicates interpretations regarding the relationship between various symptoms and their individual response to treatment. The more pragmatic approach was to design a scale covering both tics and the key BESs that cause the greatest impairment for the patient and his/her family. Therefore, a brief clinician and parent rating scale for this purpose — the Tourette Disorder Scale (TODS), a 15-item measure rating behavior, activity, attention, motor tics and vocal tics — was developed and rated. In a recent reliability and validity study, the TODS was found to have good inter-rater reliability, excellent internal consistency, and favorable levels of validity and sensitivity to change. Individual items showed good convergent and discriminant validity with existing measures available for assessment of TD illness severity.[71]

The final protocol design resulted in a multicenter, double-blind, placebo-controlled flexible-dose study assessing the safety and efficacy of mecamylamine for the treatment of TD. Eligible patients had to be between the ages of 8 and 17 and had to meet DSM-IV criteria for Tourette's disorder. In addition, the behavioral and emotional symptoms associated with Tourette's disorder had to be rated as more dominant than the tics in the presenting constellation of symptoms. All subjects had to be nonsmokers and all had to be free of psychotropic medication following appropriate washout. Following inclusion in the study, subjects were randomized to receive mecamylamine or placebo. Medication was supplied in capsules containing 2.5 mg mecamylamine or placebo; placebo and mecamylamine capsules were identical in appearance, weight, and taste. Dose was started at one capsule per day, increased to two capsules per day the second week and three per day the third week.

Subjects were evaluated at eight consecutive weekly visits to determine the efficacy and safety of the drug. Primary outcome measures were the clinician-rated Tourette's disorder scale and the 21-point clinician global improvement scale. Secondary outcome measures included the parent-rated Tourette's disorder scale (TODS-PR), the Yale global tic severity scale (YGTSS), the child and adolescent symptom inventory-4 (C/ASI-4), as well as a host of other secondary measures.

A total of 69 patients were screened and 61 enrolled in the study; 29 patients received mecamylamine and 32 received placebo. The results indicated that mecamylamine was well tolerated but no more effective than placebo in the treatment of tic symptoms using the doses employed in this study. This finding was not surprising considering that previous animal studies[28,61] and clinical experience suggested that mecamylamine may work primarily to augment neuroleptic therapy in reducing tic symptoms of TD.[62,72] Baseline assessments suggested that the mecamylamine group had more severe tic symptoms than the placebo group. This finding, coupled with the high degree of psychiatric comorbitity in this sample, makes demonstration of overall efficacy in this small sample particularly difficult. Nevertheless, exploratory post-hoc analyses revealed that patients with comorbid mood disorders may have benefited from mecamylamine monotherapy.[73] For example, an item analysis of the TODS-CR as a function of psychiatric comorbidity, revealed that patients diagnosed with major depression exhibited significant mecamylamine-related reductions in several symptoms, including irritability, sudden mood changes, difficulty paying attention, compulsions, anxiety, and depressed mood. Figures 12.3 and 12.4 illustrate the effects of mecamylamine and placebo on some of these symptoms in TD patients who were comorbid with major depression.

Of all the TODS-CR symptom items, "sudden mood changes" was most sensitive to mecamylamine related improvements. In patients with moderate to severe TD illness severity (baseline TODS-CR total score of >60), 47% of patients treated with mecamylamine (n = 17), but none of the patients treated with placebo (n = 15), had score reductions of 4 points or greater on the sudden mood changes item of the TODS-CR (data not shown). These controlled findings are consistent with recent clinical observations of mood stablizing properties of mecamylamine in two TD patients comorbid for bipolar disorder.[74]

Considering the lack of FDA approved safe and effective treatments for emotional liability in children and adolescents with neuropsychiatric disorders, future clinical studies are warranted to investigate further mecamylamine's potential mood stabilizing properties in more diverse pediatric populations.

12.4 CONCLUSIONS AND FUTURE DIRECTIONS

The available evidence thus far suggests that nicotinic receptor modulation is a reasonable therapeutic approach to the treatment of patients with TD. However, the tic reducing actions of nicotine, and most likely to mecamylamine, appear to require the concurrent use of a neuroleptic. Nevertheless, the potential mood stablizing effects of mecamylamine in the absence of nicotine-like side effects holds promise for other neuropsychiatric disorders involving dysregulation of mood.

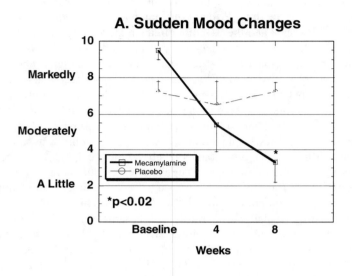

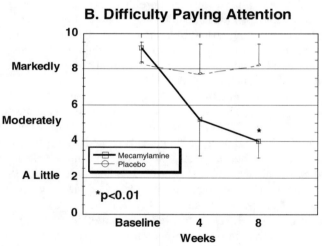

FIGURE 12.3 Reduction in (A) sudden mood changes and (B) difficulty paying attention (TODS-CR) in Tourette's disorder patients with comorbid major depression (DSM-IV) who were treated with mecamylamine (n = 4) and placebo (n = 4).

While preclinical studies have suggested that novel nicotinic agonists such as ABT-418 should have substantially improved side-effect profiles relative to nicotine, results from early phase I and phase II clinical trials indicate that these ligands still appear to have clinically significant nicotine-like side effects. For example, dizziness and nausea were reported in 41% and 16%, respectively, of adult ADHD patients treated with ABT-418 compared to a 3% incidence of each side effect when treated with placebo.[75] Similar adverse side effects were also reported recently

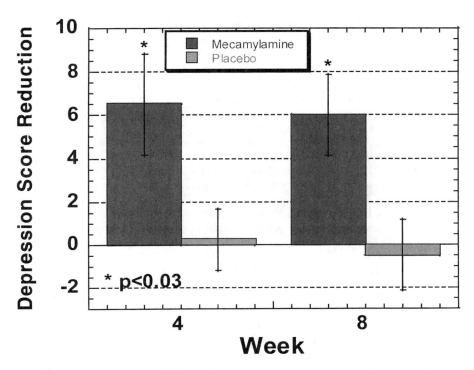

FIGURE 12.4 Improvement in depression symptom severity (C/ASI-4) in Tourette's disorder patients with comorbid major depression (DSM-IV) who were treated with mecamylamine (n = 4) and placebo (n = 4).

with SIB-1508Y, another novel nicotinic ligand in phase II trials for the treatment of Parkinson's disease.[76]

Since ABT-418, like nicotine, has mixed agonist/antagonist properties at different nicotinic receptor subtypes,[77] it is becoming less clear whether receptor activation or inactivation is responsible for the therapeutic and/or adverse side effects. For example, a recent study by the same group involved with many of the cognitive preclinical studies on ABT-418 found that low, but not high, doses of mecamylamine also improved executive cognitive function in aged primates.[78] These findings with low doses of mecamylamine replicate earlier work by Driscoll on cognition[79,80] as well as more recent studies on stress.[81] The biphasic effects of mecamylamine are important aspects for future studies, since mecamylamine may be therapeutic for some neuropsychiatric disorders at low doses, with potentially fewer side effects than that obtained with nicotinic agonist therapy.

The preclinical and clinical studies of nicotine and mecamylamine raise questions about the role of the acetylcholine system in a range of neuropsychiatric disorders, including developmental disorders and mood disorders. Growing evidence suggests that nicotine has the ability both to stimulate and inhibit nicotinic receptors, with each nicotinic receptor subunit having different sensitivities to these two processes. Further research is needed to investigate whether or not nicotinic receptor "modulation" may better characterize the pharmacological effects of nicotinic drugs.

REFERENCES

1. DSM-IV, Diagnostic and Statistical Manual of Mental Disorders, 4th ed., Washington, D.C., American Psychiatric Association, 1994.
2. Bruun, R.D. and C.L. Budman, The course and prognosis of Tourette syndrome, *Neurol. Clin.,* 15, (2) 291–8, 1997.
3. Robertson, M.M., Tourette syndrome, associated conditions and the complexities of treatment, *Brain,* 123, 425–62, 2000.
4. Pauls, D.L., et al., Gilles de la Tourette's syndrome and obsessive-compulsive disorder. Evidence supporting a genetic relationship, *Arch. Gen. Psych.,* 43, (12) 1180–2, 1986.
5. Comings, D.E. and B.G. Comings, Tourette's syndrome and attention deficit disorder with hyperactivity: are they genetically related?, *J. Am. Acad. Child Psych.,* 23, (2) 138–46, 1984.
6. Comings, D.E. and B.G. Comings, A controlled family history study of Tourette's syndrome, I: Attention-deficit hyperactivity disorder and learning disorders, *J. Clin. Psych.,* 51, (7) 275–80, 1990.
7. Coffey, B.J. and K.S. Park, Behavioral and emotional aspects of Tourette syndrome, *Neurol. Clin.,* 15, (2) 277–89, 1997.
8. Coffey, B.J., et al., Distinguishing illness severity from tic severity in children and adolescents with Tourette's disorder, *J. Am. Acad. Child Adolesc. Psych.,* 39, 5, 556–61, 2000.
9. Silver, A.A. and R.A. Hagin, Gilles de la Tourette's syndrome, in *Disorders of Learning in Childhood,* John Wiley & Sons, New York, 469–508, 1990.
10. Budman, C.L., et al., Rage attacks in children and adolescents with Tourette's disorder: a pilot study, *J. Clin. Psych.,* 59, 11, 576–80, 1998.
11. Riddle, M.A., et al., Eds. Behavioral symptoms in Tourette's syndrome, *Tourette's Syndrome and Tic Disorders: Clinical Understanding and Treatment,* Cohen, D.J., Bruun, R.D., and Leckman, J.F., Eds., John Wiley & Sons, New York, 1988, 151–162.
12. Stefl, M.E., et al., A survey of Tourette syndrome patients and their families: the 1987 Ohio tourette survey, Ed., Cincinnati, Ohio Tourette Syndrome Association, 1988.
13. Harris, D. and A.A. Silver, Tourette's syndrome and learning disorders, *Learning Disabil.,* 6, (1) 1–7, 1995.
14. Matthews, W.S., Attention deficits and learning disabilities in children with Tourette's syndrome, *Psych. Ann.,* 18, 414–416, 1988.
15. Leckman, J.F., et al., Pathogenesis of Tourette's syndrome, *J. Child Psychol. Psych.,* 38, (1) 119–42, 1997.
16. Singer, H.S., et al., Dopaminergic dsyfunction in Tourette syndrome, *Ann. Neurol.,* 12, (4) 361–6, 1982.
17. Shapiro, E.S., et al., Controlled study of haloperidol, pimozide, and placebo for the treatment of Gilles de la Tourette's syndrome, *Arch. Gen. Psych.,* 46, 722–730, 1989.
18. Jankovic, J. and J. Beach, Long-term effects of tetrabenazine in hyperkinetic movement disorders, *Neurology,* 48, (2) 358–62, 1997.
19. Singer, H.S., et al., Abnormal dopamine uptake sites in postmortem striatum from patients with Tourette's syndrome, *Ann. Neurol.,* 30, (4) 558–62, 1991.

20. Silva, R.R., et al., Causes of haloperidol discontinuation in patients with Tourette's disorder: management and alternatives, *J. Clin. Psych.*, 57, (3) 129–35, 1996.

21. Goetz, C.G., Clonidine and clonazepam in Tourette syndrome, *Adv. Neurol.*, 58, 245–51, 1992.

22. Leckman, J.F., et al., Clondine treatment of Gilles de la Tourette's syndrome, *Arch. Gen. Psych.*, 48, 324–328, 1991.

23. Chappell, P.B., et al., Guanfacine treatment of comorbid attention-deficit hyperactivity disorder and Tourette's syndrome: preliminary clinical experience, *J. Am. Acad. Child Adolesc. Psych.*, 34, (9) 1140–6, 1995.

24. Horrigan, J.P. and L.J. Barnhill, Guanfacine and secondary mania in children, *J. Affect. Disord.*, 54, (3) 309–14, 1999.

25. Sanberg, P.R., et al., Nicotine for the treatment of Tourette's syndrome, *Pharmacol. Ther.*, 74, (1) 21–5, 1997.

26. Sanberg, P.R., Haloperidol-induced catalepsy is mediated by post-synaptic dopamine receptors, *Nature*, 284, 472–473, 1980.

27. Sanberg, P., et al., The catalepsy test: is a standardized method possible?, in *Motor Activity and Movement Disorders: Research Issues and Applications*, Sanberg, P., Ossenkopp, K., and Kavaliers, M., Eds., Humana Press, NJ, 197–211, 1995.

28. Moss, D.E., et al., Evidence for the nicotinic cholinergic hypothesis of cannabinoid action within the central nervous system: exrapyramidal motor behaviors, in *Marijuana: An International Research Report*, Chesher, G., Consroe, P., and Musty, R., Eds., Australian Government Printing Service, Canaberra, Australia, 359–364, 1988.

29. Moss, D.E., et al., Nicotine and cannabinoids as adjuncts to neuroleptics in the treatment of Tourette syndrome and other motor disorders, *Life Sci.*, 44, (21) 1521–5, 1989.

30. Emerich, D.F., et al., Nicotine potentiates the behavioral effects of haloperidol, *Psychopharmacol. Bull.*, 27, (3) 385–90, 1991.

31. Emerich, D.F., et al., Nicotine potentiates haloperidol-induced catalepsy and loco-motor hypoactivity, *Pharmacol, Biochem, Behav,* 38, (4) 875–80, 1991.

32. Emerich, D., et al., Differential effect of nicotine on D_1 vs D_2 antagonist-induced catalepsy, *Soc. Neurosci. Abst., 16, 247, 1990.*

33. Elazar, Z. and M. Paz, Potentiation of haloperidol catalepsy by microinjections of nicotine into the striatum or pons in rats, *Life Sci.*, 64, (13) 1117–25, 1999.

34. Tizabi, Y., et al., Nicotine blocks quinpirole-induced behavior in rats: psychiatric implications, *Psychopharmacol.* (Berl), 145, (4) 433–41, 1999.
    35.Sanberg, P.R., et al., Nicotine potentiates the effects of haloperidol in animals and in patients with Tourette syndrome, *Biomed. Pharmacother.*, 43, (1) 19–23, 1989.

36. Sanberg, P.R., et al., Nicotine gum and haloperidol in Tourette's syndrome [letter], *Lancet,* 1, 8585, 592, 1988.

37. McConville, B.J., et al., Nicotine potentiation of haloperidol in reducing tic frequency in Tourette's disorder [published erratum appears in *Am. J. Psych.,* Sep;148(9):1282, 1991] [see comments], *Am. J. Psych.*, 148, (6) 793–4, 1991.

38. McConville, B.J., et al., The effects of nicotine plus haloperidol compared to nicotine only and placebo nicotine only in reducing tic severity and frequency in Tourette's disorder, *Biol. Psych.*, 31, (8) 832–40, 1992.

39. Devor, E.J. and K.E. Isenberg, Nicotine and Tourette's syndrome [letter], *Lancet,* 2, 8670, 1046, 1989.

40. Dimitsopulos, T. and R. Kurlan, Tourette's syndrome and nicotine withdrawal [letter], *J. Neuropsych. Clin. Neurosci.*, 5, (1) 108–9, 1993.

41. Silver, A.A., et al., Transdermal nicotine in Tourette's syndrome, in *The Effects of Nicotine on Biological Systems,* Clarke, P.B.S., Quik, M., and Thurau, K., Eds., Birkhauser Publishers, Boston, 293–299, 1995

42. Dursun, S.M. and M.A. Reveley, Differential effects of transdermal nicotine on microstructured analyses of tics in Tourette's syndrome: an open study, *Psychol. Med.,* 27, (2) 483–7, 1997.

43. Dursun, S.M., et al., Longlasting improvement of Tourette's syndrome with transdermal nicotine [letter], *Lancet,* 344, 8936, 1577, 1994.

44. Dursun, S.M., et al., Differential effects of transdermal nicotine patch on the symptoms of Tourette's syndrome., *Br. J. Clin. Pharmacol.,* 39, (1) 100P–101P, 1995.

45. Silver, A.A., et al., Case study: long-term potentiation of neuroleptics with transdermal nicotine in Tourette's syndrome, *J. Am. Acad. Child Adolesc. Psych.,* 35, (12) 1631–6, 1996.

46. Shytle, R.D., et al., Transdermal nicotine for Tourette's syndrome, *Drug Develop. Res.,* 38, (3/4), 290–298, 1996.

47. Silver, A.A., et al., Transdermal Nicotine and Haloperidol in Tourette syndrome: a double-blind placebo controlled study, *J. Clin. Psych.,* in press, 2000.

48. Silver, A.A. and P.R. Sanberg, Transdermal nicotine patch and potentiation of haloperidol in Tourette's syndrome [letter], *Lancet,* 342, 8864, 182, 1993.

49. Silver, A.A., et al., Clinical experience with transdermal nicotine patch in Tourette's syndrome, *CNS Spectrums,* 4, (2) 68–76, 1999.

50. Shytle, R.D., et al., Nicotine, tobacco and addiction [letter], *Nature,* 384, 6604, 18–9, 1996.

51. Brioni, J.D., et al., The pharmacology of (-)-nicotine and novel cholinergic channel modulators, *Adv. Pharmacol.,* 37, 153–214, 1997.

52. Lippiello, P.M., et al., RJR–2403: a nicotinic agonist with CNS selectivity II. *In vivo* characterization, *J. Pharmacol. Exp. Ther.,* 279, (3) 1422–9, 1996.

53. Sacaan, A.I., et al., Pharmacological characterization of SIB-1765F: a novel cholinergic ion channel agonist, *J. Pharmacol. Exp. Ther.,* 280, (1) 373–83, 1997.

54. Balfour, D.J., et al., Sensitization of the mesoaccumbens dopamine response to nicotine, *Pharmacol. Biochem. Behav.,* 59, (4) 1021–30, 1998.

55. Dani, J.A. and S. Heinemann, Molecular and cellular aspects of nicotine abuse, *Neuron,* 16, (5) 905–8, 1996.

56. Lukas, R.J., et al., Regulation by nicotine of its own receptors, *Drug Devel. Res.,* 38, 136–148, 1996.

57. Hulihan-Giblin, B.A., et al., Acute effects of nicotine on prolactin release in the rat: Agonist and antagonist effects of a single injection of nicotine, *J. Pharmacol. Exp. Ther.,* 252, (1) 15–20, 1990.

58. Pidoplichko, V.I., et al., Nicotine activates and desensitizes midbrain dopamine neurons, *Nature,* 390, 6658, 401–4, 1997.

59. Perkins, K.A., Baseline-dependency of nicotine effects: a review [see comments], *Behav. Pharmacol.,* 10, (6/7), 597–615, 1999.

60. Corrigall, W.A., et al., The mesolimbic dopaminergic system is implicated in the reinforcing effects of nicotine, *Psychopharmacology,* 107, (2/3), 285–9, 1992.

61. Levin, E.D. and P. Lippiello, Mutually potentiating effects of mecamylamine and haloperidol in producing catalepsy in rats, *Drug Devel. Res.,* 47, 90–96, 1999.

62. Sanberg, P.R., et al., Treatment of Tourette's syndrome with mecamylamine [letter], *Lancet,* 352, 9129, 705–6, 1998.

63. Silver, A.A., et al., Mecamylamine in Tourette's syndrome: a two-year retrospective case, *J. Child Adolesc. Psychopharmacol.,* 10, 2, 59–68, 2000.

64. Sandyk, R., Cholinergic mechanisms in Gilles de la Tourette's syndrome, *Int. J. Neurosci.*, 81, (1/2), 95–100, 1995.

65. Janowsky, D.S., et al., Acetylcholine and depression, *Psychosom. Med.*, 36, (3) 248–57, 1974.

66. Janowsky, D.S., et al., Adrenergic-cholinergic balance and the treatment of affective disorders, *Prog. Neuropsychopharmacol. Biol. Psych.*, 7, (2/3) 297–307, 1983.

67. Janowsky, D.S. and S.C. Risch, Cholinomimetic and anticholinergic drugs used to investigate and acetylcholine hypothesis of affective disorders and stress, *Drug Develop. Res.*, 4, 125–142, 1984.

68. Janowsky, D.S., et al., Is cholinergic sensitivity a genetic marker for the affective disorders?, *Am. J. Med. Genet.*, 54, (4) 335–44, 1994.

69. Silver, A.A., et al., A multicenter double-blind placebo-controlled safety and efficacy study of mecamylamine (Inversine) monotherapy for Tourette disorder, *J. Amer. Acad. Child Adolesc. Psych.*, in press, 2000.

70. Shytle, R.D., et al., Clinical assessment of motor abnormalities in Tourette's syndrome, in *Motor Activity and Movement Disorders: Research Issues and Applications,* Sanberg, P.R., Ossenkopp, K.P., and Kavaliers, M., Eds., Humana Press, NJ, 1995.

71. Shytle, R.D., et al., The Tourette disorder scale (TODS): development, reliability, and validity, *Assessment*, 2001.

72. Silver, A.A., et al., Clinical experience with mecamylamine in Tourette syndrome, *J. Child Adolesc. Psychopharmacol.*, 10, 2, 59–68, 2000.

73. Shytle, R.D., et al., Neuronal nicotinic receptor inhibition for treating mood disorders: preliminary evidence with mecamylamine, *Lancet,* submitted, 2001.

74. Shytle, R.D., et al., Comorbid bipolar disorder in Tourette syndrome responds to nicotinic receptor antagonist, mecamylamine (Inversine®), *Biol. Psych.*, 48:1028–1031.

75. Wilens, T., et al., A controlled trial of ABT-418 for attention deficit hyperactivity disorder in adults, *Am. J. Psych.*, 56, (12) 1931–1937, 1998.

76. McClure, D.E. The potential therapeutic usefulness of subtype selective neuronal nAChR agonists for the motor, cognitive and disease progression components of Parkinson's disease. in *IBC's 2nd Int. Symp. Nicotinic Acetylcholine Receptors, Adv. Mol. Pharmacol. Drug Devel.*, 1999, Annapolis, MD.

77. Papke, R.L., et al., Activation and inhibition of rat neuronal nicotinic receptors by ABT– 418, *Br. J. Pharmacol.*, 120, 3, 429–38, 1997.

78. Terry, A.V., et al., Dose-specific improvements in memory-related performance by rats and aged monkeys administered the nicotinic-cholinergic antagonist mecamylamine, *Drug Develop. Res.*, 47, 127–136, 1999.

79. Driscoll, P. and K. Battig, Cigarette smoke and behavior: some recent developments, *Rev. Environ. Hlth.*, 1, 113–133, 1973.

80. Driscoll, P., Nicotine-like behavioral effect after small dose of mecamylamine in Roman high-avoidance rats, *Psychopharmacologia*, 46, 119–121, 1976.

81. Newman, M.B., et al., Corticosterone-Attenutating and anxiolytic properties of mecamylamine in the rat, *Neuro-Psychopharmacol. Biol. Psych.*, 25:609–620, 2001.

13 Nicotine Effects on Attention Deficit Hyperactivity Disorder

Edward D. Levin

CONTENTS

13.1 INTRODUCTION

Nicotine administered by smoking and by transdermal patches has been shown to improve attention in dependent smokers, nondependent smokers, and nonsmokers significantly. Attentional improvements are also seen in patients with attention deficits including those with Alzheimer's disease, schizophrenia, and attention deficit hyperactivity disorder (ADHD). Nicotine delivered via means such as nicotine skin patches and novel nicotinic agonists holds promise for providing therapeutic treatment for attentional deficits.

Nicotine transdermal patches were originally developed by Rose and associates[36–39] to help with smoking cessation. In fact, nicotine transdermal patches have proven to be useful in helping people to quit smoking[1,14–16,21,27,37,48] and they have become one of the standard treatments for smoking cessation in clinical practice. These patches may be useful for other types of therapeutic treatment as well. As reviewed in other chapters of this book, nicotine treatment shows promise for therapeutic treatment for Alzheimer's disease, schizophrenia, and Tourette's syndrome. The use of nicotine transdermal patches is clearly superior to nicotine from tobacco because the lack the variety of carcinogenic compounds and other toxins found in tobacco smoke. Also, the abuse liability of nicotine transdermal patches appears to be low because they deliver nicotine in a slow chronic fashion.[32] Novel nicotinic ligands under development in a number of pharmaceutical companies and academic labs may further reduce the adverse effects of nicotinic therapy.

Nicotine has been reported by cigarette smokers to improve attentiveness;[29,47] however, since smoking withdrawal causes cognitive impairment,[18] there is concern about the degree to which nicotine–induced attentional improvements may merely be an attenuation of a nicotine-withdrawal induced attentional impairment. Importantly, this issue has been resolved: nicotine has been found to improve attentiveness in nondeprived smokers.[45] It has been shown that nicotine transdermal patches significantly improve choice accuracy and reduce response speed variability in normal nonsmoking subjects.[23] This provides confirmation that nicotine has attentional improving effects apart from alleviation of nicotine withdrawal symptoms.

Nicotinic treatment may be useful for treatment of ADHD. As described later, acute and chronic studies to determine the effect of nicotine transdermal treatment in adults with ADHD have been concluded. In addition, Wilens and co-workers have determined the effect of the nicotinic agonist ABT-418 in the same population.[49]

The possibility of nicotinic treatment for ADHD arises from three lines of evidence: (1) attentional improving effects in smokers and nonsmokers administered nicotine; (2) the effect of nicotine stimulating the release of dopamine,[50] — a primary mechanism of action of stimulants, such as methylphenidate and amphetamine, currently used to treat ADHD;[20] and (3) the increased smoking in adults with ADHD — about twice the societal background rate.[33] These adults with ADHD may be self-medicating with nicotine to attenuate their symptoms of ADHD. Nicotine or other nicotinic ligands may provide a promising new avenue for the treatment of ADHD, as suggested by acute and chronic studies of nicotine skin patch treatment in adults with this disorder.[7,25,26]

13.2 DRUG ADMINISTRATION

In the acute study,[7,26] smokers were given a 21 mg/day nicotine patch (Nicoderm®) after overnight abstinence, verified by endtidal CO measurements. This is the dose recommended for average cigarette smokers attempting to quit smoking. Nonsmokers were given a 7 mg/day nicotine patch (Nicoderm®) — the lowest dose of Nicoderm® available. This dose was chosen to minimize the side effects of nausea and dizziness in nonsmokers who were not tolerant of these effects of nicotine. Treatments were given in a counterbalanced order with a placebo patch in a double-blind fashion, whereby neither the subject nor the researcher was informed as to the treatment condition. One nonsmoker who had nausea and vomited while on the nicotine patch volunteered to be retested and showed this effect again. Both times the patch was removed as soon as this occurred and the nausea passed quickly. Normally the patch was removed after the final test at 1:15 p.m. The therapeutic outcome measures, starting with the profile of mood state (POMS) and ending with the interval timing test, were conducted between 1.5 and 4.5 hours after patch administration. The Nicoderm® patch was selected because it provides a rapid onset of nicotine delivery after initial administration. A pharmacokinetic study has shown that, with the 21-mg/day Nicoderm® patch, plasma nicotine levels rise to about 12 ng/ml 2 hours after application and to about 14 ng/ml 4 hours after application.[3] This is near the range of eventual steady-state nicotine levels delivered throughout the day with this patch.

In the chronic study,[25] nicotine and methylphenidate treatment was given for 4 weeks. Each subject received only one of four possible treatment combinations: control, nicotine only, methylphenidate only, or nicotine plus methylphenidate. Nicotine was administered via transdermal patches (Nicotrol®, Upjohn and Pharmacia, Helsingborg, Sweden); the 16-hour patch was used to avoid sleep disturbances. The nicotine dose was ramped up to help avoid side effects of nausea and dizziness. The nicotine dosing regimen followed was week 1: 5 mg/day; weeks 2 to 3: 10 mg/day; week 4: 5 mg/day.

The 5-mg/day patch is the lowest dose of Nicotrol® available and was chosen to minimize the side effects of nausea and dizziness in subjects who, as nonsmokers, were not initially tolerant of these effects. Methylphenidate (Ritalin®) was administered by pill PO (20 mg/day, Ritalin-SR 20®), a formulation shown to provide effective treatment of ADHD in children.[30,31] The patches and pills were administered each morning of the study. Treatments were given in a randomized, double-blind fashion with a placebo patch and pill in which neither subjects nor researchers working with them were informed as to the treatment condition.

13.3 CLINICAL ASSESSMENT

The positive diagnosis of ADHD was made by an experienced clinical psychologist using standard criteria for ADHD in adults as described in the DSM-IV manual.[10] Other useful measures for characterizing the presence of adult ADHD are the Wender Utah rating scale,[46] the Conners/Wells adolescent and adult self-report (CASS),[8] and the Barkley's adult ADHD semi-structured interview.[2] The benchmark clinical outcome measure of ADHD symptoms is the clinical global impression (CGI) scale, which is scored through a structured interview by a trained clinician. In research studies it is essential that the interviewing clinician be blind to treatment condition. The CGI is a standard seven-point scale widely used in clinical studies in which higher scores indicate more severe clinical states.[28] On the CGI scale are 1 = normal, 2 = borderline, 3 = mildly ill, 4 = moderately ill, 5 = markedly ill, 6 = severely ill, and 7 = extremely ill. It was found that the CGI was significantly reduced by acute nicotine treatment.[7,25,26]

Acute nicotine treatment[7,26] significantly diminished clinical symptoms of ADHD as assessed by CGI (see Figure 13.1). Importantly, this improvement is seen even when only nonsmokers are considered; thus, it does not seem to be merely the effect of nicotine skin patches attenuating the effects of smoking withdrawal. This was an acute study with nicotine administered over a 4.5-hour session; in the chronic nicotine study,[25] the CGI also improved, but improvement was only significant during the acute phase of treatment.

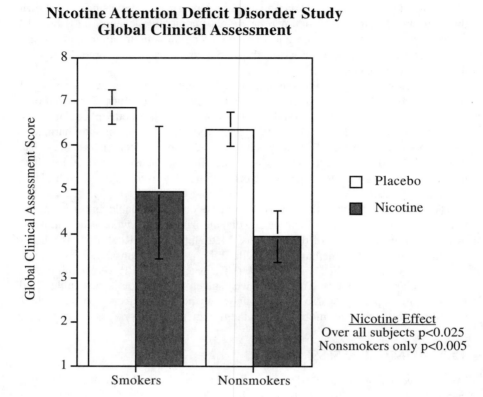

FIGURE 13.1 Nicotine skin patch effects on clinical symptoms in adults with attention deficit hyperactivity disorder.[7,26] The subjects' ADHD symptoms were rated with a modified clinical global impression (CGI) scale measuring symptom severity, treatment efficacy, and improvement. This is a standardized scale widely used in clinical studies. The CGI was completed following a brief interview regarding ADHD symptoms and drug side effects during that morning's session. The rater was blind to nicotine treatment.

13.4 SUBJECTIVE ASSESSMENT

Individual questions from the POMS battery were selected on the basis of known or hypothesized effects of nicotine on affective and cognitive state. The six questions from the POMS selected were those asking about: tension, fatigue, vigor, depression, anger, and difficulty concentrating. These items were selected on the basis of previous results showing nicotine-induced effects on these psychological domains in adults with ADHD[7,26] and hypotheses stemming from known effects of nicotine or tobacco smoking.[9,19,44]

In the acute study, smokers but not nonsmokers, reported significant self-perceived reductions in difficulty concentrating on the POMS scale. Both smokers and nonsmokers reported significant nicotine induced increases in self-perceived vigor on the POMS test.[26] In the chronic study,[25] nicotine treatment significantly improved the self-report of depressed mood during both acute and chronic phases of treatment (see Figure 13.2).

13.5 COMPUTERIZED ASSESSMENT

The principal computerized test found useful for determining nicotine effects on cognitive function in adults with ADHD is the Conners continuous performance test (CPT).[5,6] To characterize possible effects on other aspects of neurobehavioral function, components of the Automated Neuropsychological Assessment Metrics (ANAM) battery have also been used: simple reaction time, spatial mental rotation, and delayed matching to sample.[34,35]

The Conners CPT has been validated as an assessment tool for diagnosing ADHD and is sensitive to stimulant therapy;[5,6] it has previously been shown to be sensitive to the effects of acute nicotine treatment in adults with ADHD.[7,26] This is a 14-minute test in which the subject is instructed to respond as quickly as possible to a target stimulus, but to refrain from responding to a more rarely occurring nontarget stimulus. This differentiates the Conners CPT from other CPTs in which the subject must respond to rarely occurring stimuli and makes it sensitive to problems individuals may have in withholding inappropriate responses. Reaction time and variability in reaction time were measured over trial blocks during the course of the session and over different interstimulus intervals (ISI) of 1, 2, and 4 seconds. On the Conners CPT, response time variability typically increases over the course of the session and with longer ISIs. The hypothesis was that drug treatments would attenuate this degradation of performance. Errors of omission and commission were assessed, as was the composite measure of attentiveness (d'), which includes a weighted formula of response scores found to be sensitive to ADHD in adults[12] and the positive effects of therapeutic treatments including nicotine.[7,25,26]

In the acute study,[26] nicotine-induced improvement in attentiveness was discernible in terms of performance consistency on the Conners CPT. Smokers showed a significant reduction invariability of reaction time over blocks of the test session, while nonsmokers had a nearly significant reduction in variability of reaction time across different interstimulus intervals. In the chronic study,[25] there was a robust (p<0.05) nicotine-induced attenuation of the rise in CPT hit-reaction-time standard error (SE) over blocks of the session. This measure provides an index of the shift in response time variability over the course of the session. The untreated control group showed a consistent rise in variability of response time over the course of the session. Significant nicotine-induced improvements were seen during the acute (p<0.05) and chronic (p<0.025) phases of treatment. The nicotine effects on the CPT appeared to be relatively specific inasmuch as no significant effects were seen on the ANAM tests of simple reaction time, spatial mental rotation, and delayed matching to sample.

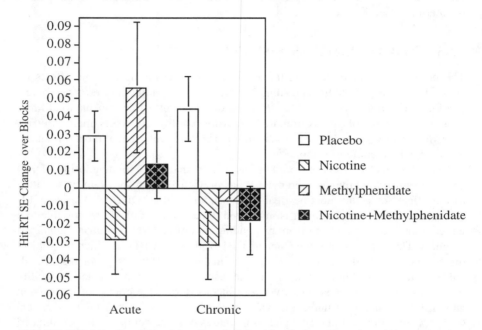

Nicotine Skin Patch and Methylphenidate Effects on Conners CPT Hit Reaction Time Standard Error Change over Session Blocks

FIGURE 13.2 Nicotine skin patch and methylphenidate-induced lowering of hit reaction time standard error over blocks of a session in the Connors CPT (mean ± SEM). Group sizes: control N = 7; nicotine N = 9; methylphenidate N = 9; and nicotine + methylphenidate N = 9.

13.6 CONCLUSIONS

Nicotinic treatment holds promise for treating attentional deficits seen in other disorders as well. In Alzheimer's disease patients, attentional performance has been found to be significantly improved by nicotine.[22,40,41] Recently, it was found that chronic nicotine patches with short-term administration — after 4 weeks of nicotine administration — nicotine caused a significant reduction in errors of omission in the Conners CPT, an attentional task.[25] Supporting a nicotine-induced improvement in attention, this reduction in errors of omission was not at the expense of increased errors of commission. Thus the nicotine effect did not seem to be a mere increase in response rate; rather, a true increase in response accuracy seemed to be caused by nicotine treatment. The consistency of attention seemed to be improved by nicotine as well. In Alzheimer's patients, nicotine caused a significant reduction in variability of response speed on the CPT.

Nicotine-induced attentional improvement has been found in schizophrenics.[24] Regardless of the haloperidol dose, nicotine caused a significant dose-related reduction in variability in response time on the Conners CPT. This contrasted with nicotine-induced improvements in short-term memory and mental processing speed, which were only seen as attenuation of deficits caused by moderate or high doses of haloperidol.

Nicotinic agonist-induced improvement in attention has been less evident in rat models than in humans. Sarter and co-workers did not find attentional improvements with either nicotine[43] or the nicotinic agonist ABT-418,[42] but did find an attentional deficit with the nicotinic antagonist mecamylamine. Bushnell also did not find any evidence of nicotine-induced improvement.[4] In contrast, nicotine-induced attentional improvements in aged rats has been found recently, using a two-choice discrimination task.[17] In this task, correct responses were signaled by the position of a light; signal duration was individually titrated so that rats were performing with a baseline accuracy between 75 to 87% correct (mean light duration = 1.8 sec). Similarly, nicotine was found to enhance performance of a five-choice "tracking" task in which lights signaled the correct choice among five possibilities.[13] It appears, then, that some of the problems in demonstrating a nicotine-induced attentional improvement in rats may be because of the structure of the task. Nicotine has been shown to increase adverse effects of proactive interference.[11] With repeated trials and alternative response sites in the tasks typically used for assessing attention in rats, nicotine's effects on proactive interference may overshadow its effects on attention.

There is clear evidence showing the attentional improvements produced by nicotine in adults with ADHD. Similar effects were seen in normal nonsmoking subjects as well as those with attentional deficits associated with Alzheimer's disease and schizophrenia. Nicotine administered via a skin patch appears to be safer than smoking inasmuch as it does not contain the components of tar and the gas phase of a cigarette. The dependence liability for the nicotine skin patch appears to be lower than for smoking. Since nicotine skin patches are currently only approved for use to aid in smoking cessation, additional evaluation is needed to demonstrate safety and efficacy for other indications, including attentional impairment. The research thus far is promising: nicotine skin patches may prove to be useful as a treatment for attentional deficits. Other novel nicotinic agonists in development may be even better, with lower side-effect profiles.

REFERENCES

1. Abelin, T., Buehler, A., Muller, P., Vesanen, K., and Imhof, P.R., Controlled trial of transdermal nicotine patch in tobacco withdrawal, *Lancet*, 7–10, 1989.
2. Barkley, R.A., Murphy, K., and Kwasnik, D., Psychological adjustment and adaptive impairments in young adults with ADHD, *J. Atten. Disord.*, 1, 41–54, 1996.
3. Benowitz, N.L. and Jacob, P., Nicotine and cotinine elimination pharmacokinetics in smokers and nonsmokers, *Clin. Pharmacol. Ther.*, 53, 316–323, 1993.
4. Bushnell, P.J., Oshiro, W.M., and Padnos, B.K., Detection of visual signals by rats: effects of chlordiazepoxide and cholinergic and adrenergic drugs on sustained attention, *Psychopharmacology*, 134, 230–241, 1997.
5. Conners, C.K., The continuous performance test (CPT): Use as a diagnostic tool and measure of treatment outcome. *Ann. Conven. Am. Psychol. Assoc.*, Los Angeles, CA, 1994.

6. Conners, C.K., The Continuous Performance Test, Multi-Health Systems, Toronto, 1995.

7. Conners, C.K., Levin, E.D., Sparrow, E., Hinton, S., Ernhardt, D., Meck, W.H., Rose, J.E., and March, J., Nicotine and attention in adult ADHD, *Psychopharmacol. Bull.,* 32, 67–73, 1996.

8. Conners, C.K., Wells, K.C., Parker, J.D., Sitarenios, G., Diamond, J.M., and Powell, J.W., A new self-report scale for assessment of adolescent psychopathology: factor structure, reliability, validity, and diagnostic sensitivity, *J. Abnorm. Child Psychol.,* 25, 487–97, 1997.

9. Decker, M.W., Brioni, J.D., Bannon, A.W., and Arneric, S.P., Diversity of neuronal nicotinic acetylcholine receptors: Lessons from behavior and implications for CNS therapeutics — minireview, *Life Sci.,* 56, 545–570, 1995.

10. DSM-IV-Task-Force, Ed., *Diagnostic and Statistical Manual for Mental Disorders: DSM-IV,* American Psychiatric Association, Washington, DC, 1994.

11. Dunnett, S.B. and Martel, F.L., Proactive interference effects on short-term memory in rats: 1. basic parameters and drug effects, *Behav. Neurosci.,* 104, 655–665, 1990.

12. Epstein, J.N., Conners, C.K., Sitarenios, G., and Erhardt, D., Continuous performance test results of adults with attention deficit hyperactivity disorder, *Clin. Neuropsychol.,* 12, 155–168, 1998.

13. Evenden, J.L., Turpin, M., Oliver, L., and Jennings, C., Caffeine and nicotine improve visual tracking by rats — a comparison with amphetamine, cocaine and apomorphine, *Psychopharmacology,* 110, 169–176, 1993.

14. Fiore, M.C., Smith, S.S., Jorenby, D.E., and Baker, T.B., The effectiveness of the nicotine patch for smoking cessation — a meta-analysis, *JAMA,* 271, 1940–1947, 1994.

15. Gawin, F., Compton, M., and Byck, R., Reduction of cigarette smoking by use of a nicotine patch, *Arch. Gen. Psych.,* 46, 288–289, 1989.

16. Glover, E.D., Varma, J.R., and Rich, J.D., Transdermal nicotine patch for smoking cessation, *N. Engl. J. Med.,* 326, 344–345, 1992.

17. Grilly, D.M., Simon, B.B., and Levin, E.D., Nicotine enhances stimulus detection performance of middle- and old-aged rats: A longitudinal study, *Pharmacol. Biochem. Behav.,* 65, 665–670, 2000.

18. Hatsukami, D., Fletcher, L., Morgan, S., Keenan, R., and Amble, P., The effects of varying cigarette deprivation duration on cognitive and performance tasks, *J. Sub. Abuse,* 1, 407–416, 1989.

19. Henningfield, J.E., Behavioral pharmacology of cigarette smoking, *Adv. Behav. Pharmacol.,* 4, 131–210, 1984.

20. Hoffman, B.B. and Lefkowitz, R.J., Catecholamines, sympathomemetic drugs, and adrenergic receptor antagonists. In J.G. Hardman, A.G. Gilman and L.E. Limbird, Eds., *Goodman and Gilman's The Pharmacological Basis of Therapeutics,* McGraw-Hill, New York, 1996, 199–263.

21. Hurt, R.D., Lauger, G.G., Offord, K.P., Kottke, T.E., and Dale, L.C., Nicotine replacement therapy with use of a transdermal nicotine patch — a randomized double-blind placebo-controlled trial, *Mayo Clin. Proc.,* 65, 1529–1537, 1990.

22. Jones, G.M.M., Sahakian, B.J., Levy, R., Warburton, D.M., and Gray, J.A., Effects of acute subcutaneous nicotine on attention, information processing and short-term memory in Alzheimer's disease, *Psychopharmacology,* 108, 485–494, 1992.

23. Levin, E., Conners, C., Silva, D., Hinton, S., Meck, W., March, J., and Rose, J., Nicotine skin patch effects on attention, *Psychopharmacology,* submitted, 1997.

24. Levin, E., Wilson, W., Rose, J., and McEvoy, J., Nicotine-haloperidol interactions and cognitive performance in schizophrenics, *Neuropsychopharmacology,* 15, 429–436, 1996.

25. Levin, E.D., Conners, C.K., Silva, D., Canu, W., and March, J., Effects of chronic nicotine and methylphenidate in adults with ADHD, *Exp. Clin. Psychopharmacol.,* in press, 2001.

26. Levin, E.D., Conners, C.K., Sparrow, E., Hinton, S., Meck, W., Rose, J.E., Ernhardt, D., and March, J., Nicotine effects on adults with attention-deficit/hyperactivity disorder, *Psychopharmacology,* 123, 55–63, 1996.

27. Levin, E.D., Westman, E.C., Stein, R.M., Carnahan, E., Sanchez, M., Herman, S., Behm, F.M., and Rose, J.E., Nicotine skin patch treatment increases abstinence, decreases withdrawal symptoms and attenuates rewarding effects of smoking, *J. Clin. Psychopharmacol.,* 14, 41–49, 1994.

28. NIMH, CGI (Clinical Global Impression) Scale, *Psychopharmacol. Bull.,* 21, 839–843, 1985.

29. Peeke, S.C., and Peeke, H.V.S., Attention, memory, and cigarette smoking, *Psychopharmacology,* 84, 205–216, 1984.

30. Pelham, W.E., Jr., Greenslade, K.E., Vodde-Hamilton, M., Murphy, D.A., Greenstein, J.J., Gnagy, E.M., Guthrie, K.J., Hoover, M.D., and Dahl, R.E., Relative efficacy of long-acting stimulants on children with attention deficit-hyperactivity disorder: a comparison of standard methylphenidate, sustained-release methylphenidate, sustained-release dextroamphetamine, and pemoline, *Pediatrics,* 86, 226–37, 1990.

31. Pelham, W.E., Jr., Sturges, J., Hoza, J., Schmidt, C., Bijlsma, J.J., Milich, R., and Moorer, S., Sustained release and standard methylphenidate effects on cognitive and social behavior in children with attention deficit disorder, *Pediatrics,* 80, 491–501, 1987.

32. Pickworth, W.B., Bunker, E.B., and Henningfield, J.E., Transdermal nicotine: reduction of smoking with minimal abuse liability, *Psychopharmacology,* 115, 9–14, 1994.

33. Pomerleau, O.F., Downey, K.K., Stelson, F.W., and Pomerleau, C.S., Cigarette smoking in adult patients diagnosed with attention deficit hyperactivity disorder, *J. Subst. Abuse,* 7, 373–8, 1995.

34. Reeves, D., Bleiberg, J., and Spector, J., Validation of the ANAM battery in multicenter head injury rehabilitation studies, *Arch. Clin. Neuropsychol.,* 8, 356, 1993.

35. Reeves, D., Kane, R., Winter, K., Raynsford, K., and Pancella, T., *Automated Neuropsychological Assessment Metrics (ANAM): Test Administrator's Guide Version 1.0,* Missouri Institute of Mental Health, St. Louis, 1993.

36. Rose, J.E., Transdermal nicotine as a strategy for nicotine replacement, in Okene, J.K., Ed., *The Pharmacologic Treatment of Tobacco Dependence: Proceedings of the World Congress.,* Institute for the Study of Smoking Behavior and Policy, Cambridge, MA, 1986, 158–166.

37. Rose, J.E., Transdermal nicotine and nicotine nasal spray as smoking cessation treatments, in *The Clinical Management of Nicotine Dependence,* Springer-Verlag, New York, 1990.

38. Rose, J.E., Herskovic, J.E., Trilling, Y., and Jarvik, M.E., Transdermal nicotine reduces cigarette craving and nicotine preference, *Clin. Pharmacol. Ther.,* 38, 450–456, 1985.

39. Rose, J.E., Jarvik, M.E., and Rose, K.D., Transdermal administration of nicotine, *Drug Alc. Depend.,* 13, 209–213, 1984.

40. Sahakian, B., Jones, G., Levy, R., Gray, J., and Warburton, D., The effects of nicotine on attention, information processing, and short-term memory in patients with dementia of Alzheimer type, *Br. J. Psych.,* 154, 797–800, 1989.

41. Sahakian, B.J. and Jones, G.M.M., The effects of nicotine on attention, information processing, and working memory in patients with dementia of the Alzheimer type, in Adlkofer, F. and Thruau, K. Eds., *Effects of Nicotine on Biological Systems,* Birkhauser Verlag, Basel, 1991, 623–230.

42. Turchi, J., Holley, L.A., and Sarter, M., Effects of nicotinic acetylcholine receptor ligands on behavioral vigilance in rats, *Psychopharmacology,* 118, 195–205, 1995.

43. Turchi, J., Holley, L.A., and Sarter, M., Effects of benzodiazepine receptor inverse agonists and nicotine on behavioral vigilance in senescent rats, *J. Gerontol. Biol. Sci. Med. Sci.,* 51, B225–31, 1996.

44. Warburton, D.M., Nicotine: an addictive substance or a therapeutic agent?, *Prog. Drug Res.,* 33, 9–41, 1989.

45. Warburton, D.M. and Arnall, C., Improvements in performance without nicotine withdrawal, *Psychopharmacology,* 115, 539–542, 1994.

46. Wender, P., *Attention-Deficit Hyperactivity Disorder in Adults,* Oxford University Press, New York, 1995.

47. Wesnes, K. and Warburton, D.M., Smoking, nicotine and human performance, *Pharmacol. Ther.,* 21, 189–208, 1983.

48. Westman, E.C., Levin, E.D., and Rose, J.E., Effectiveness of the nicotine patch in smoking cessation: a randomized controlled trial with telephone counseling, *Arch. Int. Med.,* 153, 1917–1923, 1993.

49. Wilens, T.E., Biederman, J., Spencer, T.J., Bostic, J., Prince, J., Monuteaux, M.C., Soriano, J., Fine, C., Abrams, A., Rater, M., and Polisner, D., A pilot controlled clinical trial of ABT-418, a cholinergic agonist, in the treatment of adults with attention deficit hyperactivity disorder, *Am. J. Psych.,* 156, 1931–1937, 1999.

50. Wonnacott, S., Irons, J., Rapier, C., Thorne, B., and Lunt, G.G., Presynaptic modulation of transmitter release by nicotinic receptors, in Nordberg, A., Fuxe, K., Holmstedt, B., and Sundwall, A., Eds., *Progress In Brain Research, Vol. 79,* Elsevier Science Publishers B.V., 1989, 157–163.

14 Clinical Methodologies in the Examination of Nicotinic Effects on Cognition

Alexandra Potter and Paul A. Newhouse

CONTENTS

14.1 INTRODUCTION

The aim of this chapter is to provide practical information regarding methodological issues in the study of nicotinic agents on cognition in both clinical and normal volunteer populations. The scope is rather broad, briefly reviewing clinical disorders for which nicotinic stimulation may have therapeutic uses, and then addressing several methodological issues involving subject selection, drug administration, and measures to use. The review of issues related to subject selection will address studying patient populations vs. normal volunteers, as well as issues involving characteristics of individual subjects including age, gender, and smoking status. Drug administration issues will be discussed including practical advantages and disadvantages to different dosing methods (iv, subcutaneous, transdermal, inhaled, nasal, and oral). Finally, issues involved in selecting measures for differing research purposes will be examined including tests of learning, memory, attention, behavioral ratings, electrophysiological measures, and neuroimaging techniques.

The overarching aim of nicotinic research in humans is to develop pharmaco-therapies that may alleviate cognitive and behavioral symptoms of clinical disorders. This line of investigation is based on several decades of research from many areas of neuroscience pointing to the importance of this receptor system in human behavior and cognition.[1,2] More specifically, research has demonstrated that disruptions in nicotinic functioning in several disease states exist,[3] that many of today's pharma-cotherapies for these diseases use drugs which impact nicotinic functioning, and that, in both laboratory animals and humans, nicotinic agents have measurable impact on cognitive and behavioral functioning.[2] Advances made in recent years in under-standing the structure and function of the nicotinic receptor system make human cognitive studies a viable and exciting area of experimentation.[4]

14.2 CLINICAL DISORDERS

It is helpful at this point to review briefly the disorders where nicotinic physiology and pharmacology seem relevant as a prelude to the methodological issues arising when conducting nicotinic studies in normal volunteers and patient populations. These disorders can be divided into groups based on the strength of the link to nicotinic receptor functioning.

There are two disorders for which there is an identified nicotinic pathology: autosomal dominant frontal lobe epilepsy (ADNFLE) and schizophrenia. In schizo-phrenia there is impairment in auditory sensory gating that may be related to the difficulty that these patients have in filtering extraneous sensory information.[5] Nor-mal subjects presented with a pair of auditory tones produce a smaller P50 (evoked potential at 50 ms) brain wave response to the second tone; the subjects partially inhibit or gate their electrophysiological response to the second tone. In patients with schizophrenia and their first-degree relatives, an inhibitional failure exists whereby these patients produce nearly identical P50 responses to both tones.[5] Studies have found that nicotine administration temporarily corrects this gating deficit in patients with schizophrenia. This has been linked to a polymorphism at chromosome 15, which is the locus of the alpha7 nicotinic receptor.[6,7]

ADNFLE has been found to result from a missense mutation in the alpha4 subunit gene,[8] resulting in a mutated alpha 4-beta2 nicotinic receptor unit. This is hypothesized to lead to the expression of brief seizures via effects on reduced receptor function, resulting in lower levels of GABA and glycine that may be triggered by these genetic mutations.[5]

Another classification of disorders can be considered to be those disorders where known nicotinic pathology is associated with the presence of the disease state. This includes Alzheimer's disease (AD) and Parkinson's disease (PD). Several researchers have shown that, in AD a significant loss of cortical nicotinic receptor binding in patients occurs, compared to age-matched controls.[9–11] Nordberg[12] demonstrated that there is a correlation between the level of cognitive dysfunction in AD and the loss of nicotinic receptor binding in temporal and frontal cortices and in the hippocampus using positron emission tomogrophy (PET). This link is further supported by the finding[13] that the anticholinesterase tacrine, which produces mild cognitive improvements, increases nicotinic receptor binding along with increasing cerebral blood flow. Cholinesterase inhibitors have been the primary pharmacologic treatment strategy for AD, showing moderate effects in reducing symptoms and slowing progression of the disease.[14,15] It is presumed that this therapy is partially effective through effects on the nicotinic receptor system.[5] Nicotine and nicotinic agonists have been studied in patients with AD and have significant positive effects on learning memory[16,17] and attentional functioning.[18] The nicotinic antagonist mecamylamine has been shown to cause impairment in learning and memory in normal volunteers and patients with PD and AD, with the patient groups showing increased sensitivity to nicotinic blockade, which correlates with the presumed level of nicotinic receptor loss.[19]

The primary pathology of PD is the loss of dopaminergic neurons in the substantia nigra. There is a complex relationship between the dopaminergic and nicotinic receptor systems. Nicotine stimulates dopamine release in several areas of the brain, including the striatum and substantia nigra,[20] and administration of a nicotinic antagonist has been shown to inhibit dopamine release from the striatal and mesolimbic structures.[21] Patients with PD often develop cognitive impairments that may be related to nicotinic receptor loss.[19] Whitehouse and colleagues[22] demonstrated that the loss of nicotinic receptors in PD correlates with the degree of dementia seen in the patient. Kelton and colleagues[23] administered nicotine to non-demented patients with PD and found short-term, measurable cognitive and motor benefits, which included positive effects on attention, arousal, and processing speed. These effects were inhibited when pretreatment with the nicotinic antagonist mecamylamine was administered.

In PD and AD, epidemiological studies have found tobacco smoking to be associated with reduced incidence of these diagnoses.[24] This finding, together with the finding that a loss of high-affinity nicotine binding in both AD and PD occurs, has lead researchers to hypothesize that the consequences of nicotinic cholinergic transmission may be neuroprotective.[21]

A third classification of disorders related to nicotinic function is those where there is no known loss or genetic alteration of nicotinic receptors, but there is evidence that stimulating nicotinic receptors may have therapeutic value. These disorders include ADHD, anxiety and depression, and Tourette's syndrome.

Attention deficit hyperactivity disorder (ADHD) is a clinical syndrome usually diagnosed in childhood. While to date no firm evidence of nicotinic system dysfunction in this disorder exists, current pharmacological treatments are psychostimulants, which are presumably effective via their interactions with dopamine. The primary deficits of this disorder involving attention are affected by nicotine administration in other clinical populations.[25] Pilot studies administering nicotine to patients with ADHD have shown improvements in clinical global impressions (CGI) of symptoms, and on attentional tasks.[26]

There is evidence that anxiety and depression may be linked to nicotinic functioning. Nicotine has anxiolytic effects in humans and animals,[17,27] which can be blocked by administration of mecamylamine.[28] Major depression is associated with increased rates of cigarette smoking, and a greater difficulty in quitting smoking.[29] Transdermal nicotine administration improves the mood of nonsmoking depressed patients.[30] Treatments for depression that directly impact nicotinic receptor functioning have yet to be investigated.

Tourette's syndrome (TS) is a neurological disorder characterized by motor and vocal tics, often accompanied by hyperactivity, anxiety, fear, and symptoms of obsessive-compulsive disorder. Studies have demonstrated that nicotine administration (via transdermal patch or nicotine gum) can potentiate the action of traditional neuroleptics and is effective in managing the symptoms of TS.[31,32] These effects are fairly long lasting, with improvements in symptoms being reported up to four weeks after two days of nicotine exposure.[33] It is hypothesized that this effect is related to the regulation of dopamine resulting from the desensitization of nicotinic receptors through chronic nicotine exposure. Novel nicotinic agents with a larger therapeutic index, as well as fewer side effects, offer promise for the treatment of this disorder.[5]

14.3 METHODOLOGICAL ISSUES

14.3.1 POPULATION

The determination of whom to study has many dimensions; the basic ones discussed here are (1) studying a clinical sample vs. studying normal volunteers; (2) studying smokers vs. nonsmokers; (3) studying young vs. elderly subjects; and (4) issues involving gender.

The determination of whether to use a clinical or normal volunteer population largely rests on the research question addressed. In studies that try to link a known deficit of nicotinic functioning to a specific clinical symptom, it is useful to model the receptor dysfunction using a nicotinic antagonist in a normal volunteer population.[25] Newhouse and colleagues[19] have done this, successfully demonstrating that nicotinic receptor blockade produces cognitive effects on learning and memory relevant for the study of dementia. Antagonist studies can also be conducted in clinical populations to demonstrate an increased sensitivity to the effects of receptor blockade. This type of study has been used to demonstrate the differential effects of nicotinic blockade in compromised vs. intact neurotransmitter systems using normal volunteers of varying age groups as well as clinical populations (AD and PD).[16]

While normal volunteer studies are important, studying nicotine directly in clinical populations has several advantages. One is the ability to assess the effects of the compound on compromised areas of functioning directly. For example, the novel nicotinic agonist ABT-418 has been investigated in patients with AD, using cognitive and behavioral tests of learning, memory, and anxiety.[17] This type of study can demonstrate the clinical utility of nicotinic agents, helping to determine if statistically significant results are likely to be clinical significant. This is accomplished using clinical rating scales completed by the investigator, subject, and other raters (i.e., family members, caregivers).

14.3.2 Smoking Status

Another dimension that must be considered in the determination of the subject pool is whether to include both smoking and nonsmoking subjects. Studies of cognition and nicotinic agents in clinical samples have typically used nonsmoking populations. While there are several advantages to this decision, the primary one is that these patients do not have the altered brain neurochemistry seen after chronic nicotine exposure.

Cigarette smokers have an increased number of nicotinic receptors in the brain.[34] However, these nicotinic receptors undergo rapid desensitization in the presence of even low doses of nicotine and, consequently, smokers show reduced sensitivity to the effects of nicotinic stimulation.[3] One consequence of this is that smokers show nicotine tolerance. This is manifested in the finding that repeated nicotine doses produce less effect than the first dose,[35] and that after repeated exposure to nicotine the same plasma drug level produces less effect than it did originally.[35] Investigations conducted using smokers as subjects must contend with these issues. When a nonsmoking population is studied, the issue of nicotine tolerance must also be taken into account if the study design included multiple dosing. Nicotine tolerance develops quickly[35] and thus it is important to design the dosing schedule to avoid confounding results with the development of nicotine tolerance.

Another issue arising when studying smokers is how to account for withdrawal effects from smoking abstention, which include nervousness, irritability, anxiety, and impaired concentration and cognitive functioning.[36] It is difficult to discern if the effects of the nicotinic drug being tested go beyond merely relieving withdrawal symptoms, if the subject pool includes current smokers.

Practical considerations must be addressed when studying nicotinic agents in nonsmokers. There is a higher incidence of side effects including nausea, vomiting, dizziness, and cardiovascular effects.[40] In nonsmokers, who have not developed nicotine tolerance, these side effects must be anticipated and steps taken to minimize them. Due to high rates of cigarette smoking in many psychological disorders, it can be difficult to recruit a nicotine-naïve sample of subjects.

The lay community may also raise ethical objections to the administration of nicotine in nonsmoking subjects, fearing that these subjects will be more prone to developing nicotine dependence. Theoretically the risk of this is low, particularly when the dosing method involves continuous exposure (i.e., iv or nicotine patch), which has extremely low abuse liability.[37] Other factors influencing this risk are the unpleasant side effects experienced by nonsmokers, and the short exposure to nicotine in acute studies. Hughes[37a] addressed the question of whether subjects are likely to become smokers subsequent to participation in a clinical investigation involving tobacco administration. This investigation found that, three months after completion of the study, none of the nonsmokers or ex-smokers who participated had become smokers or regular users of other nicotine-containing products.

Due to changes in brain neurochemistry, issues of tolerance and side effects are important to be aware of in selecting subjects to study. It is important to characterize subjects according to smoking status in all studies of the cognitive effects of nicotinic agents. Hughes[38] has created a classification of smoking status that divides subjects as never-smokers (who have never smoked daily for more than 2 weeks), ex-smokers (who smoked more than 10 cigarettes a day for more than 1 year and have been abstinent for more than one year), and current smokers. This system provides clear delineations to address the issue of smoking status among research subjects.

14.3.3 GENDER

Studies of the impact of gender on the pharmacodynamics of nicotine have resulted in mixed findings. One study[39] found that men had significantly greater clearance of nicotine than women, but a second study[40] found no effect of gender on nicotine clearance. As male subjects usually weigh more and have larger liver mass and partal vein capacity, it is to be expected that men will usually have greater total nicotine clearance than women. The possibility of changes in nicotine metabolism during the menstrual cycle has not been examined. There is some evidence of an increase in nicotine withdrawal symptoms and greater cigarette craving during menses;[41,42] however these data are inconclusive due to the overlap of symptoms attributable to menses and symptoms attributable to nicotine withdrawal.

14.3.4 AGE

Nicotinic agents have been studied in subjects of all ages. Children as young as 10 have participated in studies of nicotine patch in Tourette's syndrome,[43] and elderly subjects as old as 80 have participated in AD studies.[16] The question of age affecting nicotine tolerance and absorption has not been directly studied. It is obvious that the variability in body mass associated with different age groups (children vs. adults vs. elderly subjects) will impact the effect of a given dose of nicotine. However, subjects of all ages have been successfully used in nicotinic investigations when issues of dosing and side effect management have been addressed.

Newhouse and colleagues[44] demonstrated that there are varying cognitive effects associated with the age of normal volunteer subjects. This study demonstrated that elderly (60 +) subjects were more susceptible to the cognitive effects of nicotinic blockade than were young (18 to 30) subjects; presumably this was associated with the loss of nicotinic receptors, which comes with age.[19] The finding that the age of the subject impacts the magnitude of the cognitive effects in normal volunteers has implications for subject selection. One strategy for managing this finding is to use subjects in a similar age range. If this is not feasible, it will be important to examine a subset or even individual results of subjects, based upon age.

14.3.5 INDIVIDUAL VARIABILITY

One consistent finding regarding subject characterization in the study of nicotine and cognition indicates that there are individual differences in subjects' response to nicotine. Perkins[45] reviewed a large body of literature in support of the theory that baseline state may account for a significant portion of variability in results of nicotinic studies. The effects of baseline performance may explain why studies using normal volunteers show small or no improvements following nicotine administration,[46] while studies using impaired subjects (either from sleep deprivation, tobacco withdrawal, or diagnosis of AD, ADHD, Schizophrenia, etc.) show larger improvements following nicotine administration.[16,18,26,47,48] The effects of baseline performance on the magnitude and direction of the effects of nicotine administration can be seen on a number of dependant variables including mood, behavioral ratings, and tests of cognition, and have been shown in subjects using a wide variety of methodologies.[45] One investigator demonstrated that subjects who had lower initial arousal levels experienced a greater stimulant effect of nicotine than subjects whose baseline arousal was already high.[49] Although it is possible that this may represent "ceiling" or "floor" effects, due to the finding that many effects involving responding exist well within the upper and lower limits of the dependent measure,[50] this explanation does not explain away the findings relevant to baseline dependency. Individual variability has been explored in primates by individually dose-ranging subjects and their cognitive changes in order to compare individuals.[51]

The practical implications of these findings include examining data based on individuals as well as on group means. Traditionally, data from pharmacological studies are collapsed to compare means (of differing treatments, or of treatment vs. placebo). Baseline dependency may influence the response to the drugs such that a subgroup of subjects (with low baseline mood scores) may experience elevations of mood ratings, while another subgroup (with high baseline mood ratings) may experience declines of mood ratings. Collapsing the data would results in a finding of no significant difference, but examination of subgroups or even individual subjects data could provide a more complete picture of the effect of the drug.

Another important methodological implication from this body of research is to characterize the subject pool carefully, using baseline measures comparable to the dependant measures of study. This allows for post-hoc characterization of the effects of nicotinic agents on the dependent variable that accounts for the subjects' baseline performance on this variable. These findings also underscore the importance of controlling for baseline conditions (either within or between subjects) to reduce the baseline variability and, thus, potential variability in the effects seen from nicotine administration.

14.3.6 ROUTE OF DRUG ADMINISTRATION

Several considerations contribute to the decision of which preparation and route of administration to use for nicotine and/or nicotinic agents in human cognitive investigations. Nicotine is available and has been used in many forms including inhaled, iv, subcutaneous, transdermal patch, nasal spray, and gum. These have all been used successfully in human studies of performance. Issues to address in this determination include the speed of nicotinic effects, the side effects profile, the dosing accuracy required by the investigation, and practical issues of dose preparation and blinding.

IV administration of nicotine has been used successfully in cognitive studies using patients diagnosed with AD and PD.[52,53] This route of administration results in a steady rise in plasma levels over the first 30 minutes of drug administration; plasma levels decline rapidly when drug administration ceases.[54]

The primary advantage of iv administration is the precision with which a dose can be administered (based upon the weight of the subject). This allows for greater control over the drug and, presumably, plasma nicotine levels. Another advantage of the increased control with iv preparation is that drug administration can be rapidly adjusted or discontinued if side effects occur. This can reduce the discomfort experienced by patients and lead to greater compliance and higher completion rates in this type of protocol.

The main drawback to using an iv preparation of nicotine involves the creation of the pharmaceutical. IV nicotine is not commercially available and thus must be prepared by a pharmacy (in accordance with an IND from the FDA) and administered by a physician. This is in contrast to commercially available forms of nicotine (patch, gum, and nasal spray) that can be purchased over the counter and administered by other study personnel (or self-administered by the subject). IV nicotine must be prepared by a pharmacy for each subject because the doses are based upon the weight of the subject. The rapid absorption and continuous administration in iv dosing lead to the possibility of the rapid onset of nicotine toxicity. This can be managed by frequent monitoring of subjective symptoms and discontinuation of the infusion at the onset of symptoms of nicotine toxicity. IV administration of nicotine also necessitates cardiac monitoring during the infusion; cardiac telemetry has been used successfully[52,53] in several studies using iv administration of nicotine. Other drawbacks to this route of administration include subject discomfort due to the insertion of the iv, occasional arm pain during the infusion, and the need to remain stationary during the infusion. These problems are largely manageable and should not prevent subjects from completing a protocol.

Nicotine inhaled from cigarette smoking reaches the brain in 10 to 19 seconds.[55] Cigarette smokers vary individually in the actual dose of nicotine they get from each cigarette by changes in how they puff their cigarette (rate, volume, and intensity) and changes in depth of inhalation.[35] For these reasons, it is difficult to control the nicotine dosage used in smoking studies. Standard methodologies in studies using cigarette smoking as a route of nicotine delivery dictate smoking rate and length of inhalation to subjects in an attempt to control the nicotine dose; however, precise dosing is difficult to attain using inhaled cigarette smoke.

The primary advantage to cigarette smoking in nicotine research is that it is a prime model for addiction studies where subjects control their own dosing and investigators control contributing variables. Cigarette smoking is naturalistic, and these laboratory studies can provide a useful link to real-life uses of nicotine. However, for the study of cognition and clinical disorders, smoking is not typically used to deliver nicotine. The loss of precise dosing is the primary drawback of this method of drug delivery. In addition, smoking cigarettes is quite noxious to nicotine-naïve subjects, and creating placebo cigarettes with a similar taste and abrasiveness to the airways, while possible, is labor intensive and expensive.

Oral nicotine is usually administered via nicotine gum. This preparation of nicotine is absorbed through the oral mucosa and results in a slow rise of plasma nicotine levels that peak at approximately 30 minutes.[35] There is substantial variability in the amount of nicotine actually extracted from the gum due to variations in chewing between subjects.[56] Nicotine is also swallowed with this administration method, leading to inaccuracies in determining how much nicotine reaches the brain.[56]

Nicotine gum is readily available in standard doses and is easy to administer to subjects. It is portable, allowing for dosing in virtually any setting. Drawbacks include inaccuracies in dosing cited earlier, as well as difficulties in creating placebo gum due to the distinct taste of nicotine gum.

Subcutaneous administration has been used in several studies of the cognitive effects of acute nicotine administration. Pharmacokinetic studies show that, following administration, plasma levels rise steadily with peak plasma levels reached approximately 15 minutes following injection.[57] Advantages to this method of administration include the speed of administration and nicotine absorption. Unlike intravenous administration, the subjects' exposure to a needle is brief and, with the use of pH neutralization, the site of the injection is relatively painless following the injection.[57] Subcutaneous preparations of nicotine are prepared individually for each subject, allowing for precise dosing based on the weight of the subject.

Drawbacks to this route of administration include the same problems in drug manufacturing as for iv preparations. Specifically, these include the necessity of a pharmacy to create the nicotine preparation (in accordance with an IND from the FDA), and a physician to administer it. An additional drawback to subcutaneous administration is that side effects cannot be managed by adjusting or discontinuing the nicotine dose. Some subjects may experience pain and irritation at the site of the injections, and others may find the procedure itself to be unpleasant. Subcutaneous administration has not been widely used in cognitive studies of nicotine; when compared with iv administration, it appears that the drawbacks are similar, and iv administration has the added advantage of continuous administration and the ability to adjust or discontinue the drug rapidly.

Nicotine patches have been successfully used in studies of cognition and nicotine in patients with AD, PD, and ADHD; they deliver nicotine continuously while worn. The absorption of nicotine is slow, with peak plasma levels achieved approximately six hours after the patch is applied.[40] These levels remain generally steady for 7 to 8 hours and then decline over the next 6 hours.[40]

Nicotine patches are well suited for clinical studies for several reasons. They offer steady rates of nicotine delivery unlike smoking, nasal spray, or nicotine gum. This helps with the control of dosing that is problematic in many administration methods. There are many conveniences associated with use of the nicotine patch, including ease of application, and the unobtrusive nature of the transdermal patch. Nicotine patches are relatively easy to blind and are commercially available. Some brands are constructed so that it is possible to cut the patch to control the amount of nicotine administered, allowing for some flexibility in dosing.

There are several drawbacks to the use of patches as well. The onset of peak plasma levels occurs relatively slowly while the onset of side effects is often seen within 1 hour.[57] Due to the slow rise in plasma levels, nicotine continues to accumulate after the patch is first removed. This can be problematic for subjects who become symptomatic because the drug cannot be rapidly discontinued. While using the nicotine patch provides a convenience of premade doses, it is difficult to make dosage adjustments because it is available in limited dosage forms. Nicotine patches also produce skin irritation in some subjects.

14.3.7 MEASURES

Assessment of the effects of nicotine on human performance involves careful selection of appropriate cognitive, behavioral, and/or imaging measures. Increasingly, measures are combined in studies of nicotinic agents in humans in an attempt to draw correlations between cognitive and behavioral effects and neuroanatomical or neurophysiologic mechanisms.

Cognitive measures generally focus on learning, memory, and attention. Whether nicotine enhances cognitive function apart from its ability to relieve nicotine withdrawal has motivated the search for cognitive measures that would be independent of or not influenced by withdrawal. Studies of the cognitive effects of nicotine using cigarette smokers have been confounded by withdrawal effects that have obscured the effects of nicotinic stimulation alone.[58,59] Studies of nonimpaired humans and patients with a variety of disorders have suggested that nicotinic stimulation improves certain types of attentional processes, learning, and memory under conditions of significant cognitive load.[60]

Tasks that measure sustained or continuous attention performance appear to show the most robust nicotinic effects. Tasks such as the rapid visual information processing task or the Connor's continuous performance test have shown sensitivity to nicotinic stimulation in normal volunteers,[60,61] patients with ADHD, and patients with Alzheimer's disease.[62,63] Measures of selective attention have produced more mixed results. For example, the Stroop task is a measure of conflict between verbal and color processing. While some studies have demonstrated a reduction in the Stroop effect with cigarettes or nicotinic stimulation[64] others have shown inconsistent results[60] or results suggesting that nicotine enhances the Stroop effect.[23] Measures which examine attentional performance over very long periods of time have shown positive effects of nicotine,[65] especially in preventing performance decrements. Nicotinic stimulation has been shone to improve attentional performance in primates, particularly when distraction is present,[66] while little improvement was seen without the distraction. This paradigm should probably be used more frequently in studies of attentional effects of nicotinic stimulation in humans.

Within continuous or sustained performance tests, it is possible for the effects of nicotine or nicotinic stimulation to be manifest in some but not all aspects of the task. For example, in a study of adults with ADHD, nicotine patch administration reduced errors of omission but did not have any significant effect on errors of commission,[67] suggesting an effect on sustained attention without a general increase in responding or a shift in response strategy. A reduction in attentional performance errors is an effect seen in a variety of different paradigms with different populations, such as patients with Alzheimer's disease[68] and schizophrenia,[69] and suggests that error reduction in cognitive tasks is a measure which may prove particularly sensitive to nicotinic effects. This is further supported by studies of the effects of nicotinic antagonists. Newhouse and colleagues[70] showed that the nicotinic antagonist mecamylamine produced a dose-related increase in errors on a nonverbal learning task (the repeat acquisition task) that required significant and sustained attention to a computer screen. Nicotine administration improved error performance on this task in a group of Alzheimer's disease patients.[71]

In tasks which have examined divided attention, nicotine administration has shown improvement of performance on naturalistic telephone directory search task with concomitant tone counting[60] in normals. However, in this laboratory, nicotine administration appeared to produce a reduction in overall performance on verbal-visual divided attention tasks involving number string retrieval and maze completion in a group of patients with Parkinson's disease (Newhouse et al., unpublished). In a study examining the effects of nicotine on intensity and selectivity features of attention, Mancuso and colleagues[61] showed that nicotine appeared to have no effect on attentional switching or selectivity but did appear to improve intensity features of attention, suggesting a general increase in processing resources.

Measures of learning and memory have generally been focused on variants of serial learning such as words, numbers, etc. Rusted and colleagues have suggested that learning and memory tasks that involve effortful processing (as opposed to automatic processing) are more likely to demonstrate nicotinic effects or improvements. This may be due to the ability of nicotinic stimulation to enhance cognitive resources overall.[72] In addition, attentional improvement during encoding may also be responsible for an improvement in the amount of information placed in working memory. There is also evidence for nicotinic effects on memory consolidation as well.[52,72,73] An example of a useful measure in studies of learning with nicotinic stimulation with a significant effortful component is the selective reminding task (SRT). This measure involves serial list learning of unrelated words with the caveat that, from trial to trial, the only words repeated are words not recalled from the previous trial. In a study of the effects of the novel nicotinic agonist ABT-418 in Alzheimer's disease patients, significant positive effects were found on the SRT task on measures of word recall and recall failure;[17] both of these measures reflect processes of working memory. Additionally, studies of the effects of the nicotinic antagonist mecamylamine have shown that the learning rate on this task (the amount learned/the amount remaining to be learned) can be measured accurately and may be a more meaningful measure of cognitive capability.[25] Measures of verbal recall (which demand significant effort demanding) are generally used in preference to verbal recognition (presumed to be less demanding). However, Rusted and colleagues have shown significant effects of nicotine administration on recognition of Chinese characters.[73] The test subjects' lack of familiarity with these characters may have contributed to the effort required to preserve recognition of them and may have increased the effort demanded of this task. Another type of memory task that appears to show nicotinic responsivity is match-to-sample tasks, particularly in primates.[74] In these tasks, a sample item is demonstrated for a brief period of time, then a variable delay interval ensues followed by the appearance of a probe item. The subject must determine whether the probe matches the original sample. Such tasks have only occasionally been used in studies of cognitive enhancement in humans,[75] probably due to concerns about having enough unique test items and interference effects, but this task may be a useful task for nicotinic assessment, particularly when long delay intervals occur between sample and probe items. Verbal and nonverbal versions could be constructed as long as the task is made sufficiently difficult.

Measures of nonverbal learning have only rarely been used in the assessment of nicotinic drug effects. Measures of memory for spatial location have shown some positive effects in a study of the nicotinic agonist ABT-418 in Alzheimer's disease patients. Nonverbal serial learning was used in studies of the nicotinic agonist mecamylamine utilizing the repeated acquisition task,[70,76] in which subjects learn a sequence (chain) of button pushes on a key pad or button box. The subject learns a sequence prior to the beginning of the experiment and is periodically asked to reproduce that chain or sequence after drug administration. In addition, during the experiment, the subject is periodically asked to learn a new chain of button pushes. This enables the assessment of both long-term memory and retrieval as well as the acquisition of new nonverbal information and shape of the acquisition curve. Mecamylamine not only produces an increase in errors of acquisition of new information on this task but also appears to decrease the ability of working memory to hold information. By contrast, retrieval of previously learned information (original chain) was not affected. The length of the chain can be adjusted to the cognitive abilities of the subject, which makes this task widely applicable to subject groups with varying cognitive abilities. Measures of visual memory that do not involve location have not been used frequently in studies of nicotinic effects, perhaps because of concerns about lack of sufficient forms for repeated measures designs.

Neuroimaging is rapidly becoming a new area for nicotinic investigation. The availability of ligands for PET and SPECT imaging of nicotinic receptors and nicotinic receptor function has opened up the possibility of functional assessment of the effects of nicotinic drugs. Initial neuroimaging studies were directed at documenting the involvement of nicotinic cholinergic systems in AD. Nordberg[12] showed a significant correlation between the change in temporal cortex labeling of 11C-nicotine and cognitive function scores in AD patients using positron emission tomography (PET). Significant correlations were shown between cognitive dysfunction and the loss of nicotinic receptor binding in temporal and frontal cortices and hippocampus in patients using PET. Nordberg[77] also examined the effects of treatment with the anticholinesterase tacrine in AD patients using PET and showed that brain nicotinic receptor binding of 11C-nicotine increased along with cerebral blood flow after three weeks of treatment. Extensive development of novel nicotinic ligands for PET and SPECT in the past several years offers additional opportunities for functional imaging including studies where functional activation patterns in the brain during memory tasks are measured by cerebral blood flow changes with PET.[78] In addition to measuring nicotinic receptor function directly using these ligands, it may be useful to measure extracellular neurotransmitter levels such as dopamine that may be released with nicotinic stimulation. PET ligands are now available to estimate these levels. Pharmacokinetics, distribution, and pharmacodynamics of nicotinic agents at nicotinic receptors are parameters that can now be assessed with currently available or soon-to-be available PET ligands. Nicotinic SPECT ligands[79] are currently in human trials and will shortly become available. The potential advantages of SPECT imaging are the longer half-life of the isotopes and smaller technical requirements for administration and imaging. Functional magnetic resonance imaging (fMRI) of the brain has become the most widely used method for real-time assessment of cognitive-anatomical relationships and assessment of the effects of cognitive operations on cerebral activity. However applications to examine the effects of nicotinic agents on cognitive performance and dribble activity using fMRI have been minimal. The utilization of this technology may have been limited in studies of nicotinic agents because of concerns that nicotinic stimulation or blockade may directly affect cerebral blood vessels,[78] thereby confusing the effects of nicotinic stimulation or blockade on cognitive performance if blood flow becomes the proxy measure of activity (as it often is in fMRI).

REFERENCES

1. Robbe, H.W.J. and O'Hanlon, J.F., Acute and subchronic effects of paroxetine 20 and 40 mg on actual driving, psychomotor performance and subjective assessments in healthy volunteers, *Eur. Neurospsychopharmacol.,* 5, 35–42, 1995.
2. Newhouse, P.A., Potter, A., and Levin, E.D., Nicotinic system involvement in Alzheimer's and Parkinson's diseases, *Drugs Aging,* 11(3), 206–228. 97.
3. Gotti, C., Fornasari, D., and Clementi, F., Human neuronal nicotinic receptors, *Prog. Neurobiol.,* 53, 199–237, 1997.
4. Arneric, S.P. and Brioni, J.D., Neuronal nicotinic receptors: pharmacology and therapeutic opportunities, in Lukas, R.J., Ed., *Cell Lines as Models for Studies of Nicotinic Acetylcholine Receptors,* Wiley-Liss, New York, 1998, 81–97.
5. Paterson, D. and Nordberg, A., Neuronal nicotinic receptors in the human brain, *Prog. Neurobiol.,* 61, 75–111. 2000.
6. Freedman, R., Coon, H., Myles-Worsley, M., Orr-Urtreger, A., Olincy, A., Davis, A., Polymeropoulos, M., Holik, J., Hopkins, J., Hoff, M., Rosenthal, J., Waldo, M.C., Reimherr, F., Wender, P., Yaw, J., Young, D.A., Breese, C.R., Adams, C., Patterson, D., Adler, L.E., Kruglyak, L., Leonard, S., and Byerley, W., Linkage of a neurophysiological deficit in schizophrenia to a chromosome 15 locus, *Proc. Nat. Acad. Sci.,* 94, 587–592, 1997.
7. Chini B., Raimond E., Elgoyhen A., Moralli D., Balzaretti M., and Heinemann S., Molecular cloning and chromosomal localization of the human alpha-7 nicotinic receptor subunit gene (CHRNA7), *Genomics,* 19:379–81, 1994.
8. Steinlein, O.K., Mulley, J.C., Propping, P., Wallace, R.H., Phillips, H.A., Sutherland, G.R., et al., A missense mutation in the neuronal nicotinic acetylcholine receptor alpha4 subunit is associated with autosomal dominant nocturnal frontal lobe epilepsy, *Nature Genetics,* 1995; 11:201–203.
9. Whitehouse, P. J., Martino, A.M., Antuono, P.G., Lowenstein, P.R., Coyle, J.T., Price, D.R., and Kellar, K.J., Nicotinic acetylcholine binding sites in Alzheimer's disease, *Brain Res.,* 371, 146–151, 1986.
10. Flynn D.D. and Mash, D.C., Nicotine receptors in human frontal and infratemporal cortex: comparison between Alzheimer's disease and the normal, *Neurosci. Abs.,* 11:1119, 1985.
11. Aubert, I., Araujo, D. M., Cecyre, D., Robitaille, Y., Gauthier, S., and Quirion, R., Comparative alterations of nicotinic and muscarinic binding sites in Alzheimer's and Parkinson's diseases, *J. Neurochem.,* 58, 529–541, 1992.
12. Nordberg, A., Functional changes in neuronal nicotinic receptors in Alzheimer's disease, *Behav. Pharmacol.,* 7 (Suppl. 1), 77–78, 1996.

13. Nordberg A., Clinical studies in Alzheimer patients with positron emission tomography, *Behav. Brain Res.*, 57:215–224, 1993.
14. Maltby, N., Broe, G.A., Creasey, H., Jorm, A.F., Christensen, H., and Brooks, W.S., Efficacy of tacrine and lecithin in mild to moderate Alzheimer's disease: double-blind trial, *Br. Med. J.*, 308, 879–883, 1994.
15. Nordberg, A., Human nicotinic receptors -- their role in aging and dementia, *Neurochem. Int.*, 25(1), 93–97, 1994.
16. Newhouse, P., Potter, A., and Corwin, J., Effects of nicotinic cholinergic agents on cognitive functioning in Alzheimer's and Parkinson's disease, *Drug Devel. Res.*, 38(3/4), 278–289, 1996.
17. Potter A, Corwin J, Lang J, Lenox R, and Newhouse PA. Acute effects of the selective cholinergic channel activator (nicotinic agonist) ABT-418 improves learning in Alzheimer's disease, *Psychopharmacology*, 142:334–342, 1999.
18. Jones, G.M.M., Sahakian, B.J., Levy, R., Warburton, D.M., and Gray, J.A., Effects of acute subcutaneous nicotine on attention, information processing and short-term memory in Alzheimer's disease, *Psychopharmacology*, 108, 485–494, 1992.
19. Newhouse, P.A., Potter, A., Corwin, J., and Lenox, R., Acute nicotinic blockade produces cognitive impairment in normal humans, *Psychopharmacology*, 108, 480–484, 1992.
20. Benwell, M.E.M, Balfour, D.J.K., and Khadra, L.F., Studies on the influence of nicotine infustions on mesolimbic dopamine and locomotor responses to nicotine, *J. Neurochem.*, 72:233–239, 1994.
21. Court, J.A., Lloyd, S., Thomas, N., Piggott, M.A., Marshall, E.F., Morris, C.M., et al., Dopamine and nicotinic receptor binding and the levels of dopamine and homovanillic acid in human brain related to tobacco use, *Neuroscience*, 87:63–78, 1998.
22. Whitehouse, P.J., Martino, A.M., Marcus, K.A., Zweig, R.M., Singer, H.S., Price, D.L., and Kellar, K.J., Reductions in acetylcholine and nicotine binding in several degenerative diseases, *Arch. Neurol.*, 45, 722–724, 1988.
23. Kelton, M.C., Kahn, H.J., Conrath, C.L., and Newhouse, P.A., The chronic and acute effects of nicotine on Parkinson's disease, 1999.
24. Baron, J.A., Beneficial effects of nicotine and cigarette smoking: the real, the possible and the spurious, *Br. Med. J.*, 52:58–73, 1996.
25. Newhouse, P.A. and Kelton, M., Nicotinic systems in central nervous systems disease: degenerative disorders and beyond, *Pharmaceutica Acta Helvetiae*, 74, 91–101, 2000.
26. Levin, E.D., Conners, C.K., Sparrow, E., HInton, S.C., Erhardt, D., Meck, W.H., Rose, J.E., and March, J., Nicotine effects on adults with attention-deficit/hyperactivity disorder, *Psychopharmacology*, 123, 55–63, 1996.
27. Seyler, Jr., L. E., Pomerleau, O.F., Fertig, J.B., Hunt, D., and Parker, K., Pituitary hormone response to cigarette smoking, *Pharmacol. Biochem. Behav.*, 24, 159–163, 1986.
28. Curzon, P., Kim, D.J.B., and Decker, M.W., Effect of nicotine, lobeline, and mecamylamine on sensory gating in the rat, *Pharmacol. Biochem. Behav.*, 49(4), 877–882, 1994.
29. Glassman, A.H., Helzer, J.E., Covey, L.S., Cottler, L.B., Stetner, F., Tipp, J.E., et al., Smoking cessation and major depression, *J. Am. Med. Assoc.*, 264:1546–1549, 1990.
30. Salin-Pascual, R.J. and Drucker-Colin R. A novel effect of nicotine on mood and sleep in major depression, *NeuroReport*, 9:57–60, 1998.

31. Sanberg, P.R., Silver, A.A., Shytle, R.D., Philipp, M.K., Cahill, D.W., Fogelson, HM., et al. Nicotine for the treatment of Tourette's syndrome, *Pharmac. Ther*, 74:21–25, 1997.

32. McConville, B.J., Sanberg, P.R., Fogelson, M.H., King, J., Cirino, P., Parker, K.W., and Norman, A.B., The effects of nicotine plus haloperido compared to nicotine only and placebo nicotine only in reducing tic severity and frequency in Tourette's disorder, *Biol. Psych.*, 31, 832–840, 1992.

33. Dursun, S.M. and Reveley, M.A., Differential effects of transdermal nicotine on microstructured analyses of tics in Tourette's syndrome: an open study, *Psychol. Med.*, 27:483–487, 1997.

34. Benwell, M.E.M., Balfour, D.J.K., and Anderson, J.M., Evidence that tobacco smoking increases the density of nicotine binding sites in human brain, *J. Neurochem.*, 50:1243–1247, 1988.

35. Piasecki, M. and Newhouse, P., Nicotine in psychiatry: psychopathology and emerging therapeutics, in Zevin S, Benowitz NL, Eds., *Pharmacokinetics and Pharmacodynamics of Nicotine*, Washington, DC, American Psychiatric Press, Inc., 2000.

36. Hughes, J.R., Hatsukami, D.K., and Skoog, K.P., Physical dependence on nicotine in gum: a placebo substitution trial, *J. Am. Med. Assoc.*, 255(23), 3277–3279, 1986.

37. Anonymous, Behavioral toxicity of nicotine, in Hughes J.R., Ed., Dependence on and Abuse of Nicotine Replacement Medications: An update, 147–157.

38. Hughes, J.R., Observer reports of smoking status: a replication, *J. Subst. Abuse*, 4, 403–406, 1992.

39. Benowitz, N.L. and Jacob III, P., Nicotine and carbon monoxide intake from high- and low-yield cigarettes, *Clin. Pharmacol. Ther.*, 265–270, 1984.

40. Benowitz N.L., Jacob P.I., and Fong I., Nicotine metabolic profile in man: comparison of cigarette smokin and transdermal nicotine, *J. Pharmacol. Exp. Ther.*, 268:296–303, 1994.

41. Pomerleau, O.F., Nicotine and the central nervous system: biobehavioral effects of cigarette smoking, *Am. J. Med.*, 93(Supp 1A), 2S-7S, 1992.

42. Allen, S.S., Hatsukami, D.K., and Christianson, D., Symptomatology and caloric intake during the menstrual cycle in smoking women, *J. Subs. Abuse*, 8:303–319, 1996.

43. Shytle, R.D., Silver, A.A., Philipp, M.K., McConville, B.J., and Sanberg, P.R., Transdermal nicotine for Tourette's syndrome, *Drug Devel. Res.*, 38(3/4), 290–298, 1996.

44. Newhouse, P.A., Potter, A., and Lenox, R.H., The effects of nicotinic agents on human cognition: possible therapeutic applications in Alzheimer's and Parkinson's diseases, *Med. Chem. Res.*, 2, 628–642, 1993.

45. Perkins, K.A., Baseline-dependency of nicotine effects: a review, *Behav. Pharmacol.*, 10:597–615, 1999.

46. Heishman, S.J., Snyder, F.R., and Henningfield, J.E., Performance, subjective, and physiological effects of nicotine in nonsmokers, *Drug Alc. Depen.*, 34, 11–18, 1993.

47. Newhouse, P.A. and Hughes, J.R., The role of nicotine and nicotinic mechanisms in neuropsychiatric disease, *Br. J. Addic.*, 86, 521–526, 1991.

48. Fagerstrom, K.O., Pomerleau, O., Giordani, B., and Stelson, F., Nicotine may relieve symptoms of Parkinson's disease, *Psychopharmacology*, 116, 117–119. 94.

49. Parrott, A.C., Individual differences in stress and arousal during cigarette smoking, *Psychopharmacology*, 115:389–396, 1994.

50. Robbins, T.W., Behavioural determinants of drug action: rate-dependency revisited, in Cooper SJ, editor, *Theory in Psychopharmacology*, Academic Press, New York, 2–63.

51. Buccafusco, J.J., Prendergast, M.A., Terry, A.V., and Jackson, W.J., Cognitive effects of nicotinic cholinergic receptor agonists in nonhuman primates, *Drug Devel. Res.*, 38(3/4), 196–203, 1996.

52. Newhouse, P.A., Sunderland, T., Tariot, P.N., Blumhardt, C.L., Weingartner, H., Mellow, A., and Murphy, D.L., Intravenous nicotine in Alzheimer's disease: a pilot study, *Psychopharmacology*, 95, 171–175, 1988.

53. Kelton, M.C., Kahn, H.J., Conrath, C.L., and Newhouse, P.A., The effects of nicotine on Parkinson's disease, *Brain Cognit.*, 43, 274–282, 2000.

54. Benowitz, N.L., Jacob III, P., Jones, R.T., and Rosenberg, J., Interindividual variability in the metabolism and cardiovascular effects of nicotine in man, *J. Pharmacol. Exp. Ther.*, 221, 368–372, 1982.

55. Heishman, S.J., Snyder, F.R., and Henningfield, J.E., Performance, subjective, and physiological effects of nicotine in nonsmokers, *Drug Alc. Depend.*, 34(1), 11–18, 1993.

56. Benowitz, N.L., Jacob III, P., and Savanapridi, C., Determinants of nicotine intake while chewing nicotine polacrilex gum, *Clin. Pharmacol. Ther.*, 41(4), 467–473. 87.

57. Russell, M.A.H., Jarvis, M.J., Jones, G., and Feyerabend, C., Nonsmokers show acute tolerance to subcutaneous nicotine, *Psychopharmacology*, 102, 56–58, 1990.

58. Heishman, S.J., Taylor, R.C., and Henningfield, J.E., Nicotine and smoking: a review of effects on human performance, *Exp. Clin. Psychopharmacol.*, 2(4), 345–395, 1994.

59. Snyder, F.R. and Henningfield, J.E., Effects of nicotine administration following 12 h of tobacco deprivation: assessment on computerized performance tasks, *Psychopharmacology*, 97, 17–22, 1989.

60. Rusted, J.M., Newhouse, P.A., and Levin, E.D., Nicotinic treatment for degenerative neuropsychiatric disorders such as Alzheimer's disease and Parkinson's disease, *Behav. Brain Res.*, 113, 121–129, 2000.

61. Manacuso, G., Andres, P., Ansseau, M., and Tirelli, E., Effects of nicotine administered via a transdermal delivery system on vigilance: a repeated measure study, *Psychopharmacology*, 142, 18–23, 1999.

62. Conners, C.K., Levin, E.D., Sparrow, E., Hinton, S.C., Erhardt, D., Meck, W.H., Rose, J.E., and March, J., Nicotine and attention in adult attention deficit hyperactivity disorder (ADHD), *Psychopharmacol. Bull.*, 32(1), 67–73, 1996.

63. White, H.K. and Levin, E.D., Four-week nicotine skin patch treatment effects on cognitive performance in Alzheimer's disease, *Psychopharmacology*, 143, 158–165, 1999.

64. Provost, S.C. and Woodward, R., Effects of nicotine gum on repeated administration of the Stroop test, *Psychopharmacology*, 104, 536–540, 1991.

65. Wesnes, K. and Warburton, D.M., Effects of scopolamine and nicotine on human rapid information processing performance, *Psychopharmacology*, 82, 147–150, 1984.

66. Prendergast, M.A., Jackson, W.J., Terry, Jr., A.V., Decker, M.W., Arneric, S.P., and Buccafusco, J.J., Central nicotinic receptor agonists ABT-418, ABT-089, and (-)-nicotine reduce distractibility in adult monkeys, *Psychopharmacology*, 136, 50–58, 1998.

67. Levin, E.D., Conners, C.K., Sparrow, E., Hinton, S.C., Erhardt, D., Meck, W.H., Rose, J.E., and March, J., Nicotine effects on adults with attention-deficit/hyperactivity disorder, *Psychopharmacology*, 123, 55–63, 1996.

68. Sahakian, B., Jones, G., Levy, R., Gray, J., and Warburton, D., The effects of nicotine on attention, information processing, and short-term memory in patients with dementia of the Alzheimer type, *Br. J. Psych.*, 154, 797–800, 1989.

69. Levin, E.D., Wilson, W., Rose, J.E., and McEvoy, J., Nicotine-haloperidol interactions and cognitive performance in schizophrenics, *Neuropsychopharmacology*, 15(5), 429–436, 1996.

70. Newhouse, P.A., Potter, A., Corwin, J., and Lenox, R., Age-related effects of the nicotinic antagonist mecamylamine on cognition and behavior, *Neuropsychopharmacology*, 10(2), 93–107, 1994.

71. Wilson, A.L., Langley, L.K., Monley, J., Bauer, T., Rottunda, S., McFalls, E., Kovera, C., and McCarten, J.R., Nicotine patches in Alzheimer's disease: pilot study on learning, memory, and safety, *Pharmacol. Biochem. Behav.*, 51(2), 509–514, 1995.

72. Warburton, D.M. and Rusted, J.M., Cholinergic control of cognitive resources, *Neuropsychobiology*, 28, 43–46, 1993.

73. Rusted, J., Graupner, L., O'Connell, N., and Nicholls, C., Does nicotine improve cognitive function, *Psychopharmacology*, (115), 547–549, 1994.

74. Elrod, K., Buccafusco, J.J., and Jackson, W.J., Nicotine enhances delayed matching-to-sample performance by primates, *Life Sci.*, 43, 277–287, 1988.

75. Perryman, K.M. and Fitten, L.J., Delayed matching-to-sample performance during a double-blind trial of tacrine (THA) and lecithin in patients with Alzheimer's disease, *Life Sci.*, 53, 479–486, 1993.

76. Newhouse, P.A., Potter, A., Corwin, J., and Lenox, R., Modeling the nicotinic receptor loss in dementia using the nicotinic antagonist mecamylamine: effects on human cognitive functioning, *Drug Devel. Res.*, 31(1), 71–79, 1994.

77. Nordberg, A., Clinical studies in Alzheimer patients with positron emission tomography, *Behav. Brain Res.*, 57, 215–224, 1993.

78. Linville, D.G., Williams, S., Raszkiewicz, J.L., and Arneric, S.P., Nicotinic agonists modulate basal forebrain control of cortical cerebral blood flow in anesthetized rats, *J. Pharmacol. Exp. Ther.*, 267(1), 440–448, 1993.

79. Volkow, N.D., Ding, Y-S, Fowler, J.S., and Gatley, S.J., Imaging brain cholinergic activity with position emission tomography: its role in the evaluation of cholinergic treatment in Alzheimer's dementia, 49, 211–220, 2001.

Section 4

15 Nicotinic Systems: An Integrated Approach

Edward D. Levin

Nicotine is a multifaceted drug. It acts on a variety of nicotinic receptor subtypes widely distributed throughout the nervous system. Nicotine both stimulates and desensitizes these receptors to affect a variety of neurobehavioral functions. This complex of actions illustrates how nicotine can have adverse effects, such as promotion of tobacco addiction, as well as potentially therapeutic beneficial effects, such as alleviation of pain and cognitive enhancement. The chapters in this book describe the levels of investigation in nicotine research from receptor mechanisms through neural system effects to behavioral and clinical effects. With an integrated approach using molecular- and receptor-level investigation together with animal model neurobehavioral studies and human clinical studies, progress can be made to identify critical neuropharmacological mechanisms for adverse and potentially therapeutic effects of nicotine. This will aid in the effort to understand the role nicotinic systems play in neurobehavioral function and to develop novel nicotinic ligands for clinical use.

At the receptor level, nicotine has actions on a variety of nicotinic receptor subtypes and it has both stimulating and desensitizing effects on those receptors. Nicotine stimulates the release of many different neurotransmitters, and thus has cascading effects throughout the nervous system. The development of new nicotinic agonist and antagonist drugs with different receptor subtype selectivity helps in determination of the functional role of nicotinic receptor subtypes and the role of activation and desensitization in nicotine's actions.

On a systems level, nicotine and other nicotinic drugs have important effects in diverse neurobehavioral functions. The role of genetic factors in neurobehavioral response to nicotine can be elegantly studied in the mouse model of strain selective effect. The impact of null mutants with As, the primary psychoactive chemical in tobacco, nicotine plays a vital role in tobacco addiction. The role of nicotine in smoking can be studied in the rat model of self-administration. This provides an important complement to experimental human studies in which the neural bases of nicotine's reinforcing effects can be discovered. Importantly, in both animal model and human studies, the complex interaction of sensory and pharmacological aspects of nicotine addiction can be studied. This research provides an arena for developing not only a better understanding of the biobehavioral underpinnings of smoking but also a way to develop new methods of smoking cessation.

0-8493-2386-X/02/$0.00+$1.50

Clinical development of nicotinic-based therapeutics is under active investigation. Nicotinic systems are now well documented as important components in the neural substrates of cognitive and motor function. These effects may provide avenues for development of novel therapeutic agents for a variety of types of cognitive and motor dysfunction such as those seen in schizophrenia, Tourette's syndrome, Alzheimer's disease, and attention deficit/hyperactivity disorder (ADHD). The dual action on cognitive and motor action can be helpful in treating diseases such as Tourette's syndrome where there is often comorbidity of attentional deficit with the motor impairment, or schizophrenia, where there may be motor slowing (bradykinesia) resulting from classic antispychotic drug use accompanying cognitive impairment. The development of nicotinic treatment of Alzheimer's disease is quite promising. It has served as one of the prime motivators for development of nicotinic treatment for conditions other than smoking cessation. This work has been prompted by finding the cognitive enhancement caused by nicotine in other subject populations and by finding that the population of nicotinic receptors in Alzheimer's disease patients is severely decreased compared to young adults. The development of nicotinic treatment for Alzheimer's disease has been furthered considerably by positive findings that injections of nicotine or use of nicotine skin patches improves attention, learning, and memory in Alzheimer's disease patients.

The elegant work in the molecular biology of nicotinic receptor subtypes helps in the development of receptor subtype selective ligands. Identification of nicotinic subsystems and availability of specific nicotinic drugs that affect them helps in the discovery of how these neuronal nicotinic subsystems are differentially involved in the variety of behavioral functions affected by nicotine. Understanding the complex mechanisms of nicotinic actions on neurobehavioral function lights the way for exploring novel avenues for nicotinic-based therapeutics for a wide variety of disorders.

Index